THE NEW ELEMENTS
OF MATHEMATICS

Picture of C. S. Peirce with his horse.

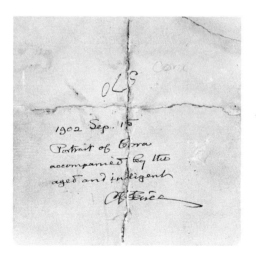

Picture of caption in Peirce's own handwriting.

(From the Charles S. Peirce Collection in the Houghton Library, Harvard University.)

THE
NEW ELEMENTS
OF
MATHEMATICS

by

CHARLES S. PEIRCE

Edited by

CAROLYN EISELE

VOLUME IV
MATHEMATICAL PHILOSOPHY

1976

MOUTON PUBLISHERS
THE HAGUE – PARIS

HUMANITIES PRESS
ATLANTIC HIGHLANDS N.J.

© Copyright 1976
Mouton & Co. B.V., Publishers, The Hague

ISBN 90 279 3045 7

The American edition published in 1976 by Humanities Press Inc.
Atlantic Highlands, N.J. 07716

Library of Congress Cataloging in Publication Data

Peirce, Charles S.
  Mathematical Philosophy

  (His The new elements of mathematics; v. 4)
  Includes index.
  1. Mathematics — Philosophy. I. Title.
QA39.2.P42 vol. 4 QA8.4   510'.8s 510'.1   76-20714
ISBN 0 391 00642 8

Printed in the Netherlands

INTRODUCTION TO VOLUME 4

My father, then, was the leading mathematician of the country in his day, — a mathematician of the school of Bowditch, Lagrange, Laplace, Gauss, and Jacobi, — a man of enormous energy, mental and physical, both for the instant gathering of all his powers and for long-sustained work; while at the same time he was endowed with exceptional delicacy of sensation, both sensuous and sentimental. But his pulse beat only sixty times in a [minute] and I never perceived any symptom of its being accelerated in the feats of strength, agility, and skill of which he was fond, although I have repeatedly seen him save his life by a hair-breadth; and his judgment was always sane and eminently cool.

Without appearing to be so, he was extremely attentive to my training when I was a child, and especially insisted upon my being taught mathematics according to his directions. He positively forbade my being taught what was then, in this country, miscalled "Intellectual Arithmetic," meaning skill in instantaneously solving problems of arithmetic in one's head. In this as in other respects I think he underrated the importance of the powers of dealing with individual men to those of dealing with ideas and with objects entirely governed by exactly comprehensible ideas, with the result that I am today so destitute of tact and discretion that I cannot trust myself to transact the simplest matter of business that is not tied down to rigid forms. He insisted that my instruction in arithmetic should be limited to exhibiting the working of an example or two under each rule and being set to do other "sums" and to having my mistakes pointed out for my correction. He preferred that I should myself be led to draw up the rules for myself, and quite forbade that I should be informed of the reason of my rule. Thus he showed me, himself, how to use a table of logarithms, and showed that in a couple of cases the sum of the logarithms was the logarithm of the product, but refused to explain why this should be or to direct me to the explanation of the phenomenon. That, he said, I must find out for myself, as I ultimately did in a more general way than that in which it is usually stated. I believe that, had he himself been acquainted with

Listing's studies in "topology," or the laws of the connections of the different parts of a place, he would have made something of that sort my introduction to geometry. But as it was, he made me study the simplest parts of "graphics," or projective geometry, including a generalized doctrine of linear perspective before I took up "metrics;" i.e. the geometrical doctrine taught in Euclid's Elements. . . .

As a child, I had opportunities at table and elsewhere, of observing a great number of distinguished people, statesmen and reformers, lawyers, divines, philosophers, historians, philologists, purely literary writers in verse and prose, and scientific men in all departments; —all of them, however, either of strong intellect or of strong character, mostly both. For it was such that were attracted to the house; and they were of various nationalities and stripes. It is really surprising how few commonplace people I saw much of. It must have been in the year 1851, when I would have been twelve years old, that I remember picking up Whateley's Logic in my elder brother's room and asking him what logic was. I see myself, after he had told me, stretched on his carpet and poring over the book for the greater part of a week for I read it through and I must have done so and pretty thoroughly mastered the greater part of it, as tests showed. From that day to this logic has been my passion although my training was chiefly in mathematics, physics, and chemistry. In 1887, being 48 years old and having attained a respectable, though not particularly eminent place among American scientists by sundry contributions mainly to exact logic, but also to several parts of physics and to other subjects, I took the decisive step largely to escape distractions from my studies in logic, of retiring to the wildest country, that is at once south of the Adirondacks and east of the Alleghenies, in the Northern United States; and I have since remained, publishing very little but incessantly working as hard as I possibly could to advance the science of logic.

                              (Notebook fragments from MS. 905)

[The] matter [is] of so much concern to me to know just what my comparative powers in logic are that I have taken the utmost pains to estimate them correctly, and neither too high nor too low. Now I am well-acquainted and deeply read in the whole literature of the subject in the widest sense and have so carefully studied the question that personally concerns me, that I feel sure I can have made no great mistake about it; and the only writers known to me who are in the same rank as I are Aristotle, Duns Scotus, and Leibniz, the three greatest logicians in my estimation, although some of the most important points escaped each. Aristotle was a marvellous man in many other directions; a great writer, a great zoologist, a great psychologist, a profound sociologist, and a very able practical politician. I only compare myself to him in respect to logical powers. Leibniz, too, was a sublime mathematician, as well as very able in all that concerns politics and jurisprudence. But I consider him only as a logician.

The problem of rating one's own intellectual powers is, quite apart from self-tenderness, one that is full of pitfalls. But I have studied them all with care; and I think that most men of 50 manage to appraise their own powers pretty accurately.

(From fragments L 482)

A Comparative and Critical Synopsis of the Considerable Methods of Logical Representation

The methods have to be noticed as follows:
I The traditional method by words
II Aristotle's (Conjectured) diagrams
III Hamilton's Notation
IV De Morgan's Spicular Notation
V Boole's Algebra of Logic
VI Jevon's Modification of Boole's Algebra
VII Peirce's Algebra of 1870
VIII Peirce's Universal Algebra
IX Peirce's Algebra of Dyadic Relations

(From s-44)

The circumstances that at the age of 12 I did not know the meaning of the word "logic" even vaguely, although my father was a mathematician, leads me to mention that he had a low opinion of the utility of formal logic, and wished to draw directly upon the geometric instinct rather than upon logical deductions from the smallest number of axioms. This disposition of his is manifest in his "Analytic Mechanics" of 1852 and still more in his interesting text-book of Elementary Geometry (of 1837). This was one of a number of allied points in which I have long disagreed with my father, mainly because I do not believe instinct is an absolutely infallible source of truth. On the contrary, it must develop like every feature of creation. The truth of this emerges from every topic of argumentation that can be applied to it, and in these days especially is as little likely to be disputed as anything I can think of.

(From letter to F. A. Woods, 14 October 1913)

It concerns the reader of a treatise upon logic to know just what the author's acquaintance with science is. I will therefore make an explicit confession on this point.

I have a pretty fair acquaintance with most of the branches of mathematics, although I am weak in the theory of forms, or "higher algebra," as it is often called. I have made various investigations of my own, though none of them have been very large.

My reading in the philosophical sciences, ethics, metaphysics, and especially logic, has been extended and careful; but I am not as well acquainted with Hegel as I ought to be. My original work in logic is well-known.

I have a general acquaintance with all branches and sects of psychology, and have made various researches of my own. Yet I do not call myself a psychologist.

I have made some studies of a pretty wide variety of languages, having looked into the logical structure of dozens of all families. I have read more or less Arabic and Ancient Egyptian, getting some lessons in the former from my friend E. H. Palmer. I have constantly used Latin and Greek; and have paid some attention to comparative philology, so-called. Of the commoner of the modern European languages I have a decent acquaintance, including English. I have quite no natural talent in this direction, nor at all a good memory. Most of what I once knew, I have lost all ready hold of. I am not infrequently at a loss even for a genuine English or French word, which are the languages I am least ignorant of.

I have always taken the deepest interest in archeology, although my sole attempt at original study is too slight to be mentioned. The methods of the sciences have been a matter of careful study with me. I have carefully studied ancient metrology.

As to the other branches of ethnology, it is probably because I know so little of them that they hardly appear to me to be sciences. But if political economy is to be put under this head, I must except this, in which I have read and carefully considered, some of the chief works of the different schools of the nineteenth century, with a preference for the more logical and analytical ones, such as Ricardo.

Of modern political history my knowledge is next to nothing; but I think I can perceive that most of it is not scientific. I have studied with great pains those branches of history in which the testimonies are few and inadequate; since these I could get to the bottom of without too much labor. In these directions, I have studied with critical attention the methods of different historians. The history of the human intellect has been one of my chief interests; especially in reference to the physical and philosophical sciences. Here, I have always gone to the sources.

Biography has always been a favorite subject with me. When I am fatigued with my day's work, I always take up a biography, if I have one at hand. Of course, the mass of mendacity which clusters here is so great that to acknowledge oneself the author of a biography is almost tantamount to confessing oneself a liar. It certainly supports that view that many presidents of American colleges have written biographies; mendaciousness being, of course, the first quality looked for in filling such a position. The individual man has always interested me particularly, as have those generalizations which have been confined to classes of men. There is a great deal of this sort of lore in novels; but I am rather an impatient student of novels. Too many pages have to be devoured; and then after all perhaps the guarantee of its truth is not always sufficient, although scepticism on this point may be carried to excess. From many works which, while I read them, seem to be foolishness mainly, I afterwards seem to have brought away something of value. It must be that there is a future science here. All those general works about great men and eminent men, especially those of Galton, have been of great interest to me; and I have learned a great deal about methodeutics from my own extensive studies of great men.

In physics, I have done serious work in gravitation and optics. Chemistry is the profession for which I was specially educated.

I am ignorant of biology except for some lessons by Asa Gray, six months under Agassiz, some little practice under other teachers. Reading textbooks is hardly to be counted.

In astronomy I have had considerable observatory experience and am the author of a book of Photometric Researches.

Of geology I am more completely ignorant than of any other branch of science.

(From MS. 427)

Whatever the philosopher thinks that every scientific intelligence must observe, will necessarily be something which he himself observes, or seems to observe. I have several times argued, at some length, that the unity of personality is in some measure illusory, that our ideas are not so entirely in the grasp of an *ego* as we fancy that they are, that personal identity differs rather in degree than in kind from the unity of "public opinion" and gregarious intelligence, and that there is a sort of identity of dynamic continuity in all intelligence. Accepting this opinion, a man is not radically devoid of the power of saying what every scientific intelligence must observe, if he has the power of saying what he observes himself. If he is in dynamic continuity with his whole self, he is in the same kind of continuity, albeit less intimate, with the whole range of intelligence. He can observe, in a fallible, yet genuine, observation what it is that every scientific intelligence must observe. Such observations will, however, require correction; because there is a danger of mistaking special observations about intelligences peculiarly like our own for observations that are open to every "scientific intelligence," by which I mean an intelligence that needs to learn and can learn (provided there be anything for it to learn) from experience. I would here define experience as the resultant of the mental compulsions from the course of life; and I would define learning as the gradual approximation of representations toward a limiting definite agreement. My theory has to be that not only can *man* thus observe that certain phenomena are open to every scientific intelligence, but that this power inheres essentially in every scientific intelligence.

The doctrine of *exact philosophy*, as I understand that phrase, is, that all danger of error in philosophy will be reduced to a minimum by treating the problems as mathematically as possible, that is, by constructing some sort of a diagram representing that which is supposed to be open to the observation of every scientific intelligence, and thereupon mathematically,—that is, intuitionally,–deducing the consequences of that hypothesis.

(Unidentified fragment)

Philosophy requires exact thought, and all exact thought is mathematical thought. Especially, it behooves a sentimentalist to take double and triple pains to make his thought rigidly exact. For it is the nature of his cast of philosophical thought to be exceedingly dangerous if it is not bound down to a logic at least as rigid as that of Euclid. Whether my thought really is so, or not, of course without expounding it at length, I am unable to show. I can only say that I have been bred in the lap of the exact sciences and I know what mathematical exactitude is, that is as far as I can see the character of my philosophical reasoning.

(From MS. 438)

My special business is to bring mathematical exactitude,—I mean *modern* mathematical exactitude into philosophy,—and to apply the ideas of mathematics in philosophy. . . .

I don't mean to shackle anybody with any condition other than that

they shall work at the rendering of philosophy mathematically exact
and scientifically founded on positive experience of some kind.

(To F. Russell, 23 September 1894)

But it is time to consider the idea of continuity, or unbrokenness, which
is the leading idea of the differential calculus and of all the useful
branches of mathematics, which plays a very great part in all scientific
thought, the more so the more exact that thought is, and which is the
master-key which in the opinion of adepts must unlock the arcanes of
philosophy.

(Unidentified fragment)

Insofar as human thought has been [tychistically] developed in the succession of generations, its source has been as follows. There is, as we
know, a ruling tendency for son to think like father, and for any one man
to think like as he has thought. When I say "like," I do not mean merely
to think the same thing, but to apply analogous ideas to different subjects. For instance, it is natural and usual for a political economist to
adopt the theory of natural selection; it is natural for a physicist to be
a necessitarian; it is natural for a moralist, who is always insisting on the
gulf between right and wrong, to be a dualist in metaphysics. At the same
time, people are not exactly consistent, and their views accidentally
vary a little this way and that way. Some of these modifications are
improvements, in that they lead to practical solutions of problems, and
in that way, their further propagation is favored. Others are not propagated, because they are not adapted to meeting the facts which are
continually impressed upon the mind from without, and they do not,
consequently, become fixed. The peculiarity of this account of mental
development is that it represents the different new ideas which are
started as mere freaks, upon the principle that an idea in its first suggestion has not been critically examined, — cannot have been so, — and
therefore can be little more than a whimsy.

(From unidentified fragments)

While I am speaking of deduction, I ought to allude to the meaning of
infinitesimal. The ordinary explanation by the doctrine of limits seems to
me to leave the difficulty just where it was, since the conception of a
limit, even when this conception is not narrowed as mathematicians

narrow it, involves precisely the same difficulty as "infinitesimal" does. But I think that my father (though he would have *scorned* to be called a logician) put his finger on the solution of the question in the Preface to his Geometry (A.D. 1837) [where] he speaks of "infinitely small, that is, as small as we choose" or as I would define it: An infinitesimal is a finite numerical quantity whose maximum possible value is indefinite but is regarded as being so small *in each numerical* computation that it may be safely neglected, provided it is never multiplied by an indefinitely large number in a similar sense. This is plainly what B.P. meant; and it is not only far simpler than the doctrine of limits, but it really solves the difficulty, which *that* merely diverts attention from.

(From a letter to F. A. Woods, 14 October 1913)

That we ought to experiment without preconceived ideas is one of those vague logical maxims which characterize the loose reasoner. An experiment has been well called the putting of a question to Nature, which she answers mostly by "no" or "yes." To have no preconceived idea in experimentation is to take an interrogatory position without putting any definite question. To be wedded to a preconceived idea is [to] replace interrogation by affirmation, and not to listen to what Nature has to say. There is a game called "Twenty Questions," in which one party thinks of something well-known to the other, who may then ask at most twenty questions answerable by yes or no, after which he has a right to make three guesses. If there are no more than 3145728 objects that [are] well-known to both parties, and if each question is so skilfully devised as precisely to bisect the number of possibilities, the answer to the twentieth question will leave just three possibilities open. But if the questioner, becoming possessed of a fancied anticipation, allows such preconceived notion to hurry the laying of the parallels of his siege, he will be pretty sure to fail. That illustrates how much value the vague maxim has. Unless there be some truth in our hereditary metaphysics, unless the nature of the mind is such that the right idea will be developed at last, unless the number of utterly wrong guesses that we should ever make is finite, no number of questions, however large, will bring us measurably nearer the truth than we were at the outset.

The principle of pragmatism is applicable here. In reference to any particular investigation that we may have in hand, we must *hope* that, if it is persistently followed out, it may ultimately have some measure of success; for if it be not so, nothing that we can do can avail, and we might as well give over the inquiry altogether, and by the same reason stop applying our understanding to anything. So a prisoner breaks through the ceiling of his cell, not knowing what his chances of escape may be, but feeling sure there is no other good purpose to which he can apply his energies.

The success for which we hope is that we shall attain some rule which further experience will not force us to repeal.

(From MS. 1519)

Logic requires us, with reference to each question we have in hand, to hope some definite answer to it may be true. That *hope* with reference to *each case* as it comes up is, by a *saltus*, stated by logicians as a *law* concerning *all cases*, namely, the law of excluded middle. This law amounts to saying that the inverse has a perfect reality. It is, therefore, inapplicable to universes of fiction, concerning which we may, nevertheless, reason. Whether Hamlet had or had not a strawberry mark on his left shoulder, admits of no other answer than that Hamlet is indeterminate in that respect. Now what pure mathematicians reason about is always a universe of their own creation; and therefore the doctrine of some logicians that an individual is whatever is determinate in every respect must be banished from the logic of mathematics.

(From MS. 140)

$$x^2 \equiv x \,(\text{mod. } 2)$$

A congruence is but an equation, and if we write this in the form

$$x^2 = x$$

it expresses the principle of excluded middle. For solving this quadratic equation, we find it signifies that $x = 0$ or $x = 1$.

(From unidentified fragments)

It is to be remarked, to begin with, then, that this philosophizing about mathematics cannot itself be purely mathematical, since it involves categorical statements of what is true of mathematics. Undoubtedly the problem may be put into conditional form, thus: "If we are to discover that anything would be true were something else true, then what can we say will be true of the procedure?" But in order to solve this problem we must make use of some universally and unconditionally true remarks about any such procedure. We have, for example, to notice that any such discovery involves making an assertion if not to others at least to oneself. This is not a proposition in mathematics. We assure ourselves of its truth by performing a variety of experiments in our imagination.

(From unidentified fragments)

I exclaim it is not he who is teaching, it is the genius, the intuitions of the pupils which is relieving him of his duty, and he gets so accustomed to this that when he meets a boy who really does require teaching and not mere instruction, he loses his patience and seems to think that the boy is in some way at fault, when it is really the teacher who does not know what the logic of mathematics is. If the teacher feels the force of a demonstration but doesn't [know] why, which happens in 999,99[9] cases out of a million, he cannot make another person feel the force of that demonstration who does not do so already. . . .

A mathematician has to be very particular too. Accuracy and temper go together. Musicians are also of uncertain temper. It is one of the many respects in which mathematicians and musicians have a certain degree of resemblance. Many mathematicians love instrumental music: why should they not? For the delight equally of the science and the art consists in the contemplation of complicated systems of relationship. It really needs explanation if a mathematician is not musical. The intelligent listening to a fugue of Bach is certainly more like reading a piece of higher mathematics than the lesson of the schoolboy in elementary geometry is like the higher geometry. Musicians are irascible but there is an enormous difference between the bad temper of the mathematician and the irascibility of the musician. Take away from a composer his excessive susceptibility and substitute a stronger will power and abstraction and you will produce a geometer. All these characters of mathematicians ought to be taken account of and and explained by the man whose office it is to teach mathematics especially to average pupils. Such a teacher has a hard task. The proportion of persons is small to whom mathematics comes easy. He has therefore a problem in psychology before him. What are the faculties he has to call into play? Into what state of mind is he to put his pupil? These are questions requiring study. But besides these psychological questions, logical questions of extreme difficulty present themselves to him. What is the real nature of mathematical demonstration? If he cannot answer that how can he expect to get the demonstration into the head of his pupil? Many pupils have genius; they have intuitions; they relieve the teacher of the hardest part of his task. He gets to expect this of his pupils, and when they haven't got it he is at a loss, and rails at the stupidity of his scholars, as if it were not precisely boys dull in mathematics that it is his business to know how to instruct. That he doesn't know how to do this, is in great measure because he doesn't comprehend himself the real nature of the demonstration the force of which he feels he doesn't know why. Feeling, learning, willing, these are the elements of mental life.

(From MS. 748)

The truths of mathematics are truths about ideas merely. They are all but certain. Only blundering can introduce error into mathematics. Questions of logic are questions of fact. Can the premise be true and the conclusion be false at the same time? But the logician, as such, knows nothing about the truth of nature. He only hopes that a few assumptions he makes may be near enough correct to answer his purpose in some measure. These assumptions are, for instance, that things are sufficiently steady for something to be true, and what contradicts it false, that nothing is true and false at once, etc.

The assertion that mathematics is purely ideal requires some explanation. Thomson and Tait (Natural Philosophy §438) wisely remark that it is "utterly impossible to submit to mathematical reasoning the exact conditions of any physical question." A practical problem arises, and the physicist endeavors to find a soluble mathematical problem that resembles the practical one as closely as it may. This involves a logical analysis of the problem, a putting of it into equations. The mathematics begins when the equations or other purely ideal conditions are given. "Applied Mathematics" is simply the study of an idea which has been constructed so as to be more or less like nature. Geometry is an example of such applied mathematics; although the mathematician often makes use of space imagination to form icons of relations which have no particular connection with space. This is done, for example, in the theory of equations and throughout the theory of functions. And besides such special applications of geometrical ideas, all a mathematician's diagrams are visually imagined, and involve space. But space is a matter of real experience; and when it is said that a straight line is the shortest distance between two points, this cannot be resolved into a merely formal phrase, like 2 and 3 are 5. A straight line is a line that viewed endwise appears as a point; while length involves the sense of muscular action. Thus the connection of two experiences is asserted in the proposition that the straight line is the shortest. But 2 and 3 are 5 is true of an idea only, and of real things so far as that idea is applicable to them. It is nothing but a form, and asserts no relation between outward experiences. If to a candle, a book, and a shadow,—three objects, is joined a book, one, the result is 5, because there will be two shadows; and if five more candles be brought, the total will be only 8, because the shadows are destroyed. But nobody would take such facts as violations of arithmetic; for the propositions of arithmetic are not understood as applicable to matters of fact, except so far as the facts happen to conform to the idea of number. But it is quite possible that if we could measure the angles of a triangle with sufficient accuracy, we should find they did not sum up to 180°, but either exceeded it or fell short. There is no difficulty in conceiving this; although, owing to numerous associations of ideas, it is necessary to devote some weeks to a careful study of the matter before it becomes perfectly clear. Accordingly, geometrical propositions and arithmetical propositions stand upon altogether different ground. The footing of logical principles is intermediate between these. The logician does not assert anything, as the geometrician does; but there are certain assumed truths which he hopes for, relies upon, banks upon, in a way quite foreign to the arithmetician. Logic teaches us to expect some residue of dreaminess in the world, and even self-contradictions; but we do not expect to be brought face to face with any such phenomenon, and at any rate are forced to run the risk of it. The assumptions of logic differ from those of geometry, not merely in not being assertorically held, but also in being much less definite.

(From MS. 409)

We must not say the sum of the three angles of a triangle is two right angles, but that it equals that $\pm$ some very small quantity. We must not say that space has three dimensions, but that bodies cannot move to any hitherto perceptible extent in more than three orthogonal directions. We must not say that phenomena are perfectly regular, but that their degree of regularity is very high indeed. We must not say that material and psychical phenomena are entirely distinct,—whether as properties of different substances or as different sides of one shield,— but we must say that all phenomena are of one character, but some more psychical and free, others more material and purposeless, yet all more or less purposive, i.e. mixedly regular and spontaneous.

I have no room here to argue the truth of synechism: the reasoning is of the highest difficulty. What I want to ask here is, supposing this *synechism* should be generally accepted, as I think it must be as soon as logic is more accurately studied, what will be the influence of it in respect to the onement of religion and science?

(Unidentified fragment)

All mathematics is, in its purity, a study of an ideal system.

(From s-25)

. . . .The evidence thus seems to be in favor of a hyperbolic space. This might be due to chance; but a rough calculation of the chances shows that that is improbable. It may be objected that it may be due to the motion of the stars being systematic. The answer to that is, that it is a legitimate first result, and further inquiry in the same line will correct the error from any such source, if there be any such error. I have not published my work in detail because, at present, the number of stars used is too small; although the result is so decided as to be exceedingly encouraging. I have not hitherto had the funds to get the necessary computations made from all the stars of Auwers even without entering into any calculations of general constants. Should any help be extended to me, I am ready to do what I can; but this seems so unlikely that I think it best to make this record, so that when the decisive discovery is made, as my results make me confident it someday will, it may be seen that I had come so near to it as means would permit. The question is not without significance in regard to a general scientific conception of nature.

(From fragments L 484)

There are other cases where the absolute upon a line of variation consists of two points. Infinite heat is perhaps not the same as absolute cold, though the difference seems trifling. In the former case, the molecules would be stationary, in the latter case they would appear at all portions of their paths at the same instant. A clearer example of a double absolute is that of probability which ranges from absolute certainty for a proposition to absolute certainty against it. Both these degrees of probability are absolutely unattainable and therefore truly constitute an absolute, although they are commonly denoted by the numbers 0 and 1. But if probability be measured by the logarithm of the ratio of chances in favor to chances against (which is the natural way of estimating it), these degrees are denoted by $\log \frac{0}{1} = -\infty$ and $\log \frac{1}{0} = +\infty$, and these two infinities, being of the zero order of infinity, do not coincide.

(From unidentified fragments)

The *calendar* of an almanac is chiefly looked at for the sake of finding where to open one's Prayer Book, and with reference to moon light. The meridian passage of the sun is also important in the country.

(From unidentified fragments)

As a further indication that the mechanical philosophy is on the wane, it may be mentioned that various attempts have been made by scientific men to enlarge this hypothesis by the supposition of other agencies which should explain the facts that seem to the authors of these hypotheses to militate strongly against the pure mechanical theory, if not to refute it. Most of these attempts have been made by biologists who urge that all observed facts point to Evolution in one direction and that nothing whatever in experience goes to support Mr. Spencer's theory that the universe during half the time is undergoing a reverse operation of Devolution. I will add that all mathematicians are in accord in holding that Spencer's attempts to connect either Evolution or Devolution with the conservation of energy by mathematical reasoning is simply beneath all criticism as puerile nonsense.

(Unidentified fragment dated "1898")

We measure all physical quantities. We measure the mass or amount of bodies, we measure time, we measure speed, we measure pressure, we measure energy, we measure heat, we measure heat capacity, we measure the pitch and intensity of sounds, we measure the brightness of light, we measure colors, we measure many electrical quantities, we measure wants by prices. For many of these quantities we have some rude means of direct measurement; but only in the case of mass is it very precise. In most cases high precision of measurement depends upon directly measuring a spatial quantity. Thus, we measure temperature by a long thermometer-scale, light by the distance at which shadows will match, etc. Thus, space measure, without being the only direct measure, is of fundamental importance for all measurement.

(Unidentified fragment)

I think I ought to say a few words more concerning the advantage of the sextal system, which I seriously hold to be so superior to the decimal or any other that were I a radical, by which I understand one of those minds who hold that it is every man's duty to endeavor to improve the condition of humanity in every way, and bestir himself to bring about every condition of society which he is quite sure would be the best, I would certainly agitate in favor of revolutionizing numeration.

(Fragment in MS. 201, ca. 1908)

So, in dynamics, at least, simple theories have a tolerable chance of being true. Observe, that the character of a theory as to being simple or complicated depends entirely on the constitution of the intellect that apprehends it. Bodies left to themselves move in straight lines, and to us straight lines appear as the simplest of all curves. This is because when we turn an object about and scrutinize it, the line of sight is a straight line; and our minds have been formed under that influence. But abstractly considered, a system of like parabolas similarly placed, or any one of an infinity of systems of curves, is as simple as the system of straight lines. Again, motions and forces are combined according to the principle of the parallelogram, and a parallelogram appears to us a very simple figure.

(Unidentified fragment)

Science consists in the sincere and thorough search for truth according to the best available methods. Its only quite indispensable condition is the absolute single hearted energy with which it works to ascertain the truth, regardless of what the character of that truth may be. It is not science if it is not an intelligently directed research. But it will come to be so if it is absolutely sincere and highly energetic. These dispositions will generate the intelligence required. It is not science, if it is not a well-informed research. But sincerity and energy will bring about study. It is not science if it does not invent cunning ways. . . .

(From unidentified fragments)

With sufficient accuracy for ordinary purposes, we may say that the law of habit is the law of the mind. To state it so, has this convenience, that habit taking is a recognized property of protoplasm, and it is not easy in that field to resolve it into any higher property. . . .

Mr. A. B. Kempe published in the Philosophical Transactions . . . a Memoir on Mathematical Forms. He illustrates these by what the mathematician calls *graphs*, that is, diagrams consisting of spots connected with lines. . . . This is what I mean by an icon; and the ideas attach themselves to the dots. We must not, however, exclude other kinds of connective diagram. Algebraical arrays are another example, probably even more valuable. Here the connecting lines are replaced by repetitions of the letters together with the connecting signs, to which significance is given by algebraical rules. Even non-visual connective forms are useful, as in speech. But the visual sort of thinking is by far the most perspicuous, and powerful. That is to say it gives us facile mastery of far more complicated forms of connection.

I say that the whole law of mind, in the department of science, of art, and of practical life, — whether it be what we call knowing, or emotion, or reaction with the world, consists in this that ideas connect themselves with iconical ideas, so as to make up sets.

(Fragment from MS. 1008)

Closely allied to the principle of excluded middle is that of the identity of indiscernibles, that is, that different things must differ in some general and specifiable respect. Different things, being constituted by different acts or reactions, certainly must differ in their relations. But it is not necessary that they should differ in respect to any of those qualities which belong to each independently of any other individual things. They may be subject to precisely the same laws. In a mathematical universe, though different things must differ in some respects, it may be quite indeterminable what those respects are.

A *collection* is not a thing, but an *ens rationis*, since its distinctive identity is constituted not only by an arbitrary act, but by the distinctive identities of other things, namely of those that it comprises. At the same time, it so far partakes of the nature of a thing, that it is constituted not at all by anything of a general nature but by individual things, irrespective of their characters. This gives the collection individuality.

A collection is not a thing, and therefore a collection which comprises a single thing and excludes everything else is not identical with that thing. Whatever things may be given, there is just one collection which comprises them all and nothing else. According to this rule, there is a collection which comprises nothing. This collection subsists in every universe.

Characters and other objects of a general or continuous nature may be reckoned as things and comprised, as such, in collections, provided they involve arbitrary elements which serve to make their distinctive identities absolute. Colors, for example, are not, in themselves, susceptible of enumeration. We cannot reckon up the probability that a color drawn at random from all possible colors should be yellow. But Newton, by attaching arbitrary names to them, made the colors of the spectrum seven. So areas of a sphere, unless marked off, cannot be enumerated.

But by reference to an arbitrary square the surface of the sphere may be measured. In all such cases, the counting consists in counting distinct acts, or things that are constituted by distinct acts.

Admitting that everything has some distinctive character (though it may involve something arbitrary, or may not be determinate), it follows that every collection has a distinctive character. Hence, everything comprised in a collection stands to the other things comprised in the same collection in a relation to which it stands to no other set of things and to which nothing else stands to anything. Such a relation is an individual relation. An individual relation, like the collection to which it belongs, though an *ens rationis*, possesses distinctive identity. Two individual relations do not necessarily differ in respect to any character not involving a reference to anything individual; but they must differ in respect to their relative characters.

A *sign* is a thing which is the representative, or deputy, of another thing for the purpose of affecting a mind. . . . The utility of icons is evidenced by the diagrams of the mathematician, whether they involve continuity, like geometrical figures, or are arrays of discrete objects like a body of algebraical formulae, all of which are icons. Icons have to be used in all thinking.

(From MS. 142)

The only use of ideas at all is to bind facts into one great continuum, by a logic at once subjective and objective.

As for topics of vital importance, all sensible talk about them must be commonplace, all reasoning about them unsound, and all study of them sordid. Reason, anyway, is a faculty of secondary rank. Cognition is but the superficial film of the soul, while sentiment penetrates its substance. Reason is divine insofar as it fills up the gaps of the discrete and displays a continuum. But it is of just as much worth in this respect or even more when it is applied for example to the theory of numbers of which a great mathematician said that its glory was that it never had been nor ever could be prostituted to any practical application whatsoever, or when it counts the imaginary inflections of a curve of which only a small fraction can correspond to anything in nature, as when it is applied to the study of real phenomena. But reasoning has no monopoly of the process of generalization —Sentiment also generalizes itself; but the continuum which it forms instead of being like that of reason merely cognitive, superficial, or subjective . . . penetrates through the whole being of the soul, and is objective or to use a better word exstant, and more than that is exsistant.

(From fragment of Detached Ideas, MS. 438)

The commonest relations to be considered are dyadic relations. A dyadic relation is any one of a collection of facts about pairs of individuals. We distinguish between the two individuals, calling one the *relate*, the other the *correlate*. We often denote a relation by a letter, as r; and use the expression A is r to B, or in the same sense, B is r'd by A. We mean by that that there is a collection of collections of two objects each, which two objects are distinguished grammatically as the active and the passive subjects. We denote this collection of collections by the letter r to show that we do not refer to any particular individual collection of collections, but to any one of a collection of collections of collections.

Two collections, say the As and the Bs, may be such that there is a possible relation, r, such that every A is r to a B to which no other A is r, while there is a possible relation, l, such that every B is l to an A to which no other B is l. When that is the case, the two collections are said to be *equal*, or to have the *same multitude*.

Again, two collections, the As and the Bs, may be such that there is a possible relation, r, such that every A is r to a B to which no other A is r, while there is no possible relation such that every B is l to an A to which no other B is l. In that case the Bs are said to be *more* than the As, or the collection of Bs to be greater than that of the As, or to have a *higher multitude* than that of the As. The As are said to be *fewer* than the Bs, the collection of As to be *less* than that of the Bs, and to have a *lower multitude*.

This last definition is, however, redundant. For if there is no relation, l, whatsoever, such that every B is l to an A to which no other B is l, it can be shown that that very fact constitutes a relation, r, such that every A is r to a B to which no other A is r. This proposition ought to be called the *fundamental theorem of multitude*. Like other fundamental theorems, it has justly been deemed excessively difficult of proof. Cantor after studying multitude profoundly and with great power for many years, only produced a demonstration of this theorem in one of his last papers. It is a very roundabout way, anything but clear. To my mind it is clearly fallacious. This is not said to depreciate Cantor, whom I greatly admire, but because I wish to show that the occasion is a good one for putting the power of the logic of relatives to a test.

(From MS. 114)

What has, however, particularly attracted me to the subject is that there is no recognized method of proof in topical geometry, and that it is beset with logical difficulties. In order to set it upon a proper basis, it seems first necessary to produce a workable definition of continuity, an extremely difficult task of logic. In order to effect this, it is indispensable to study the nature of multitude. Here I am aided by the great papers of Cantor republished in the second volume of the *Acta Mathematica*. I have not had the opportunity of studying two later memoirs by him but I do not regret it, since my own researches were made before I knew of their existence, and having gone so far on a road cut out for myself, it is probably better that I continue in it. The subject of multitude requires great logical caution, and leads to a paradox which cannot be satisfactorily treated without a thorough logical preparation. For this reason I have been obliged to begin with a section on logic.

In order to set a definite term to my inquiries, I determined to carry them far enough to prove that four colors suffice to distinguish confine regions on any map drawn on a spheroidal surface, a demonstration which has been attempted by several mathematicians of great strength.

(From MS. 141)

Accordingly Cantor, finding no contradiction involved, assumes that just as the denumeral multitude follows after the enumerable multitudes as the multitude of an aggregate of collections of enumerable multitudes of which there is no greatest, so following after all the alephs, or abnumerable multitudes, there is a multitude which is that of an aggregate of abnumerable collections among which there is no greatest; and that there is thereupon another endless series of multitudes and so on, unceasingly. But the present writer is in possession of what seems to be a clear demonstration that a contradiction is involved of a kind against which Cantor was not upon the watch. Namely, the contradiction lies in supposing the members of such an aggregate to be independent of one another, so that any one is present whether the others are present or not, and any one might be removed without affecting the others. Such independence is essential to the theory of multitude which reposes upon the idea that each member of one collection may have a relation to a single one of another collection. This is precisely what would not be possible for the aggregate considered. It is, however, of secondary importance at what particular place in the series of multitudes this phenomenon occurs. It must occur somewhere; otherwise our ordinary conception of a terminated line would not be possible. Cantor, indeed, conceives a line as an ordered collection of independent points. He talks, for example, of removing the terminal point, and so leaving the line without a terminal point. But according to our ordinary conception the single points of a line have no separate identity. If a single point flies off from the end of a line, the line continues to have a terminal point. . . .

(From unidentified fragments)

Cantor's theory . . . is a perfectly reasonable supposition. But why limit ourselves so to two *classes* of *infinity*? This limitation springs from no more being needed in mathematics, nothing else. Rational and irrational numbers.

(From MS. 1574)

The universe is to become a more and more perfect mirror of that system of ideas which would result from the indefinitely continued action of objective logic. The universe is, as it were, an awaking Mind. Now just as we say this man has such and such a character, not because of any ideas he has this minute present, but because under suitable circumstances such ideas are bound to be evolved by him, so the universe may be said to be governed by a God insofar as it is bound more and more to conform to the ultimate result of the evolution of pure ideas. But the sole Ancient of Days is Continuity in the abstract, a spontaneity which might be assumed to be very slight, though it is probably enormous.

As to the continued existence of the soul after death, the general idea of Continuity, if unreservedly accepted, hardly permits us to doubt it. The difficulty is, that there is no positive evidence in favor of it. It seems, however, easier to account for such defect in other ways than by a breaking off of consciousness. That the second element of consciousness, the reactive consciousness, ceases when external stimulation ceases, is certain. We see it in sleep. But the mind does not cease to exist in sleep; and many persons perform their most difficult operations of thought best in their sleep. They wish to "sleep upon" a difficult question. There is no more reason to suppose that death at once causes the annihilation of mind. Whatever may happen later, at first it can be nothing but a sleep. To be awakened, the soul must in some way be acted upon. But there our information ceases.

(Unidentified fragment)

Now when I die, I want proper justice done to my memory as to these things. Not at all that they are any credit to me, but simply that, by being made to appear considerable, they may invite attention and study, when I think they will do considerable good. For logic and exact reasoning are a good deal more important than you are able to see that they are. So I hope that some account of my work may appear in some publication that people will look into, and not solely in the Biographical memoirs of the National Academy of Sciences.

(From a draft letter to William James,
ca. 25 December 1909, identified
by Max Fisch)

That a man like me cannot possibly in thirty years evolve any thought that is worth a sensible man's two dollars and a half to read will, of course, be the reflection of the great majority.

Good bye great and dear majority. I waft you a farewell sigh, —one of the downward kind.

(Unidentified fragment)

CONTENTS

Introduction to Volume 4   v

1 Logic of History (691)   1

2 [Parts of Carnegie Application] (L 75)   13

3 Fermatian Inference and DeMorgan's Syllogism of
   Transposed Quantity
   A. The Conception of Infinity (819)   77
   B. Fermatian Inference (820)   79
   C. The Critic of Arguments (589)   82

4 Logico-Mathematical Glosses (812)   95

5 Qualitative Logic: Preface; The Modus Ponens;
   The Logical Algebra of Boole (736)   101

6 Meaning (Pragmatism) (622)   117

7 Abstracts of Eight Lectures [Topological Basis of
   Philosophy of Continuity] (942)   127

8 Lectures on Pragmatism, Lecture II (302, 303)   149

9 Types of Reasoning (441)   167

10 Reason's Conscience; A Practical Treatise on the Theory
   of Discovery Wherein Logic Is Conceived as Semeiotic
   (parts of six notebooks in 693)   185

11 An Appraisal of the Faculty of Reasoning (616, 617) . .   217

12 [Necessary Reasoning] (760)   225

13 Of the Place among the Sciences of Philosophy and
   of each Branch of it (from 328)   227

14 Καινὰ στοιχεῖα (517)   235

15 On Quantity with Special Reference to Collectional
   and Mathematical Infinity (15)   265

16   A. Sketch of Dichotomic Mathematics (4)   285
     B. On the Number of Dichotomous Divisions;
         a Problem in Permutations (74)   301

17   The Categories (717)   307

18   (PAP) (293)   313

19   Detached Ideas Continued and the Dispute Between
     Nominalists and Realists (439)   331

20   [The Problem of Map Coloring] (154)   347

21   How to Reason: A Critick of Arguments (397)   353

22   On Physical Geometry (257)   359

23   Methods of Reasoning (748)   363

24   Sketch of a New Philosophy (928)   375

25   [Conceptions of Modern Mathematics] (from 950)   . . .   381

Key to Greek Terms   387
Index of Names   389
Subject Index   391

1

LOGIC OF HISTORY (691)[1]

It was because those logicians who were mathematicians saw that the notion that mathematical reasoning was as rudimentary as that was quite at war with its producing such a world of novel theorems from a few relatively simple premisses, as for example it does in the theory of numbers, that they were led,—first Boole and DeMorgan, afterwards others of us,—to new studies of deductive logic, with the aid of algebras and graphs. The non-relative logic having soon been exhausted, we went into the study of the logic of relatives, first the dyadic, and subsequently I, almost alone, into polyadic relations. These studies threw a great deal of light upon logic; but still they did not really explain mathematical reasoning, until I opened up the subject of abstraction. It now appears that there are two kinds of deductive reasoning, which might, perhaps, be called *explicatory* and *ampliative*. However, the latter term might be misunderstood; for no mathematical reasoning is what would be commonly understood by *ampliative*, although much of it is not what is commonly understood as *explicative*. It is better to resort to new words to express new ideas. All readers of mathematics must have felt the great difference between *corollaries* and *major theorems*, although these words are not sharply distinguished. It is needless to say that the words come to us, not from Euclid, but from the editions of Euclid's elements. The great body of the propositions called corollaries (all but 27 in the whole 13 books) are due to commentators, and are of an obvious kind. Kant's characterization of all deductive reasoning is true of them: they are mere explications of what is implied in previous results. The same is true of a good many of Euclid's own theorems; probably the numerical majority of the whole 369 of them are of this character. But many of them are of a different nature. We may call the two kinds of deduction *corollarial* and *theorematic*. Let us examine a

---

[1] A note in Peirce's hand on a folder containing these pages of MS. 691 reads "these pages are to be used in the Chapter of the Logic treating of Deductive Reasoning. But the theory needs completion."

simple example of reasoning of each kind, in order to ascertain whether there is or is not any essential difference between the two procedures. Both shall relate to the doctrine of multitude. In order to illustrate corollarial inference, I will prove that $(x + y) + z = x + (y + z)$. For ordinal numbers nothing can be easier. We have merely to note that by saying that two numbers are equal we mean no more than that either is at least as great as the other and if A is at least as great as B, and B as C, then A is at least as great as C. Now any number is at least as great as itself, and if N is at least as great as M then GN any number next greater than N is at least as great as GM and as M. And whatever is true of *zero* and which if true of any number, N, is also true of GN the ordinal number next greater than N, is true of all numbers. Now then what do we mean by the sum of two numbers? The definition consists of three propositions, as follows:

$$0 + 0 = 0$$
$$GM + N = G(M + N)$$
$$M + GN = G(M + N).$$

We ask then whether

$$(x + 0) + z = x + (0 + z)$$

Now it is true when $x = 0$ that $x + 0 = x$ and if it is true that $x + 0 = x$ then it's true that $Gx + 0 = Gx$. Therefore it is true of all numbers that $x + 0 = x$ and in like manner that $0 + z = z$. Hence

$$(x + 0) + z = x + z$$
$$x + (0 + z) = x + z$$

and therefore $(x + 0) + z$ is at least as great as $x + z$ and $x + z$ as $x + (0 + z)$. So that $(x + 0) + z$ is at least as great as $x + (0 + z)$, and in like manner the converse is proved so that $(x + 0) + z = x + (0 + z)$. But now suppose it true that $(x + y) + z = x + (y + z)$, is it true that

$$(x + Gy) + z = x + (Gy + z)$$

since $x + Gy = G(x + y)$, $(x + Gy) + z = G(x + y) + z = G[(x + y) + z]$ and since $Gy + z = G(y + z)$, $x + (Gy + z) = x + G(y + z) = G[x + (y + z)] = G[(x + y) + z]$. Whence, we readily conclude that in all cases the proposition holds. But this does not prove that it holds for all multitudes since there are multitudes which are infinite and have no count. Nor are we supposed to know even that every multitude of which the count can terminate always terminates and with the same number. We consider multitudes, then, and ask what it means to say

$$(x + y) + z = x + (y + z)$$

when x, y, z, are multitudes. Now the sum of a multitude is a multitude;

and to say that the multitude, m is equal to the sum of the multitudes x and y, means two things; first that m is as small as $x + y$ and second that $x + y$ is as small as m. To say that m is as small as $x + y$ means that if the Ms (say M, M, M, etc.) the Xs, and the Ys are three collections whose multitudes are respectively m, x, y, then there are two relations α and β, such that every M is either in the relation α to some X to which no other M is in the relation α or is in the relation, β, to some Y to which no other M is in the relation β. In order to show that that is what we mean, let us suppose all the crosses in the diagram are Xs and all the circles Y. Where a circle surrounds a cross, the object is supposed to be at once X and Y.

Fig. 1

Here are four Xs [in Fig. 1], four Ys, and seven Ms. There are two relations α and β conforming to the conditions; for the relation which consists of being possibly connected with a full line of the system of full lines I have drawn may be taken as the relation α, and the relation of being possibly connected with a dotted line of the system of dotted lines that I have drawn may be taken as the relation β. We see that every M is either connected by a full line to some X with which no other M is connected by a full line, or is connected by a dotted line to some Y to which no other M is connected by a dotted line. If we were to add one more M, a slight change in the way of drawing the lines would show that relations fulfilling the conditions exist. But if there were nine Ms there would be no such relations, since in that case even if every one of the Xs were joined by a full line, each to a separate M, and the four Ys by dotted lines to four several and separate Ms there would still remain an M unconnected, and it could not be connected by a full line to any X to which some other M was not already connected nor by a dotted line to any Y to which some other M was not connected. Next, by saying that $x + y$ is as small as m, we mean that there are two relations γ and δ such that every X is in the relation γ to some M to which no other X is in that relation and to which no Y is in the relation δ, while every Y is in the relation δ to some M to which no other Y is in that relation nor any X in the relation γ [Fig. 2].

Fig. 2

Now, then, having these definitions, we proceed to apply them.

Let the *L*s be a collection of multitude $(x + y) + z$
Let the *M*s be a collection of multitude $(x + y)$
Let the *N*s be a collection of multitude $y + z$
Let the *P*s be a collection of multitude $x + (y + z)$

Then the *L*s are as few as the sum of the *M*s and the *Z*s. The *M*s are as few as the sum of the *X*s and *Y*s. The *N*s are as many as the sum of the *Y*s and *Z*s. The *P*s are as many as the sum of the *X*s and *N*s. That is to say there are eight relations, ϵ, ζ, η, θ, ι, κ, λ, μ such that every *L* is either ϵ to an *M* to which no other *L* is ϵ (I thus abbreviate rather than to say "every *L* stands in the relation ϵ to an *M* to which no other *L* stands in that relation") or is ζ to a *Z* to which no other *L* is ζ. Every *M* is either η to an *X* to which no other *M* is η or is θ to a *Y* to which no other *M* is θ. Every *Y* is ι to some *N* to which no other *Y* is ι and to which no *Z* is κ, while every *Z* is κ to an *N* to which no other *Z* is κ and to which no *Y* is ι. Every *X* is λ to some *P* to which no other *X* is λ and to which no *N* is μ, while every *N* is μ to some *P* to which no other *N* is μ and to which no *X* is λ. Let us call the relation of being an *X* that is λ to a *P* to which no other *X* is λ and to which no *N* is μ the relation of being λ' to that *P*. Thus, nothing but an *X* is λ' to anything and no two *X*s are λ' to the same *P*. Let us call the relation of being an *N* that is μ to a *P* to which no other *N* is μ and to which no *X* is λ the relation of being μ' to that *P*. Then nothing but an *N* is μ' to anything nor are any two *N*s μ' to the same *P*; nor is anything or any two things at once λ' and μ' to the same *P*. Let us call the relation of being a *Y* that is ι to an *N* to which no other *Y* is ι and to which no *Z* is κ the relation of being ι' to that *N*. Let us call the relation of being a *Z* which is κ to an *N* to which no other *Z* is κ and to which no *Y* is ι, the relation of being κ' to that *N*. Then only *Y*s are ι' to anything and no two *Y*s to the same *N*; and only *Z*s are κ' to anything and no two to the same *N*; nor is anything or any two things at once ι' and κ' to the same *N*. Let us call the relation of being an *M* that is η to an *X* to which no other *M* is η, the relation of being η' to that *X*. So no two *M*s are η' to the same *X*. Let us call the relation of being an *M* that is θ to a *Y* to which no other *M* is θ the relation of being θ' to that *Y*. So no two *M*s are θ' to the same *Y*. Let us call the relation of being an *L* that is ϵ to an *M* to which no other *L* is in that relation, the relation of being ϵ' to that *M*; so that no two *L*s are ϵ' to the same *M*. Let us call the relation of being an *L* that is ζ to a *Z* to which no other *L* is ζ the relation of being ζ' to that *Z*; so that no two things are ζ' to the same *Z*. No *P*, then, has two different things λ' to it; and whatever is λ' to it is an *X*; and no *X* has two different things η' to it; and whatever is η' to it is an *M*; and no

M has two things ϵ' to it; but whatever is ϵ' to it is an L. Hence, no P has two things ϵ' to an η' to a λ' to it; and whatever is so is an L. In like manner no P has two things that are ϵ' to a θ' to an ι' to a μ' to it; and whatever is so is an L. And no P has two things that are ζ' to a κ' to a μ' to it; and whatever is so is an L. Moreover, no P has one thing λ' to it and another thing or the same thing μ' to it. Hence a P which has an L which is ϵ' to an η' to a λ' to it neither has an L ϵ' to a θ' to an ι' to a μ' to it nor has it an L that is ζ' to a κ' to a μ' to it. In like manner, no P has anything μ' to it but a single N and no N has at once anything ι' to it and κ' to it. Hence no P has two different things either ϵ' to a θ' to an ι' to a μ' to it or ζ' to a κ' to a μ' to it. Thus no P has two Ls in the relation of being either ϵ' to an η' to a λ' to it or ϵ' to a θ' to an ι' to a μ' to it or ζ' to a κ' to a μ' to it. But every L is in this compound relation to some P. Hence, the Ls are at least as few as the Ps; and similar argument would show the Ps to be at least as few as the Ls. Q.E.D.

This is perfectly straightforward although a little complicated. I may add that by defining the product of the multitudes of the Xs and Ys as the multitudes of possible pairs of an X with a Y, and by defining the multitude of the involution of x to the y power as the multitude of ways in which every Y can be joined to an X, all the elementary rules for those operations could be deduced by similar corollarial inference.

I will now give an example of *theorematic* inference by proving that every multitude is less than a multitude. That is given any multitude, as that of the Xs, there is a collection, which we may call the Ξs such that there is a relation, ρ, such that every X is ρ to a Ξ to which no other X is ρ; but any relation, σ, whatsoever is either such that some Ξ is not σ to any X or else is such that two different Ξs are σ to the same X. I do not think it is worth while to prove that this proposition could not be proved by straightforward application of the definition of fewer. On the contrary it seems, at first sight, to be manifestly false; for if the Xs are all self-identical objects, the collection of them must be at least as multitudinous as any (it would seem), since the relation of identity is a relation which every member of any other collection bears to some member of this collection. Nevertheless, I argue as follows; and observe that the question is not whether my argument is sound but whether it is a species of deductive argument different from the corollarial kind. I say then that every describable and distinguishable subject, we will say, for a wish or a fancy, is a collection; and if any thing or things be added to or taken from the collection it becomes a different collection. If, therefore, the Xs form a collection, the collection of all possible Xs in regard to which a wish is conceivable, as for example that they and they only were in Jericho, is a collection. And so, in particular nothing, or no

Xs is to be reckoned among the possible collections of Xs. If, therefore, the collection of Xs were such that there was none of greater multitude, either there would be no relation in which every X would stand to a collection of Xs to which no other X stood in the same relation, or else there would be a relation in which every collection of Xs would stand to some X to which no other collection of Xs should stand in the same relation. But the former alternative is readily disproved since there are various relations in each of which every X stands to a collection of Xs to which no other X stands in the same relation. For example, one such relation is being the only X excluded from the collection. There is one collection and but one collection to which each X is in this relation. We are thus driven to the second alternative if there is no collection greater than the whole collection of the Xs. Namely, we are obliged to suppose, if this true at all, that there is some relation, call it ρ, in which each collection of Xs stands to some X to which no other collection of Xs stands in the same relation. But now, I care not what relation ρ may be, I will describe a collection of Xs which is not the sole collection of Xs that is ρ to any X. Namely, this collection shall include every X (if there be any) of which a collection of Xs not containing it is the sole ρ, and it shall exclude every X (if there be any) of which a collection containing it is the sole ρ. As for other Xs, if there be any, I care not whether they are included or not. I say that this collection is not the sole ρ of any X; for if it were, it must be either of an X it includes, or of an X it excludes. But it is not the sole ρ of any X that it includes; for if so it would include an X of which a collection containing it is its sole ρ, contrary to the hypothesis. Nor can it be the sole ρ of any X that it excludes; for if so it would exclude an X of which a collection not containing it was the sole ρ, again contrary to the hypothesis. Hence this collection is not the sole ρ to any X; that is, there [is] no relation, ρ, such that every possible collection of Xs is sole ρ to some X; the Xs are fewer than the possible collections of Xs; and every collection is smaller than some other. This proof might be attacked on the ground that ρ is the relation of identity, and that therefore the collection which is not in the relation ρ to any X excludes every X and is nothing, which does not seem to be a collection. But the answer to that is that *nothing* must be considered as a collection. For a collection is whatever may form the exclusive subject of a wish or fancy. Now we may wish that no X shall go to Jericho. I might have argued that there is no contradition in supposing that each of the Xs wishes every X to go to heaven or to hell, and all possible wishes of that kind form a collection, among them being the wish that all Xs go to heaven and the wish that no Xs go to heaven. But if each X has but one wish, that possible wish which should consist in wishing that every one of those wishes should be disappointed insofar as it refers to the wisher's self is a wish that cannot be included among the wishes. In point of fact, this is the form in which the argument first suggested itself to me. The theorematic proof is, therefore, only strengthened by the study of the objection. On the other hand, the counter argument that every member of every collection is a member of the entire collection of all possible self-identical things is fallacious. The fallacy is this. Identity is an ordinary relation in this respect that it only exists when its cor-relates exist, and can only be supposed to exist when its correlates are supposed to exist. Now all possible things cannot be supposed to exist; for no matter what exists, other possibilities can be specified. It is, therefore, absurd to speak of all possible self-identical things. One cannot do so without falling into absurdities innumerable. It is in this way that Dr. Georg Cantor has been led into the absurdity of supposing the terminal point of a line to be removed, and the line to be left without any terminal point. An oval line has no points upon it: it has only *room* for points; that is, possibility of points.. It is therefore absurd to speak of the *collection* of points upon a line; because a collection consists of units, which a possibility being essentially general in its nature, has no self-identity, or, as we say in the terminology of logic, no numerical identity (ἔν ἀριθμῷ). When the oval is cut at any point, this operation creates two terminal points. Those two terminals were before the cutting identical, in the sense that they would have been so, if the point of cutting had been marked in advance. For this marking would have made a possible point actual and so would have conferred self-identity, as well as other relations, upon it. Now if, with Cantor, we suppose the terminal point flies away, a similar thing happens. That is to say, what has been one point becomes two points; for the terminal place remains, and the fact that this is terminal and so different from all other places near it, gives it existence and identity. I think that these considerations ought to make the absurdity of talking of the collection of all possible self-identical things clear. Any collection of self-identical things, being a collection in which each one is absolutely, and not merely compara-tively, distinct from every other, always leaves room, that is to say, possibility, for accretions to it. The aggregate of all that is possible is like a line, which is the aggregate of all possible points which a moving point might occupy in the course of a lapse of time. It can only be an aggregate possibility, divisible indefinitely into more particular possibil-ities, but as long as it has not actually been divided an aggregate of possibilities which have a mere comparative and not any absolute distinction from one another, and therefore have no numerical self-identity, and not constituting what we mean by a collection.

Having now placed before ourselves a definite example of theorematic reasoning, let us ask ourselves whether its procedure differs in any essential respect from that of corollarial reasoning, which consists merely in carefully taking account of the definitions of the terms occurring in the thesis to be proved. It is plain enough that this theorematic proof we have considered differs from a corollarial proof from a methodeutic point of view, inasmuch as it requires the invention of an idea not at all forced upon us by the terms of the thesis. We remark the same thing in regard to many of Euclid's theorems. If he had not happened to think of drawing certain subsidiary lines in his diagram, he could not have constructed his proof. He derives the right to draw these lines from his postulates. A postulate, indeed, in the sense in which the word is used by the Germans is an indemonstrable practical proposition; which amounts to saying that it is the affirmation of a possibility; while an axiom or definition is the denial of a possibility. The very few explicit references which Euclid seems to make to postulates occur in problems. Yet Euclid's list of postulates and his general procedure, together with what Aristotle (who was no mathematician) says, leads us to think that the idea in Euclid's mind, however confused, was that a postulate was a geometrical proposition of so fundamental a kind that it must be assumed, and not deduced; so that it would be a clause in the definition of space; so that postulates would be of two kinds, statements of the possibilities for which space gives room, such as Euclid's first three, and statements of what space cuts off from the realm of possibility, such as the fifth. The fourth appears to be a clause of the definition of geometrical equality. It appears in advance as if we might draw up, in any deductive study, a regular definition of what we would consider possible as well as of what we would consider impossible, and that thus we might reduce *theorematic* proofs to *corollarial* proofs. I suppose it has been true of all students of the logic of relatives, as I know it has of me, that all our labors on that subject have been largely motivated by a desire to clear up the paradox, that on the one hand it seems quite impossible to admit that all the theorems of higher arithmetic are mere explications of what was virtually stated in the few and simple premisses, while on the other hand, one cannot see how a first premiss of mathematics can have any other origin than as the definition of a term or the definition of what shall be regarded as involved in the idea of the general subject of discussion. It thus seemed as if there could be but one real set of circumstances from which the necessity of any theorem flows; and that there ought to be a direct corollarial manner of deducing it from those circumstances. In order to test this idea, let us define a collection, and see what help that definition will bring to the theorematic proof just considered. A collection is not precisely of all things the very easiest in the world to define. We may say that a collection is an object distinguished from everything which is not a collection by the circumstance that its existence, if it did exist, would consist in the existence of certain other individual objects, called its members, in the existence of these, and not in that of any others; and which is distinguished from every other collection by some individual being a member of the one and not a member of the other; and furthermore every fact concerning a collection will consist in a fact concerning whatever members it may have.

This definition exhibits certain remarkable peculiarities which I will not stop to discuss. It turns out to be of very little assistance, if any, toward proving that there is no maximum multitude. So that if that proof is to be made corollarial, it can only be by a deeper study of possibility. I have not neglected this study; but I have not as yet reached sufficiently matured conclusions; and I will drop this side of the question for the present. Meantime, I will note some results of the study of the logic of relatives which are novel and put a somewhat different face upon the matter; and after having mentioned them, I will call attention to another characteristic of the greater steps in mathematical reasoning.

Ordinary syllogistic, as it has generally been expounded since Kant, maintains that two premisses are necessary in every reasoning, except in certain trivial inferences to which Kant refuses the name of *Schluss*. This dogma, taken strictly, is false. But it is true that there is an inferential step which enters into almost all reasoning which involves two premisses or two parts of the same premiss. Take the syllogism

> All who are wise are happy,
> Some poor men are wise;
> ∴ Some poor men are happy.

One step of this, is of this form:

> All who are wise are happy,
> Some poor men are students
> ∴ Some poor men are students who unless they are unwise
> are happy.

This transformation, however, is precisely of the same form as the following:

> Some woman is adored by all catholics,
> Hence, every catholic adores a woman.

One proceeds from everything is either happy or unwise and something is a poor student to something poor is either happy or is an unwise student. The other proceeds from everything is either a noncatholic or adores something and that is a woman, to everything is either non-catholic or adores something and that is a woman. Although this is a small remark, it inflicts a pretty serious wound upon the ordinary logic.

In the next place, ordinary syllogistic gives us to understand that when a premiss has once been used to draw a conclusion, it is impossible to apply it again to draw an ulterior conclusion. If that were true, the whole number of conclusions which could be drawn from the few premisses of the theory of numbers would be very moderate, and would be capable of exact calculation. It is, of course, not true. The same premiss can be used over and over again; and considering only corollarial conclusions, we find that, up to a certain point, the importance and interest of the conclusions increases the oftener the same premiss is applied; but, after a certain time, no further corollarial inferences are of any particular significance. This is very interesting as tending to show that if theorematic inference is, at bottom, of the same nature as corollarial, then such a subject as the theory of numbers must ultimately become substantially exhausted, no further consequences of great interest remaining to be drawn.

In the third place, ordinary syllogistic leads us to talk of *the* conclusion from given premisses as if there could be but one, and it represents this one as drawn according to rigid rules; so that machines have been constructed which would not only perform all the steps of the traditional logic, but a good many others beside. But the study of logic of relatives has shown conclusively that it is not true that only one corollarial conclusion can be drawn from given premisses, even when the number of times that each is to be used is fixed. On the contrary, there is very frequently considerable range of choice, which must be determined by the consideration of the purpose that the reasoner has in view, and which puts a machine at defiance. And furthermore, it is shown that even in corollarial reasoning, no matter how simple, observation is called for, and a machine cannot perform even the simplest inference until the premisses have been prepared for it by the exercise of this observation, leaving relatively little for the machine to do. I may add, though it is not the logic of relatives which proves this, that even in the performance of the mechanical part of the work, it is a matter of difficulty to make a machine which will commit as few blunders as a well-trained man. That is my experience with various forms of calculating machines. There is, no doubt, a kind of extreme regularity about machine-work, dreary and purposeless. But when it comes to making one machine perform processes very varied at different times, requiring frequent rapid changes of adjustment, the accuracy of the machine thus mixed with human interference is in my experience [more] theoretical than practical. Considering that the brain is a million times as complicated as any machine and has no positive movements at all in it, the infrequency with which it goes wrong is the most amazing thing about it.

I now come to a point more germane to the question in hand. The logics of today mostly confound abstraction with generalization. Not that they draw no distinction between them; but they consider them as closely connected, and abstraction is in their eyes mostly the hand-maid of generalization. It is chiefly a detail of economy. Nothing to the ordinary notion of logic is more ridiculous than attaching any serious value to abstraction. "Why does opium always put people to sleep? Because it has a soporific virtue." This question and answer were invented in order to put abstraction in a laughable light; and therefore I cannot be expected to find much that is serious here. Yet even here, the answer does say something pertinent. It asserts there is some explan-ation behind the fact. Theorematic reasoning, at least the most efficient of it, works by abstraction; and derives its power from abstraction. I proved that there is no maximum multitude by considering the collec-tion of all possible collections of the members of a collection. Now a *collection* is an abstraction. For what is an abstraction but an object whose being consists in facts about other things? The soporific virtue of opium is an object whose being consists in the fact not merely that opium has, in sufficient doses, always put people to sleep, but that it always will, and always would unless some other influence counteracted it. *Number* is another product of abstraction; and how useful number is in demonstration I need not say. That which abstraction does is to take a circumstance and regard it as a subject acting, suffering, and being. When the mathematician regards an operation as itself a subject of operations, he is using abstraction in one of its most abstract forms. We all know what a valuable weapon the calculus of operations has proved to be. A particle is somewhere quite definitely. It is by abstraction that the mathematician conceives the particle as occupying a *point*. The mere place is now made a subject of thought. The particle moves; and it is by abstraction that the geometer conceives it as describing a line. This *line* which, like the soporific virtue of opium, is merely a fact turned into a substantive, is regarded as so substantial that we talk of the line as moving, and as generating a *surface*, which is a new abstraction; and even the surface is made to move. Now you will need no argument to persuade you that geometry which should translate everything about points, lines, and surfaces, into terms of particles and their motions, especially a geometry of three dimensions, would be hard to read. But that does not begin to state the truth, which is that the reasoning of many,

if not most, of the theorems, would completely disappear, and could not
be performed at all.

PARTS OF CARNEGIE APPLICATION (L 75)

Application
of C. S. Peirce   Milford, Pa., 1902 July 15.

To the Executive Committee of the Carnegie Institution,

Gentlemen:

 I have the honor respectfully to submit to you herein an application for aid from the Carnegie Institution in accomplishing certain scientific work. The contents of this letter are as follows:

Page
[in Peirce's Holograph]

§1. Explanation of *what work* is proposed 2
Appendix containing a fuller statement 6
§2. Considerations as to its *Utility* 50
§3. Estimate of the *Labor* it will involve 60
§4. Estimate of *Other Expense* involved 62
§5. Statement as to the *Need* of aid from the Carnegie Institution 63
§6. Suggestion of a *Plan* by which aid might be extended . 64
§7. Estimate of the *Probability of Completion* of the work, etc. 66
§8. Remarks as to the *Probable Net Cost* to the Carnegie Institution, in money and in efficiency 71
§9. Statement of my apprehension of the *Basis* of my claim for aid 73

EXPLANATION OF WHAT THE PROPOSED WORK IS

Some personal narrative is here necessary. I imbibed from my boyhood the spirit of positive science, and especially of exact science; and early became intensely curious concerning the theory of the methods of science; so that, shortly after my graduation from college, in 1859, I determined to devote my life to that study; although indeed it was less a resolve than an overmastering passion, which I had been for some years unable to hold in check. It has never abated. In 1866, and more in 1867, I ventured upon my first original contributions to the science of logic, and have continued my studies of this science ever since, with rare interruptions of a few months only each. Owing to my treating logic as a science, like the physical sciences in which I had been trained, and making my studies special, minute, exact, and checked by experience, and owing to the fact that logic had seldom before been so studied, discoveries poured in upon me in such a flood as to be embarrassing. This has been one reason why I have hitherto published but a few fragments of outlying parts of my work, or slight sketches of more important parts. For logic differs from the natural sciences and, in some measure, even from mathematics, in being more essentially systematic. Consequently, if new discoveries were made in the course of writing a paper, they would be apt to call for a remodelling of it, a work for mature reconsideration. Still, as far as I remember, no definitive conclusion of importance to which I have ever been led has required retraction, such were the advantages of the scientific method of study. Modification in details, and changes (very sparse) of the relative importance of principles are the greatest alterations I have ever been led to make. Even those have been due, not to the fault of the scientific method, but chiefly to my adherence to early teachings. But what has, more than that cause, prevented my publishing has been, first, that my desire to teach has not been so strong as my desire to learn, and secondly, that far from there having been any demand for papers by me, I have always found no little difficulty in getting what I wrote printed; and when the favor was accorded, it was usually represented to me that funds were sacrificed in doing so. My first papers, which have since been pronounced good work, were sent to almost every logician in the world, accompanied in many cases with letters; but for ten years thereafter I never could learn that a single individual had looked into them. Since then, I have had little ardor about printing anything. Now, however, being upon the threshold of old age, I could not feel that I had done my best to do that which I was put into the world to do, if I did not spend all my available forces in putting upon record as many of my logical results as I could.

Therefore, what I hereby solicit the aid of the Carnegie Institution to enable me to do is *to draw up some three dozen memoirs, each complete in itself, yet the whole forming a unitary system of logic in all its parts*, which memoirs shall present in a form quite convincing to a candid mind the results to which I have found that the scientific method unequivocally leads, adding in each case, rational explanations of how opposing opinions have come about; the whole putting logic, as far as my studies of it have gone, upon the undeniable footing of a science. . . .

I here insert an *Appendix* to this section of my letter. I give it that title to signify that I do not ask that members of the Executive Committee should read more of it than they may think it concerns them to read. It is a list of the proposed memoirs, their titles, and in most cases, brief indications of the nature of their contents. The brevity of the explanations precludes any hint of the scientific procedure, except where it is of such a nature that some notion of it can readily be conveyed in general terms. Fuller sketches can be furnished, if desired. No important changes in this scheme will be made. Lines of division between memoirs may be slightly shifted; and some of the introductory memoirs may be transposed.

LIST OF PROPOSED MEMOIRS ON LOGIC

No. 1. *On the Classification of the Theoretic Sciences of Research.*

This will be a natural classification, not of possible sciences, but of sciences as they exist today; not of sciences in the sense of "systematized knowledge," but of branches of endeavor to ascertain truth. I shall not undertake to prove that there is no other natural classification of the sciences than that which I give; and this being merely an introductory memoir, cannot have the same convincing character as the others. Every unitary classification has a leading idea or purpose, and is a natural classification in so far as that same purpose is determinative in the production of the objects classified. The purpose of this classification is nearly the same as that of Comte, namely, so to arrange a catalogue of the sciences, as to exhibit the most important of the relations of logical dependence among them. In fact, my classification is simply an attempt to improve upon that of Comte; first, by looking less at what has been the course of scientific history, and more at what it would have been if the theoretically best methods had been pursued; secondly, by supplying the shocking omissions which Comte's rage against nonsense led him to commit; and thirdly, by carrying down the subdivision as far as my knowledge enables me to do. It was necessary for me to determine what I should call one science. For this purpose I have united under one science studies such as the same man, in the present state of science, might very well pursue. I have been guided in determining this by noting how scientists associate themselves into societies, and what contributions are commonly admitted into one journal; being on my guard against the survival of traditions from bygone states of science. A study to which men devote their lives, but not, in the present stage of development of science, so numerously as to justify exclusive societies and journals for it, I call a *variety* of science. That which forms the subject of the narrowest societies and journals, so that any student of any part of it ought to be pretty thoroughly informed about every part, I call a *species* of science. That branch, of which the student of any part is well qualified to take up any other part, except that he may not be sufficiently acquainted with the facts in detail, I call a *genus* of science. If the only new training necessary to pass from one part to another is a mere matter of skill, the general conceptions remaining the same, I call the department a *family* of science. If different sorts of conceptions are dealt with in the different families of a department, but the general type of inquiry is the same, I call it an *order* of science. If the types of inquiry of the different orders of a department are different, yet these orders are connected together so that students feel that they are studying the same great subject, I call the department a *class* of science. If there are different classes, so that different students seem to live in different worlds, but yet there is one general animating motive, I call the department a *branch* of science. Of course, there will be sub-branches, sub-classes, etc., down to sub-varieties; and even sometimes, sub-sub-divisions. To illustrate, I call Pure Science and Applied Science different branches; I call Mathematics and the Special Sciences, different classes, I say that general physics, biology, and geology belong to different orders of science. Astronomy and geognosy are different families. Thermotics and Electrics are different families. Optics and electrics are now different genera. Entomology and Ichthyology are different species of one genus. The study of Kant and the study of Spinoza are different varieties of one species.

Of course, the execution of this useful but ambitious design can, in the first instance, notwithstanding all the labor on my part that seemed economically recommended, be but a sketch. It will have fully attained all I hope for, if it is respectable enough to merit serious picking to pieces in its smaller and in its larger divisions. Indeed, I may say of all these memoirs that what I most desire is that their errors should be exposed, so long as they lead to further scientific study of the subjects to which they relate. The relation of this present memoir to those which follow it in the series is that it gives, from a general survey of science, an idea of the place of logic among the sciences. I will here set down the larger divisions of the scheme as well as I remember it (not having the notes in my possession). But it will be the *discussion* which will form the chief value of the memoir, not the scheme itself. Nearly a hundred schemes given hitherto will be criticized.

A. Theoretical Science.

   I. Science of Research.

     i.   Mathematics.

     ii. Philosophy, or Cenoscopy.
        1. Categorics.
        2. Normative Science.
          a. Esthetics.
          b. Ethics.
          c. Logic.
        3. Metaphysics.

     iii. Idioscopy, or Special Science.
        1. Psychognosy.
          a. Nomological, or General Psychology.
          b. Classificatory.
            · Linguistics.
            · Critics.
            · Ethnology.
          c. Descriptive.
            · Biography.
            · History.
            · Archeology.
        2. Physiognosy.
          a. Nomological, or General Physics.
            · Dynamics.
              · Of particles.
              · Of Aggregations.
            · Elaterics and Thermotics.
            · Optics and Electrics.
          b. Classificatory.
            · Crystallography.
            · Chemistry.
            · Biology.
          c. Descriptive.
            · Astronomy.
            · Geognosy.

   II. Science of Review, or Synthetic Philosophy (Humboldt's Cosmos; Comte's Phil. Positive).

B. Practical Science, or the Arts.

No. 2. *On the Simplest Mathematics.*[1]

This is that mathematics which distinguishes only two different values, and is of great importance for logic.

No. 3. *Analysis of the Conceptions of Mathematics.*

Such are Number, Multitude, Limit, Infinity, Infinitesimals, Continuity, Dimension, Imaginaries, Multiple Algebra, Measurement, etc. My former contributions, though very fragmentary, have attracted attention in Europe, although in respect to priority justice has not been done them. I bring the whole together into one system, defend the method of infinitesimals conclusively, and give many new truths established by a new and striking method.

No. 4. *Analysis of the Methods of Mathematical Demonstration.*

I shall be glad to place early in the series so unquestionable an illustration of the great value of minute analysis as this Memoir will afford. The subjects of Corollarial and Theorematic Reasoning, or the Method of Abstraction, of Substantive Possibility, and of the method of Topical Geometry, of which I have hitherto published mere hints, will here be fully elaborated.

No. 5. *On the Qualities of the Three Categories of Appearance.*

An analysis and description of three irreducibly different kinds of elements found in experience and even in the abstract world of pure mathematics. This memoir rests upon Observation of the experience of every day and hour; this observation being systematized by thought. It is proved, beyond doubt, that there are no more than the three categories. The list was first published by me in May, 1867; but has since been repeatedly subjected to the severest criticism I could bring to bear upon it, with the result of making it far more evidently correct. The categories were originally called Quality, Relation, and Representation. The question of names and other terminology for them still somewhat perplexes me. I am inclined to call them Flavor, Reaction, and Mediation.

No. 6. *On the Categories in their Reactional Aspects.*

No. 7. *Of the Categories in their Mediate Aspects.*

These two memoirs develop and render clear a considerable number

---

[1]   MS. 1 is so named (see Vol. 3, 17, a).

of conceptions of which I shall make constant use in the remaining memoirs, and which are of constant use in all parts of philosophy and even in mathematics.

No. 8. *Examinations of Historical Lists of Categories.*

My list differs from those of Aristotle, Kant, and Hegel, in that they never really went back to examining the Phenomenon to see what was to be observed there; and I do not except Hegel's *Phänomenologie* from this criticism. They simply took current conceptions and arranged them. Mine has been a more fundamental and more laborious undertaking; since I have worked up from the percepts to the highest notions. I examine those systems as well as some others.

No. 9. *On the Bearing of Esthetics and Ethics upon Logic.*

I begin by explaining the nature of the normative sciences. They have often been mistaken for Practical Sciences, or Arts. I show that they are at the opposite pole of the sphere of science, and are so closely allied to mathematics, that it would be a much smaller error to say that like Mathematics they were simply occupied in deducing the consequences of initial hypotheses. Their peculiar dualism, which appears in the distinctions of the Beautiful and the Ugly, Right and Wrong, Truth and Falsity, and which is one cause of their being mistaken for Arts, is really due to their being on the border between mathematics and positive science; and to this, together with their great abstractness, is due their applicability to so many subjects, which also helps to cause their being taken for arts. Having analyzed the nature of the precise problems of the three, and given some considerations generally overlooked, I show that Ethics depends essentially upon Esthetics, and Logic upon Ethics. The latter dependence I had shown less fully in 1869 (*Journal of Speculative Philosophy*, Vol. II, pp. 207 et seq.). But the methods of reasoning by which the truths of logic are established must be mathematical, such reasoning alone being evident independently of any logical doctrine.

No. 10. *On the Presuppositions of Logic.*

I here show that much that is generally set down as presupposed in logic is neither needed nor warranted. The true presuppositions of logic are merely *hopes*; and as such, when we consider their consequences collectively, we cannot condemn scepticism as to how far they may be borne out by facts. But when we come down to specific cases, these hopes are so completely justified that the smallest conflict with them suffices to condemn the doctrine that involves that conflict. This is one

of the places where logic comes in contact with ethics. I examine the
matter of these hopes, showing that they are, among other things which
I enumerate, that any given question is susceptible of a true answer,
and that this answer is discoverable, that *being* and *being represented*
are different, that there is a reality, and that the real world is governed
by ideas. Doubt and everyday belief are analyzed; and the difference
between the latter and scientific acceptance is shown. Other doctrines
are examined.

No. 11. *On the Logical Conception of Mind.*

This memoir is here placed, or perhaps better before No. 9, for the
sake of perspicuity of exposition. The matter of it will have to be somewhat transformed at a later stage. If the logician is to talk of the operations of the mind at all, as it is desirable that he should do, though it is
not scientifically indispensable, then he must mean by "mind" something quite different from the object of study of the psychologist; and
this logical conception of mind is developed in this memoir and
rendered clear.* (*My order of arrangement of the first eleven memoirs
is subject to reconsideration. The Categories are applicable to the
logical analysis of Mathematics. It is even a question whether this
fact does not derange my classification, although I have carefully
considered it, and have provisionally concluded that it does not. It
further seems to me better to let the Categories first emerge in the
mathematical memoirs before explicitly considering them. This is a
question of methodeutic, which is not so exact in its conclusions as is
critical logic. I think the arrangement I here propose is favorable to
the reception of the Categories. But if I were to decide to postpone the
mathematical memoirs until after the categories, they might better be
placed last among the first eleven memoirs. In that case, also, and
indeed in any case, it might do well to place the memoir on the logical
conception of mind before that upon Esthetics and Ethics. The present
arrangement has been pretty carefully considered; and the last trans-
position is the only one that I think there is much likelihood of my
deciding upon. After No. 12, the only changes possible in the list are
shifts of boundaries in order to equalize the lengths of memoirs.)

No. 12. *On the Definition of Logic.*

Logic will here be defined as *formal semiotic*. A definition of a sign
will be given which no more refers to human thought than does the
definition of a line as the place which a particle occupies, part by part,
during a lapse of time. Namely, a sign is something, A, which brings

something, *B*, its *interpretant* sign determined or created by it, into the same sort of correspondence with something, *C*, its *object*, as that in which itself stands to *C*. It is from this definition, together with a definition of "formal," that I deduce mathematically the principles of logic. I also make a historical review of all the definitions and conceptions of logic, and show, not merely that my definition is no novelty, but that my non-psychological conception of logic has *virtually* been quite generally held, though not generally recognized.

No. 13. *On the Division of Logic.*

By an application of Categoric, I show that the primary division of logic should be into Stechiology, Critic, and Methodeutic. There is a cross-division into the doctrines of Terms, Propositions, and Arguments, to which three kinds of signs, however, stechiology, critic, and methodeutic, are quite differently related. The various historical divisions of logic are considered.

No. 14. *On the Methods of Discovering and of Establishing the Truths of Logic.*

I shall here show that no less than thirteen different methods of establishing logical truth are in current use today and mostly without any principle of choice and in a deplorably uncritical manner. I shall show that the majority of these methods are quite inadmissible, and that of the remainder all but one should be restricted to one department of logic. The one universally valid method is that of mathematical demonstration; and this is the only one which is commonly avoided by logicians as fallacious. I shall show in the clearest manner that this notion is due to a confusion of thought, which I shall endeavor to trace through all its metamorphoses. I hope to give this its quietus.

The methods of *discovering* logical truth can naturally not be numerous when discovery is pretty nearly at a standstill. I explain my own method.

No. 15. *Of the Nature of Stechiologic.*

This will contain especially a discussion of *Erkenntnisslehre*, what it must be, if it is an indispensable preparatory doctrine to critical logic.

No. 16. *A General Outline of Stechiologic.*

No. 17. *On Terms.*

This memoir will be based on my paper of November 1867. Practice has shown that that paper needs extension in several directions. Besides,

account has to be taken of important classes of terms there barely
mentioned. The historical part, too, needs great amplification. My very
conception of what a term is has been much improved by studies subsequent to that paper, and altogether original. The study of "agglutina-
tive" languages has been an aid to me.

No. 18.   *On Propositions.*

The question of the nature of the Judgment is today more actively
debated than any other. It is here that the German logicians are best
worthy of attention; and I propose to take occasion to give here an
account of modern German logic. Although this seems rather the
subject for a book than for a single paper, yet I think by stretching this
memoir, I can bring into it all that it is necessary to say about these
treatises, which belong to near a dozen distinct schools.

I shall then show how my own theory follows from attention to the
three categories; and shall pass to an elaborate analysis, classification,
symbolization, and doctrine of the relations of propositions.

This will probably be the longest of all the memoirs, and will balance
No. 16, which will be short. I think I shall treat No. 16 as a supplement
to No. 15 and divide No. 21 into two parts to be handed in separately.

No. 19.   *On Arguments.*

I first examine the essential nature of an argument, showing that it
is a sign which separately signifies its interpretant. It will be scrutinized
under all aspects.

I shall then come to the important question of the classification of
arguments. My paper of April 1867 on this subject divides arguments
into Deductions, Inductions, Abductions (my present name, which will
be defended), and Mixed Arguments. I consider this to be the key of
logic. In the following month, May 1867, I correctly defined the three
kinds of simple argument in terms of the categories. But in my paper on
Probable Inference in the Johns Hopkins "Studies in Logic," owing to
the excessive weight I at that time placed on formalistic considerations, I
fell into the error of attaching a name, the synonym I then used for
Abduction, to a probable inference which I correctly described, forgetting that according to my own earlier and correct account of it, abduction is not of the number of probable inferences. It is singular that I
should have done that, when in the very same paper I mention the
existence of the mode of inference which is true abduction. Thus, the
only error that paper contains is the designation as Abduction of a mode
of induction somewhat resembling abduction, which may properly be
called abductive induction. It was this resemblance which deceived me,

and subsequently led me into a further error contrary to my own previous correct statement. Namely, continuing to confound Abduction and Abductive Induction, in subsequent reflections upon the rationale of Abduction, I was led to see that this rationale was not that which I had in my Johns Hopkins paper given of Induction; and in a statement I published in the *Monist*, I was led to give the correct rationale of Abduction as applying to Abductive Induction and so, in fact, to all Induction. All the difficulties with which I labored are now completely disposed of by recognizing that Abductive Induction is quite a different thing from Abduction. It is a very instructive illustration both of the dangers and of the strength of my heuretic method. Similar errors may remain in my system. I shall be very thankful to whoever can detect them. But if its errors are confined to that class, the general fabric of the doctrine is true. I at first saw that there must be three kinds of arguments severally related to the three categories; and I correctly described them. Subsequently, studying one of these kinds, I found that besides the typical form, there was another, distinguished from the typical form by being related to that category relation to which distinguishes abduction. I hastily identified it with abduction, not being clear-headed enough to see that, while related to that category, it is not related to it in the precise way in which one of the primary divisions of arguments ought to be, according to the theory of the categories. This is the form of error to which my method of discovery is peculiarly liable. One sees that a form has a relation to a certain category, and one is unable, for the time being, to attain sufficient clearness of thought to make quite sure that the relation is of the precise nature required. If only one point were obscure, it would soon be cleared up; but the difficulty is at first that one is sailing in a dense fog through an unknown sea without a single landmark. I can only say that if others, after me, can find some way of making as important discoveries in logic as I have done while falling into less error, nobody will be more intensely delighted than I shall be. My gratitude to the man who will show me where I am wrong in logic will have no bounds. Thus far, I have had to find out for myself as well as I could. Meantime, be it observed that the kind of error which I have been considering can never amount to anything worse than a faulty classification. All that I asserted about probable inference in my Johns Hopkins paper and in my *Monist* paper was perfectly true.

In this paper, besides very important improvements in the subdivision of the three kinds of simple arguments, with several hitherto unrecognized types, and far greater clearness of exposition, I shall have much that is new to say about mixed arguments, which present many points of importance and of interest that have never been remarked. I shall give a new classification of them based, not upon the nature of their elements, but upon their modes of combination. Besides setting forth my own doctrine of the stechiology of argument, I shall examine the most important of those which are opposed to it.

No. 20. *Of Critical Logic, in General.*

A thorough discussion of the nature, division, and method of critical logic.

No. 21. *Of First Premisses.*

My position on this subject comes under the general head of sensationalism; but I contend that criticism is inapplicable to what is not subject to control. Consequently, not sensation nor even percepts are first premisses, but only perceptual judgments. I subject what goes under the title of the test of inconceivability to an elaborate examination, bringing out various useful truths. I also examine the tests of universality and necessity, first adding certain other characters which as much prove a priority as do those. These tests have been taken in two senses, and there is a third more advantageous than either.

No. 22. *The Logic of Chance.*

I here discuss the origin and nature of probability, by my usual method; also the connection between objective probability and doubt; the nature of a "long run"; in what sense there can be any probability in the mathematical world; the application of probability to the theory of numbers. I show that it is not necessary that there should be any definite probability that a given generic event should have a given specific determination. It is easy to specify cases where there would be none. There appears to be no definite probability of a witness's telling the truth. I also show that it is quite a mistake to suppose that, for the purposes of the doctrine of chances, it suffices to suppose that the events in question are subject to unknown laws. On the contrary, the calculus of probability has no sense at all unless it in the long run secures the person who trusts to it. Now this it will do only *if there is no law*, known or unknown, of a certain description. The person who is to trust to the calculus ought to assure himself of this, especially when events are assumed to be independent. The doctrine of chances is easily seen to be applicable to the course of science. Its applicability to insurance companies and the like is not in any case to be assumed off-hand. When it comes to the case of individual interests, there are grave difficulties.

The rules of probability are stated in a new way, with the application to high numbers and method of least squares according to several different theories. Pearson's developments examined. Inverse probabilities are shown to be fallacious.

There are many matters here under dispute; more than I here set down. In all these cases, I take pains to state opposing arguments in all their force, and to refute them clearly.

This memoir is intended to form a *complete vade mecum* of the doctrine of chances, and to be plentifully supplied with references. It will be somewhat long, but I hope not of double length.

No. 23. *On the Validity of Induction.*

This restates the substance of the Johns Hopkins paper; relegating formalistic matters to separate sections, taking account of types of induction with which I was not acquainted twenty years ago, and rendering the whole more luminous. Other views will be considered more at large.

No. 24. *On the Justification of Abduction.*

The categories furnish the definition of abduction, from which follows its mode of justification, and from this again its rules. The various maxims which are found in different books are passed in review and, for the most part, are found to sin only in vagueness. One question not very commonly studied is what is the character of a phenomenon which makes it call for explanation. The theory of Dr. Carus that it is *irregularity*, and that of Mr. Venn that it is *isolation*, though the latter is defended with some power, are positively refuted. This refutation does not apply to the theory that the character sought is that of being *surprising*. This, however, is open to another kind of objection. The true doctrine is nearly this, however.

No. 25. *Of Mixed Arguments.*

This is a highly important memoir upon a subject of singular difficulty, although at first blush one would not anticipate any difficulty or interest in it.

No. 26. *On Fallacies.*

There would be no advantage in devoting a special memoir to a strictly scientific treatment of fallacies in general. It would be like a chapter in a treatise on trigonometry which should treat of possible errors in trigonometry. But since my purpose is that these memoirs should not only be scientific but that they should also be useful, I propose to devote this to Fallacies, because I think, though it is not an attractive subject for a logician, that I can make the discussion very useful. I shall not attempt a strict theoretical development, but shall discuss fallacies under five heads, according to their causes, showing under each head how they come about, how we can avoid them in original reasoning and in controversy, how [we can] detect them and reply to others who fall into them. The five heads are; 1st, slips; 2nd, misunderstandings; 3rd, fallacies due to bad logical notions; 4th, fallacies due to moral causes; and 5th, sophisms invented to test logical rules, etc. This will thus be of *an entirely exceptional character* among the memoirs, more so even than the first.

No. 27. *Of Methodeutic.*

The first business of this memoir is to show the precise nature of methodeutic; how it differs from critic; how, although it considers, not what is admissible, but what is advantageous, it is nevertheless a purely theoretical study, and not an art; how it is, from the most strictly theoretical point of view, an absolutely essential and distinct department of logical inquiry; and how upon the other hand, it is readily made useful to a researcher into any science, even mathematics itself. It strongly resembles the purely mathematical part of political economy, which is also a theoretical study of advantages. Of the different classes of arguments, abductions are the only ones in which after they have been admitted to be just, it still remains to inquire whether they are advantageous. But since the whole business of heuretic, so far as its theory goes, falls under methodeutic, there is no kind of argumentation that methodeutic can pass over without notice. Nor is methodeutic confined to the consideration of arguments. On the contrary, its special subjects have always been understood to be the definition and division of terms. The formation of systems of propositions, although it has been neglected, should also evidently be included in methodeutic. In its method, methodeutic is less strict than critic.

No. 28. *On the Economics of Research.*

In all economics the laws are ideal formulae from which there are large deviations, even statistically. In the economics of research the "laws" are mere general tendencies to which exceptions are frequent. The laws being so indefinite, at best, there is little advantage in very accurate definitions of such terms as "amount of knowledge." It is, however, possible to attach a definite conception to one increment of knowledge being greater than another. To work this out will be the first business of the memoir. I also establish a definite meaning for the amount of an increment in diffusion of knowledge. I then consider the relation of each of these to the expenditure of energy and value required to produce them in varying conditions of the advancement or diffusion of knowledge already attained. Comparing knowledge with a material commodity, we know that in the latter case a given small increment in the supply is very expensive, in most cases, when the supply is very small, that as the supply increases, it sinks to a minimum, from which it increases to a very large but finite value of the supply where no further increment would be possible at any finite cost. Putting instead of supply, the amount of knowledge attained, we find that there is a "law," or general tendency, subject to similar large irregularities as in the case of the supply of a material commodity, but here even greater. The final increase of cost of an increment with the increase of attainment already achieved is marked, on the whole, in almost all cases, while in many cases, at least, there is a point of attainment where the cost of an increment is at a minimum. The same general tendency appears in reference to the diffusion of knowledge; but there is this striking difference, that attainments in advance of sciences are very commonly actually on the upward slope where increments are costing more and more, while there are few branches of knowledge whose diffusion is already so great that a given increment of the diffusion will cost more and more, as the diffusion is increased.

I shall next pass to a study of the variation of the *utility* (meaning, generally, the *scientific* utility) of given small increments of scientific knowledge and of the diffusion of knowledge in varying states of attainment. This is to be compared with the variation of the total amount that will be paid for a commodity for a fixed small increment of the *demand*, or amount thrown upon the market to fetch what it will, with varying amounts of that demand. Here, the additional total amount that will be paid for the small increment of amount sold will correspond to the utility of the small fixed increase of scientific knowledge or of the diffusion of knowledge; while, the demand being equal to the supply, this demand, or total amount that is sold, will correspond as before to the amount of attainment in scientific knowledge or in the diffusion of knowledge. For a material commodity we know that if it is given away people will only carry home a finite amount. One would have to pay them to carry away more. On the other hand there is probably some maximum price for most things, above which none at all would be sold. It necessarily follows that beyond a certain amount thrown upon the market, a small increment in that amount would actually diminish the total receipts from the sale of it, while for any smaller amount the increment of receipts for a given small increment of amount sent to market would be less and less. With regard to the scientific utility of a small fixed advance of knowledge, the "law" is certainly very different from that. In the first place, there is no degree of knowledge of which a small increase would be worse than useless, and while the general tendency is that the utility of such fixed increase becomes less and less, yet the curve is rather saw shaped, since like Rayleigh's small addition to our knowledge of the density of nitrogen, now and then a small increment will be of great utility and will then immediately sink to its former level. The scientific advantage of the diffusion of knowledge is difficult to determine. It cannot be believed that any increment of diffusion is positively unfavorable to science. It is favorable in two ways; first, by preparing more men to be eminent researchers; and secondly, by increasing general wealth, and therefore the money bestowed on science. I am inclined to think that the general tendency is that a given increment of diffusion is less and less advantageous to science the greater the attained diffusion. But I am not confident that this is so, at any rate without very important deflexions. The general effect, however, is nearly the same for the advancement as for the diffusion of knowledge. Namely, beginning with dense ignorance, the first increments cost more than they come to. That is, knowledge is increased but scientific energy is spent and not at once recovered. But we very soon reach a state of knowledge which is profitable to science, that is, not only is knowledge increased, but the facility of increasing knowledge gives us a return of more available means for research than we had before the necessary scientific energy was spent. This increases to a maximum, diminishes, and finally, there is no further gain. Yet still, in the case of energy expended upon research, if it is persisted in, a fortunate discovery may result in a new means of research. I shall analyze as far as I can the relative advantages, *for pure science* exclusively, of expending energy (which is of such a kind as to be equally capable of being directed either way) to the direct advancement of knowledge and to the diffusion of knowledge. I find the latter so overwhelmingly more important (although all my personal sympathies are the other way) that it appears to me that, for the present, to give to research, in money, one or two per cent of what is spent upon education is enough. Research must contrive to do business at a profit; by which I mean that it must produce more effective scientific energy than it expends. No doubt, it already does so. But it would do well to become conscious of its economical position and contrive ways of living upon it.

Many years ago I published a little paper on the Economy of Research, in which I considered this problem. Somebody furnishes a fund to be expended upon research without restrictions. What sort of researches should it be expended upon? My answer, to which I still adhere, was

this. Researches for which men have been trained, instruments pro-
cured, and a plant established, should be continued while those conditions subsist. But the new money should mainly go to opening up new
fields; because new fields will probably be more profitable, and, at any
rate, will be profitable longer.

I shall remark in the course of the memoir that economical science
is particularly profitable to science; and that of all the branches of
economy, the economy of research is perhaps the most profitable: that
logical methodeutic and logic in general are specially valuable for
science, costing little beyond the energies of the researcher, and helping the economy of every other science. It was in the middle of the 13th
century that a man distinguished enough to become Pope opened his
work on logic with the words, "Dialectica est ars artium et scientia
scientiarum, ad omnium methodorum principia viam habens." This
memorable sentence, whose gothic ornamentation proves upon scrutiny
to involve no meaningless expression nor redundant clause, began a
work wherein the idea of this sentence was executed satisfactorily
enough for the dominant science of the middle ages. Jevons adopted
the sentence as the motto of his most scientific contribution to logic;
and it would express the purpose of my memoirs, which is, upon the
ground well prepared by Jevons and his teacher DeMorgan, and by the
other great English researchers, especially Boole, Whewell, Berkeley,
Glanvill, Ockham, and Duns Scotus, to lay a solid foundation upon
which may be erected a new logic fit for the life of twentieth century
science.

No. 29. *On the Course of Research.*

Comparing the two wings of the special sciences, i.e. Psychognosy
and Physiognosy, and taking the history of their development as a basis,
but correcting the history, as well as we can, in order to make it conform
to what good logic and good economy would have made it, we get the
idea of rational courses of development which those branches might
have followed. Between these two there is a striking parallel; so that
we can formulate a general rational course of inquiry. Now passing to
the study of the history of special sciences, also modified by the same
process, we find some traces of the same law; or to express it more
clearly, it is as if the special science showed us one part of the general
scheme under a microscope. By successively examining all the sciences
in this way (or all I am sufficiently able to comprehend), we can fill in
details, and make the general formula more definite. We find here a
succession of conceptions which we can generalize in some measure,
but which we find it difficult to generalize very much without losing their

peculiar "flavors." These I call the Categories of the course of research. They have not the fundamental character of the Categories of Appearance; but appear, nevertheless, to be of importance,

No. 30. *On Systems of Doctrine.*

Singularly enough, it seems to have been left to me to make a first attempt to formulate in detail what a system of doctrine ought to be. I follow the same general heuretic method as in the memoir, No. 29, taking some of the most perfect systems extant, and imagining how they might be more rational. In this way, I work out a series of conceptions which I term the Categories of Systems.

No. 31. *On Classification.*

I study classification, after some general considerations, by actually drawing up a number of classifications of the only sort of objects which we can sufficiently comprehend; that is to say, different classes of objects of human creation; such as, contrivances for keeping the skin warm, languages, words, alphabets, sciences, etc. From these, I endeavor to elicit a general series of Categories of Classification.

No. 32. *On Definition and the Clearness of Ideas.*

In January, 1878, I published a brief sketch of this subject wherein I enunciated a certain maxim of "Pragmatism," which has of late attracted some attention, as indeed, it had when it appeared in the *Journal Philosophique.* I still adhere to that doctrine; but it needs more accurate definition in order to meet certain objections and to avoid certain misapplications. Moreover, my paper of 1878 was imperfect in tacitly leaving it to appear that the maxim of pragmatism led to the last stage of clearness. I wish now to show that this is not to the case and to find a series of Categories of clearness.

No. 33. *On Objective Logic.*

The term "objective logic" is Hegel's; but since I reject Absolute Idealism as false,,"objective logic" necessarily means *more* for me than it did for him. Let me explain. In saying that *to be* and *to be represented* were the same, Hegel ignored the category of Reaction (that is, he imagined he reduced it to a mode of being represented) thus failing to do justice to *being*, and at the same time he was obliged to strain the nature of *thought*, and fail in justice to that side also. Having thus distorted both sides of the truth, it was a small thing for him to say that *Begriffe* were concrete and had their part in the activity of the world;

since that activity, for him, was merely represented activity. But when I, with my scientific appreciation of objectivity and the brute nature of reaction, maintain, nevertheless, that ideas really influence the physical world, and in doing so carry their logic with them, I give to objective logic a waking life which was absent from Hegel's dreamland. I undertake in this memoir to show that so far from its being a metaphorical expression to say that Truth and Right are the greatest powers in this world, its meaning is just as literal as it is to say that when I open the window in my study, I am really exercising an agency. For the mode of causation in the one case and in the other is precisely the same. In fact, there are two modes of causation, corresponding to Aristotle's efficient and final causation, which I analyze and make clear, showing that both must concur to produce any effect whatever. The mind is nothing but an organism of ideas; and to say that I can open my window is to say that an idea can be an agent in the production of a physical effect. This naturally looks toward a special metaphysics of the soul; but I pass this by, as not germane to my present subject, and go on to examine the logic of ideas in their physical agency. Herein I find the key to the different series of categories which the studies of memoirs Nos. 29, 30, 31, 32, developed.

The remaining three memoirs are of the nature of elucidations of sound methodeutic by applying it in practice to the solution of certain questions which, although they do not belong to logic, are of special interest in the discussion of logic.

No. 34. *On the Uniformity of Nature.*

The vagueness of the language with which men commonly talk of the uniformity of nature at once masks the diversity of a number of distinct questions which are wrapped up together in that phrase, and at the same time masks the great diversity of opinions that are very commonly held upon these questions. I have discussed these different questions in half-a-dozen different papers; but there is none of them upon whose statement of my argumentation cannot be much amplified and improved, and to which new historical matter cannot bring considerable light. Moreover, I wish to bring all the different questions to one focus, and consider them together. This, I am sure, will cause thinkers to be more favorable to the views which I have at different times defended. Among the questions is that of nominalism and realism, in connection with which I shall show that all modern philosophy, by an accident of history, has been blind to considerations of the greatest evidence and moment.

No. 35. *On Metaphysics.*

The great distinction between Aristotelian philosophy and all modern philosophy is that the former recognized a germinal mode of being inferior to existence, which hardly Schelling does; certainly no other modern philosopher. This question is considered in the light of the methodeutic developed in previous memoirs. The result is applied to all the questions of high metaphysics.

No. 36. *On the Reality and Nature of Time and Space.*

This applies my methodeutic to the discussion of a question which will have repeatedly emerged during the course of the memoirs. I may say briefly that I defend the well-known opinion of Newton. But other questions are considered. I do not think any theory satisfactory which does not offer some explanation (a mathematically exact and evident one) of why space should have three dimensions.

ESTIMATE OF THE UTILITY OF THE WORK

To my apprehension, any man over sixty years of age, who is endowed with reason, is a better judge of his own powers and of the utility of his performances than other people can be expected to be. Particularly is this true when the man has accumulated a large fund of unpublished results. Yet as soon as such a man assumes the attitude of seeking recognition for the utility of his work, suspicions as to the candour of his appreciations may be suggested by those who, for any reason, are unfavorable to the action he desires.

For that reason, I shall confine myself to asserting in a general way my profound conviction of the utility of publishing my results, as likely to influence some sciences, but still more as themselves stimulating a most important branch of science, that of logic, which is at present in a bad way. The latter kind of utility is not much diminished if I have fallen into some errors. Beyond averring that conviction, I do not offer myself as a witness to the utility of the work. I should, indeed, not have gone so far as I have done, were I not persuaded that the Executive Committee ought to require, as one of the first conditions of extending aid to any work, that the person who was to do it should be saturated with faith in its utility and value.

I will indicate certain lines of thought which, if pursued by the Executive Committee, may determine an opinion in regard to the utility of the work I propose. These lines of thought are two. The one bears upon the value of my researches considered as contributions to pure science;

the other relates to their probable influence, direct or indirect, upon the progress of other sciences. I will first venture upon a few suggestions along the latter line.

What would be the degree of utility of a really good and sound methodeutic, supposing that it existed, for the other sciences? I am not of opinion that a science of logic is altogether indispensable to any other science; because every man has his instinctive *logica utens*, which he gradually corrects under the influence of experience. Indeed, instinct, within its proper domain, is generally less liable to err and is capable of greater subtlety than is any human theory. Perhaps it may sound like a contradiction to talk of "instinctive logic." It may possibly be thought that instinct is precisely that which is not logic or reason. But think of a man whose business it is to lend out money. The accuracy of his cool reason is what he relies upon; and yet he is not guided by a theory of reasoning, but much rather upon an intense love of money which stimulates his faculties of reasoning. That is what I call his *logica utens*.

There are many fields in which few will maintain that any theoretical way of reaching conclusions can ever be so sure as the natural instinctive reasoning of an experienced man. Yet let instinct tread beyond its proper borders but by ever so little, and it becomes the most helpless thing in the world, a veritable fish out of water. Sciences do often go wrong: that cannot be denied. Their history contains many a record of wasted time and energy that a good methodeutic might have spared. Think of the Hegelian generation in Germany! Is reasoning the sole business whose method ought not to be scientifically and minutely analyzed? To me, it is strange to see a man like Poincare (whom I mention only as a most marked case among many) who, in his own science, would hold it downright madness to trust to anything but the minutest and most thorough study, nevertheless discussing questions of the logic of science in a style of thought that seems to imply a deliberate disapproval of minute analysis in that field, and a trust to a sort of "On to Richmond" cry,—I mean a cry that those who have not closely studied are better judges than those who have.

Many will say that all that may be true, but that, as a matter of fact, we are already in possession of a scientific system of logic, that of Mill. Now it is displeasing to me to be forced to decry Mill's Logic; because, looking at it in certain very broad outlines, I approve of it. The book has unquestionably done much good, especially in Germany, which needed it most. But I must declare that quite no deep student of logic entertains a very high respect for it. If, however, that book, though written by a literary and not a scientific man, by a mere advocate of a shallow metaphysics, has had so beneficial an influence as unquestionably it has, would it not seem to be desirable that the same subject should be pursued, I do not say by me, but by scientific students of it? Surely, enough has been done to make it manifest that there is such a thing as strictly scientific logic. For instance, the doctrine of chances is nothing else. The doctrine of chances has been called the logic of the exact sciences; and as far as it goes, so precisely it is. Its immense service to science will not be disputed by any astronomer, by any geodesist, nor probably by any physicist. Pearson and Galton have shown how useful it may be in biological and psychognostic researches. The utility of truly scientific logic, then, is indisputable. But that general logic is today in a bad way would seem to be sufficiently shown by the fact that it is pursued by thirteen different methods, and mostly by a confused jumble of those methods, of which I, a very fallible person of course but still a scientific man who has carefully weighed them, pronounce but one, — and that one, in bad odor, — to be alone of general validity. Is it, then, not desirable that an interest in pursuing logical inquiries in a true scientific spirit and by acknowledged scientific methods should be aroused? If it be so, is not the publication of my researches, even if they contain some errors, as likely to stimulate such studies as anything that could be suggested? Slight and fragmentary as my publications have been, dealing with the less important of my results, have they not in some appreciable degree stimulated the production of such work? I point to the third volume of Schröder's *Logik*. Look at it, or ask him, and I think you will say that I have exercised some stimulating agency. Everybody admires (nobody more than I) the beautiful presentation by Dedekind of the logic of number; and Dedekind, by the way, pronounces all pure mathematics to be a branch of logic. Read his *Was sind und was sollen die Zahlen*, and then read my paper on the *Logic of Number*, published six years earlier, and sent to Dedekind, and ask yourselves whether there is anything in the former of which there is not a plain indication in the latter. Let me not be misunderstood. I am simply arguing that my papers have stimulated the science of logic. I wish with all my heart that the Executive Committee could have in view some other student of logic of vastly greater powers than mine. But even if they had, considering how much energy has been spent in obtaining my results, would it not be a pity not to have them presented to the world?

It is my belief that science is approaching a critical point in which the influence of a truly scientific logic will be exceptionally desirable. Science, as the outlook seems to me, is coming to something not unlike the age of puberty. Its old and purely materialistic conceptions will no longer suffice; while yet the great danger involved in the admission of any others, ineluctable as such admission is, is manifest enough. The influence of the conceptions of methodeutic will at that moment be decisive.

Vast, however, as the utility of logic will be in that direction, provided that logic shall at the critical moment have developed into that true science which it is surely destined someday to become, yet the pure theoretical value of it is greater yet. No doubt, it is possible, while acknowledging, as one must, that logic produces useful truths, to take the ground that it is a composite of odds and ends, a crazy-quilt of shreds and patches, of no scientific value in itself. But seeing that pure mathematics is so close to logic, that eminent mathematicians class it as a branch of logic, it is hard to see how one can deny pure scientific worth to logic and yet accord such worth to pure mathematics. Probably there are naturalists of culture so narrow that they would deny absolute scientific value to pure mathematics. I do not believe the Committee will embrace such views. And then, there is Metaphysics to be considered. Everybody must have his *Weltanschauung*. It certainly influences science in no small measure. But metaphysics depends on logic, not merely as any science may occasionally need to appeal to a logical doctrine, but, according to the greatest metaphysicians, the very conceptions of metaphysics are borrowed from the analyses of logic. Now if there is any such thing as pure scientific value, as distinguished from the admiration one might have for a newly discovered dye, in what can it consist if not in intellectual relations between truths? If so it be, then, in view of the relation of logic to metaphysics, and that of metaphysics to all science, how can it be said that logic is devoid of scientific value, if there be such thing as scientific value? If logic is the science which my memoirs go to show that it is, it is the very keystone in the arch of scientific truth.

Little known as my papers have been, I believe that there are some men whose judgments must command respect in the world of science, who will testify to the utility of the work that I have done, and to the probable utility of that which I am about to do.

[EXCERPTS FROM EARLIER DRAFTS OF CARNEGIE APPLICATION]

Chapter 3 will examine the nature of mathematical reasoning. Logic can pass no judgment upon such reasoning, because it is evident, and as such, beyond all criticism. But logic is interested in studying how mathematical reasoning proceeds. Mathematical reasoning will be analyzed and important properties of it brought out which mathematicians themselves are not aware of.

But what would be the contents of my three ponderous volumes of logic? I answer, in the first place, in reference to the expectations which would be roused in uninstructed minds by the word "logic," that it would contain a theory of scientific reasoning and also a theory of the reasoning of practical men about every day affairs. These two would be shown to be governed by somewhat different principles, inasmuch as the practical reasoning is forced to reach some definite conclusion promptly, while science can wait a century or five centuries, if need be, before coming to any conclusion at all. Another cause which acts still more strongly to differentiate the methodeutic of theoretical and practical reasoning is that the latter can be regulated by instinct acting in its natural way, while theory of how one should reason depends upon one's ultimate purpose and is modified with every modification of ethics. Theory is thus at a special disadvantage here; but instinct within its proper domain is generally far keener, surer, and above all swifter, than any deduction from theory can be. Besides, logical instinct has, at all events, to be employed in applying the theory. On the other hand, the ultimate purpose of pure science, as such, is perfectly definite and simple; the theory of purely scientific reasoning can be worked out with mathematical certainty; and the application of the theory does not require the logical instinct to be strained beyond its natural function. On the other hand, if we attempt to apply natural logical instinct to purely scientific questions of any difficulty, it not only becomes uncertain, but if it is heeded, the voice of instinct itself is that objective considerations should be the decisive ones.

The methodeutic utility of logic is still further limited by the fact that the reasonings of pure mathematics are perfectly evident and have no need of any separate theory of logic to reinforce them. Mathematics is its own logic.

Furthermore, the three normative sciences, esthetics, ethics, and logic itself, although they do not come under that branch of science called practical, that is, the arts, are nevertheless so far practical that instinct in its natural operation, is perfectly adapted to their reasonings after the subtle analyses of which these sciences themselves take cognizance have prepared the premisses.

It follows that the only reasonings for which a science of logic is methodeutically useful are those of metaphysics, and the special theoretical sciences, of the physical and the psychical wing. Physical science has hitherto done well enough without any appeal to a science of logic. But at this moment questions of a logical nature have arisen which nothing but a scientific logic are likely to settle. Witness the controversy between those who are about Poincaré and those who are about Boltzmann. Witness the still more difficult question of the constitution of matter. To my prevision physics seems to be entering a period when such questions will be multiplied.

There are three different ways in which a method may be calculated to lead to the truth, these three senses constituting three great classes of reasonings. *Deduction* is reasoning which professes to pursue such a method that if the premisses are true the conclusion will in every case be true. Probable deduction is, strictly speaking, necessary; only, it is necessary reasoning concerning probabilities. *Induction* is reasoning which professes to pursue such a method that, being persisted in, each special application of it (when it is applicable) must at least indefinitely approximate to the truth about the subject in hand, in the long run. *Abduction* is reasoning, which professes to be such, that in case there is any ascertainable truth concerning the matter in hand, the general method of this reasoning though not necessarily each special application of it must eventually approximate to the truth.

Of these three classes of reasonings Abduction is the lowest. So long as it is sincere, and if it be not, it does not deserve to be called reasoning, Abduction cannot be absolutely bad. For sincere efforts to reach the truth, no matter in how wrong a way they may be commenced, cannot fail ultimately to attain any truth that is attainable. Consequently, there is only a relative preference between different abductions; and the ground of such preference must be economical. That is to say, the better abduction is the one which is likely to lead to the truth with the lesser expenditure of time, vitality, etc.

Deduction is only of value in tracing out the consequences of hypotheses, which it regards as pure, or unfounded, hypotheses. Deduction is divisible into sub-classes in various ways; of which the most important is into Corollarial and Theorematic. *Corollarial deduction* is where it is only necessary to imagine any case in which the premisses are true in order to perceive immediately that the conclusion holds in that case. All ordinary syllogisms and some deductions in the logic of relatives belong to this class. *Theorematic deduction* is deduction in which it is necessary to experiment in the imagination upon the image of the premiss in order from the result of such experiment to make corollarial deductions to the truth of the conclusion. The subdivisions of theorematic deduction are of very high theoretical importance. But I cannot go into them in this statement.

Induction is the highest and most typical form of reasoning. In my essay of 1883, I only recognized two closely allied logical forms of pure induction, one of which is undoubtedly the highest. I have since discovered eight other forms which include those almost exclusively used by reasoners who are not adepts in logic. In fact, Norman Lockyer is the only writer I have met with who in his best work, especially his last book, habitually restricts himself to the highest form. Some of his work, however, as for example, that on the orientation of temples, is logically poor.

Besides these three types of reasoning there is a fourth, Analogy, which combines the characters of the three, yet cannot be adequately represented as composite. There are also composite reasonings where an argument of one type is joined to an argument of another type. Such for example, is an induction fortified by the consideration of some known uniformity. Uniformities are of four principal kinds of which Mill distinctly recognizes only a single one.

There is one point which I have so far passed over without notice which is of great importance for the solidity of the foundation of my method of ascertaining whether a reasoning is good or bad. My position, in opposition to almost all the German logicians and those who blindly worship them in this country, is that in the science of the logic of science it will not do to rely upon our instinctive judgments of logicality merely; it is necessary to prove that, from the nature of things, the given method of reasoning will conduce to the truth in the sense in which it professes to do so. But here two questions arise: First, Have you not, after all, to rely upon the veracity of the logical instinct in judging of the validity of this proof? and Secondly, Where do you get the premisses from which this proof proceeds; and how do you know they are true? Have you not to rely upon instinct again, here? These questions are pressed by the German and other subjectivist logicians; and their pressing them with confidence in their unanswerable difficulty is a good example of a characteristic of those writers; namely that they look at everything through the spy-glass of logical forms and metaphysical theories, and often overlook plain facts before their faces. In order to answer those questions, it is necessary to recognize certain very plain and easy distinctions which the German logicians habitually overlook. In the first place, it is necessary to distinguish between a proposition and the assertion of it. To confound those two things is like confounding the writing of one's name idly upon a scrap of paper, perhaps for practice in chirography, with the attachment of one's signature to a binding legal deed. A proposition may be stated without being asserted. I may state it to myself and worry as to whether I shall embrace it or reject it, being dissatisfied with the idea of doing either. In that case, I doubt the proposition. I may state the proposition to you and endeavor to stimulate you to advise me whether to accept or reject it: in which I put it interrogatively. I may state it to myself; and be deliberately satisfied to base my action on it whenever occasion may arise: in which case I judge it. I may state it to you: and assume a responsibility for it: in which case I assert it. I may impose the responsibility of its agreeing with the truth upon you: in which case I command it. All of these are different moods in which that same proposition may be stated. The German word *Urtheil* confounds the proposition itself with the psychological act of assenting to it. This confusion is a part of the general refusal of idealism, which still considerably affects almost all German thought, to acknowledge that it is one thing to *be* and quite another to *be presented*. I use the word *belief* to express any kind of holding for true or acceptance of a representation. Belief, in this sense, is a composite thing. Its principal element is not an affair of consciousness at all; but is a habit established in the believer's nature, in consequence of which he would act, should occasion present itself, in certain ways. However, not every habit is a belief. A belief is a habit with which the believer is deliberately satisfied. This implies that he is aware of it, and being aware of it does not struggle against it. A third important characteristic of belief is that while other habits are contracted by repeatedly performing the act under the conditions, belief may be, and commonly if not invariably, is contracted, by merely imagining the situation and imagining what would be our experience and what our conduct in such a situation; and this mere imagination at once establishes such a habit that if the imagined case were realized we should really behave in that way. Take for example, the way in which ninety-nine ordinary men, not sharps, out of every hundred would form the belief that the sum of the angles of a triangle is two right angles. Any one of them would probably imagine himself to be in a field facing the north. He would imagine himself to march some distance in that direction, turn through an angle, march again, turn again so as to face his original position and then turn again so as to face as he did in the first place. Then he would say, I should, in effect, have turned through four right angles; for if I had stood at one point or hardly moved I should have had to make a complete rotation before the North star would again be in front of me; and therefore the sum of my three turnings would have been four right angles (there would be his fallacy). Therefore the sum of the exterior angles of a triangle is four right angles. But the sum of exterior and interior angle at each angle is two right angles; and since there are three angles, the sum of the sums of exterior and interior angles is six right angles. Subtracting from these six the four right angles equal to the sum of the exterior angles, I find two right angles left as the sum of the interior angles of a triangle. Thereupon, a habit would have been formed so that he would thereafter always act on the theory that the sum of the interior angles of a triangle is two right angles. This habit would have been the consequence of what he imagines would be forced upon his experience in that situation; this imagination being due to another habit which like every belief affects imagination, unless there is a special inhibition, just as it does real conduct. If anybody says that in this description of belief I make too much of conduct, I admit it frankly. It will not be so in the book itself; but in the present statement I do so in order to counteract the effect of the neglect of a certain point the statement of which would be too long. Such, then, roughly, is what a belief, or holding for true is. A *doubt* is of a very different nature. A belief is chiefly an affair of the soul, not of the consciousness; a doubt, on the contrary, is chiefly an affair of consciousness. It is an uneasy feeling, a special condition of irritation, in which the idea of two incompatible modes of conduct [is] before the doubter's imagination, and nothing determines him, indeed he feels himself forbidden, to adopt either and reject the other. Of course it is not necessary that the degrees of dissatisfaction with the opposite alternatives should be equal. Like irritations generally, doubt sets up a reaction which does not cease until the irritation is removed. If we accept this account of the matter, doubt is not the direct negation or contrary of belief; for the two mainly affect different parts of the man. Speaking physiologically, belief is a state of the connections between different parts of the brain, doubt an excitation of brain-cells. Doubt acts quite promptly to destroy belief. Its first effect is to destroy the state of satisfaction. Yet the belief-habit may still subsist. But imagination so readily affects this habit, that the former believer will soon begin to act in a half-hearted manner and before long the habit will be destroyed. The most important character of doubt is that no sooner does a believer learn that another man equally well-informed and equally competent doubts what he has believed, than he begins by doubting it himself. Probably the first symptom of this state of irritation will be anger at the other man. Such anger is a virtual acknowledgment of one's own doubt. Such doubt, at first of a purely external nature, that is to say, not a genuine doubt, or feeling of uneasiness but a sense that it is possible we may come to doubt it, sets up as reaction an effort to enter into the doubt and to comprehend it. Indeed, it is not necessary that one should actually meet with a man who doubts; for such is the influence of imagination in such matters that as soon as a believer can imagine that a man, equally well-informed and equally competent with himself, should doubt, doubt actually begins to set in, in his own state of feeling. From this follows the important corollary that if a man does not himself really doubt a given proposition he cannot imagine how it can be doubted, and therefore cannot produce any argument tending to allay such doubt. It thus appears that it is one thing to *question* a proposition and quite another to *doubt* it. We can throw any proposition into the interrogative mood at will; but we can no more call up doubt than we can call up the feeling of hunger at will. What one does not doubt one cannot doubt, and it is only accidentally that attention can be drawn to it in a manner which suggests the idea that there might be a doubt. Thence comes a critical attitude, and finally, perhaps, a genuine doubt may arise. It is this critical attitude that must next be examined. I regret very much the necessity of entering into such details; but the two questions I am preparing to answer are of such fundamental importance in regard to the value of the methodeutical part of my book, that the briefest account of what is to characterize must necessarily dwell upon these matters. The word *criticism* carries a meaning in philosophy which has so little resemblance to the criticism of literature, that the latter meaning throws no light on the former. Philosophical criticism is applied to an idea which we have already adopted, but which we remark that we have not deliberately adopted. The mere fact that it has been adopted, as if hastily, that is, without deliberation though it does not necessarily create a doubt, suggests the idea that perhaps a doubt might arise. The critical attitude consists in reviewing the matter to see in what manner corrections shall be made. This is what one does when one reads over a letter one has written to see whether some unintended meaning is suggested. The criticism is always of a process, the process which led to the acceptance of the idea. It supposes that this process is subject to the control of the will; for its whole purpose is correction, and one cannot correct what one cannot control. Reasoning, in the proper sense of the word, is always deliberate and, therefore, is always subject to control.

A third class of judgments not open to criticism are judgments concerning objects created by one's own imagination. Imagine, for example, an endless succession of objects. Then there will be there two distinct endless sequences; namely that of the objects in the oddly numbered places, and that of the objects in the evenly numbered places. That this is so is not to be discovered by merely analyzing what one had in mind. The judgment is the result of a physical process of experimentation, considerably like an induction. But it differs from any kind of reasoning in not being subject to control. It is true that after one has once lit up the idea that there are two endless series whose members so alternate, the analysis of *that* idea does show that it will be applicable to any endless series; and this analysis can be thrown into the form of a proof that it will be so. Yet this proof will rest on some proposition which is simply self evident. But as long as one only has the idea of the simple endless series, one may think forever, and not discover the theorem, until something suggests that *other* idea to the mind. What I call the *theorematic* reasoning of mathematics consists in so introducing a foreign idea, using it, and finally deducing a conclusion from which it is eliminated. Every such proof rests, however, upon judgments in which the foreign idea is first introduced, and which are simply self-evident. As such, they are exempt from criticism. Judgments of this kind are the very foundation of logic except insofar as it is an experiential science. If a proposition appears to us, after the most deliberate review, to be quite self-evident, and leave no room for doubt, it certainly cannot be rendered *more* evident; for its evidence is perfect already. Neither can it be rendered less evident, until some loophole for doubt is discovered. It is, therefore, exempt from all criticism. True, the whole thing may be a mistake. The sixteenth proposition of the first book of Euclid affords an example. The second postulate was that every terminated right line can be continuously prolonged. Καὶ πεπερασμένην εὐθεῖαν κατὰ τὸ συνεχὲς ἐπ' εὐθείας ἐκβαλεῖν. This is by no means saying that it can be prolonged to an indefinitely great length. He, however, virtually has proved (in prop. 2) that from the extremity of a straight line can be drawn continuously with that line a line of any given length. He imagines, then, a triangle ΑΒΓ. He prolongs the side ΒΓ a little beyond Γ to a point Δ; and he then

proposes to prove that $\angle A \Gamma \Delta > \angle \Gamma BA$. For that purpose, he bisects
$A\Gamma$ in E, draws BE and prolongs it, through E to Z, making EZ = BE.
He then joins Z to Γ by a straight line, and argues that $\angle E\Gamma\Delta >$
$\angle E\Gamma Z$ because the whole is greater than its part. He is thinking of
Fig. 1; but he has not proved that Z cannot fall as in Fig. 2; so that the

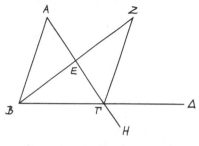

Fig. 1

whole demonstration falls to the ground. He ought to have appealed to
the third postulate to show that E would be the centre of a circle
passing through B and Z, and therefore by definition 15 *within* the circle
(why else than for such purposes should the centre's being within have
been so emphatically insisted on?), to prove that Fig. 2, was inadmis-

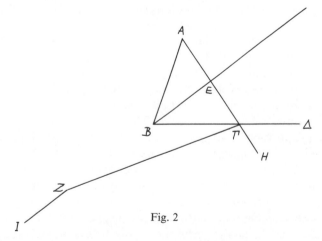

Fig. 2

sible. It is curious that there is not one down-right fallacy in the first
Book of the Elements (the only part of the work drawn up with supreme
circumspection) into which Euclid is not drawn by that axiom that the
whole is greater than its part; nor is this axiom ever appealed to without
resulting in a fallacy. We know now, as Euclid himself half-knew, that

the axiom is false. Yet it is not its falsity which causes Euclid's fallacies. It is always as here because it tempts him to draw a figure and judge by the looks of it what is part and what whole. Though this first book of Euclid has been for twenty centuries under a fire of criticism in comparison with which the strictures of professed logicians are blank cartridges printed by babes, yet to this day its real faults have escaped detection, as [have] its real merits, which are phenomenal. It simply shows how rare a thing correct reasoning is among men.

If Euclid had not been able to save his sixteenth proposition by means of the third postulate about the circle, he could not have saved it at all. For his first postulate is not that only one right line can join two points as its terminals, but merely that there is a right line from one point to the other. Ἡιτήσθω ἀπὸ παντὸς σημείον ἐπὶ πᾶν σημεῖον εὐθεῖαν γραμμὴν ἀγαγεῖν. He does not postulate that only one straight line can be drawn through two points, but only that all right angles are equal. That postulate would have enabled him to prove that only one unlimited straight line cannot be drawn between two points (a proposition he does not give because he deals only with what is limited), but not that there were not two limited straight lines having those points as terminals. Had he omitted from his definition of the circle the clause represented in our language by the single word "within" (τῶν ἐντὸς τοῦ σχήματος κειμένων) as some moderns would have had him do, he would have had no logical way of proving that the sum of the angles of a plane triangle do not exceed two right angles. Still, as long as he continued to overlook the possibility of Fig. 2, his proof would have appeared convincing, and there would have been no criticism to make upon it. In all these cases, of whatever class, it is only the act of judging that is exempt from criticism in the strict sense of an inquiry whether an operation has been performed rightly or wrongly. There is nothing to prevent the resulting proposition from being confronted with objections showing that there is something wrong somewhere. For example, though the act of judging that the sky looks blue is itself exempt from criticism, yet one can imagine a person to be so thoroughly persuaded of the falsity of Goethe's theory of colors that, not seeing any other way of accounting for the skies seeming blue that he might suspect that it does not seem blue. Again if a man analyzing his idea of matter deliberately judges that he means by "matter" something which in its nature is not a representation of anything, his judgment would, as an act, not be open to any criticism; but still, that would not prevent a Berkeley from raising the difficulty that since we can have no experience or imagination of anything but representations, there does not seem to be any possible way in which a man ever could attach such a meaning to a word consistently.

So again, certain saints have declared that they would go voluntarily and deliberately to Hell, if such were the will and good pleasure of the Lord; but a Hobbes would not be prevented from suspecting that they had deceived themselves, since Hell means a state of utter dissatisfaction, and it is absurd to say that a person could find any satisfaction in complete dissatisfaction. So in the present case, had Euclid omitted the word "within," or rather the corresponding Greek phrase, from his definition of the circle, it might have occurred to him that he was provided with no postulates about straight lines in a plane that were not equally true of great circles on a sphere, and therefore, since a spherical triangle may have the sum of its angles anything up to six right angles, or even ten, if you please, there must be something wrong with the proof that this is impossible in a plane.

This third class of judgments exempt from criticism coincides with that of evident judgments or judgments of evident propositions. For "evident" means manifest to any mind who clearly apprehends the proposition, no matter how lacking in experience he may be. The truth of a perceptual judgment, analysis of meaning, or declaration of intention, is manifest only to the one person whose experience it concerns. It is only when we judge concerning creatures of the imagination that all minds are on a par, however devoid of experience some of them may be.

When a mathematical demonstration is clearly apprehended, we are forced to admit the conclusion. It is evident; and we cannot think otherwise. It is, therefore, beyond all logical criticism; and the forms of syllogism cannot lend it any support. Pure mathematics, therefore, stands in no need of a science of logic. Methodeutically, mathematics is its own logic; and the notion that a calculus of logic can be of any help to mathematics, unless merely as another mathematical method supplying a speedier process of demonstration (which is just what a logical calculus rather opposes) is futile. Mathematics, however, is of great aid to logic. The reasoning of mathematics is also an instructive subject for logical analysis, teaching us many things about the nature of reasoning. But although a mathematical demonstration, once completely apprehended, is evident, indubitable, beyond control, and beyond criticism, yet the process of arriving at it is certainly a matter of skill and art, subject to criticism, and controlled by anticipated criticism. This control implies that different ways of proceeding are considered hesitatingly; and until the demonstration is found there is doubt of the conclusion. The theorem is not self-evident or it could not really be proved. But over what elements of the process is the control exercised? Over two: the invention of the proof, and the acceptance of the proof.

But the process of invention of the proof is not of the nature of that demonstrative reasoning which we call mathematical. There is nothing evident about it except that, as it turns out, it evidently answers the purpose. It is, in fact, a piece of probable reasoning in regard to which a good logical methodeutic may be a great aid. As to the acceptance of the proof, after it is framed, all the artifices which may be employed to assist it are of the nature of checks. That is to say, they are merely equivalent to a careful review of the proof itself in which some minor details may be varied in order to diminish the chances of error. In short, this is an operation by which the proof is brought fully and clearly before the mind. That the proof is absolute is evident and beyond criticism. The theorem which was not evident before the proof was apprehended, now becomes itself entirely evident, in view of the proof. Such reasoning forms the principal staple of logic. It is not itself amenable to logic for any justification; and although logic may aid in the discovery of the proof, yet its result is tested in another way. This disposes of the German objection that to use reasoning in order to determine what methods of probable reasoning will lead to the truth begs the whole question, so that the only way is to admit that the validity of reasoning consists in a feeling of reasonableness.

I now pass to a rough statement of my results in regard to the heuretic branch of mathematical thought. At the outset, I set up for myself a sort of landmark by which to discern whether I was making any real progress or not. Cayley had shown, while I was, as a boy, just beginning to understand such things, that metric geometry, the geometry of the *Elements*, is nothing but a special problem in projective geometry, or perspective, and it is easy to see that projective geometry is nothing but a special problem in topical geometry. Now mathematicians are entirely destitute of any method of reasoning about topical geometry. The 25th proposition of the 7th book of the *Éléments de Géométrie* of Legendre, which is strictly all that is known of the subject except some extensions of it, of which the chief is Listing's Census-theorem, was demonstrated with extreme difficulty by Legendre, having exceeded the powers of Euler. Really the proof is not satisfactory, nor is Listing's. The simple proposition that four colors suffice to color a map on a spheroid has resisted the efforts of the greatest mathematicians. If, then, without particularly attending to that proposition or to topical geometry I find that my studies of the method of discovering heuretic methods leads me naturally to the desired proof of the map-problem I shall know that I am making progress. From time to time, as I advanced, I have tried my hand at that problem. I have not proved it yet; although the last time I tried, I thought I had a proof, which closer examination proved to contain a flaw. Since then, I have made, as it seems to me, a considerable advance; but I have not been induced to reexamine that subject, as I certainly should do if I were quite confident of being able to solve it with ease. I have, however, applied my logical theory directly with success to the demonstration of several other propositions which had resisted powerful mathematicians; and I have greatly improved upon Listing's theory; so that I am confident that what I have found out is of value; and I believe the same method has only to be pushed a little further to solve the map-problem.

I can show that numbers whether integral, fractional, or irrational, have no other use of meaning than to say which of two things comes earlier, which later, in a serial arrangement. To ask, How much does this weigh? is answered as soon as we know what things among those which concern us it is heavier than what it is lighter than. A system of measurement has no other purpose than that; and it appears to be the best artificial device for that purpose.

But all relations whatever can be reduced to relations of serial order; so that every mathematical question can be looked upon as a metrical question in a broad sense; and perhaps the best and readiest way to get command of a branch of mathematics is to find what system of measurement is best adapted to it. Thus, the barycentric calculus applies to projective geometry a sort of measurement; and in fact modern analytic geometry results from just that application. But it evidently labors under the difficulty of not being a sufficiently flexible and well-adapted system of measurement. Hermann Schubert's Calculus of Geometry gives some hint of what is wanted.

No. 2. *The Reasoning of Mathematics.* No science of logic is needed for mathematics beyond that which mathematics can itself supply, unless possibly it be in regard to mathematical heuretic. But the examination of the methods of mathematical demonstration sheds extraordinary light upon logic, such as I, for my part, never dreamed of in advance, although I ought to have guessed that there must be unexpected treasures hidden in this quite unexplored ground. That the logic of mathematics belonged to the logic of relatives, and to the logic of triadic, not of dyadic relations, was indeed, obvious in advance; but beyond that, I had no idea of its nature. The first things I found out were that all mathematical reasoning is diagrammatic and that all necessary reasoning is mathematical reasoning, no matter how simple it may be. By diagrammatic reasoning, I mean reasoning which constructs a diagram according to a precept expressed in general terms, performs experiments upon this diagram, notes their results, assures itself that similar experiments performed upon any diagram constructed according to the same precept would have the same results, and expresses this in general terms. This was a discovery of no little importance, showing, as it does, that all knowledge without exception comes from observation.

At this point I intend to insert a mention of my theory of grades of reality. The general notion is old; but in modern times it has been forgotten. I undertake to prove its truth, resting on the principle that a theory which is adapted to the prediction of observational facts, and which does not lead to disappointment, is *ipso facto* true. This principle is proved in No. 1. Then my proof of grades of reality is inductive, and consists in often turning aside in the course of this series of memoirs to show how this theory is adapted to the expression of facts. This might be mistaken for repetitiousness; but in fact is logically defensible, and it also has the advantage of leading the reader, step by step, to the comprehension of an idea which he would not be able to grasp at once, and to the appreciation of an argument which he could not digest at one time. I will not here undertake to explain what the theory is in detail. Suffice it to say that since reality consists in this, that a real thing has whatever characters it has in its being and its having them does not consist in its being represented to have them, not even in its representing itself to have them, not even if the character consists in the thing's representing itself to represent itself; since, I say, that is the nature of reality, as all schools of philosophy now admit, there is no reason in the nature of reality why it should not have gradations of several kinds; and in point of fact, we find convincing evidences of such gradations. It is easy to see that according to this definition the square root of minus 1 possesses a certain grade of reality, since all its characters except only that of being the square root of minus one are what they are whether you or I think so or not. So when Charles Dickens was half through one of his novels, he could no longer make his characters do anything that some whim of a reader might suggest without feeling that it was false; and in point of fact the reader sometimes feels that the concluding parts of this or that novel of Dickens is false. Even here, then, there is an extremely low grade of reality. Everybody would admit that the word might be applied in such cases by an apt metaphor; but I undertake to show that there is a certain degree of sober truth in it, and that it is important for logic to recognize that the reality of the Great Pyramid, or of the Atlantic

Ocean, or of the Sun itself, is nothing but a higher grade of the same thing.

But to say that the reasoning of mathematics is diagrammatic is not to penetrate in the least degree into the logical peculiarities of its procedure, because all necessary reasoning is diagrammatic.

My first real discovery about mathematical procedure was that there are two kinds of necessary reasoning, which I call the Corollarial and the Theorematic, because the corollaries affixed to the propositions of Euclid are usually arguments of one kind, while the more important theorems are of the other. The peculiarity of theorematic reasoning is that it considers something not implied at all in the conceptions so far gained, which neither the definition of the object of research nor anything yet known about could of themselves suggest, although they give room for it. Euclid, for example, will add lines to his diagram which are not at all required or suggested by any previous proposition, and which the conclusion that he reaches by this means says nothing about. I show that no considerable advance can be made in thought of any kind without theorematic reasoning. When we come to consider the heuretic part of mathematical procedure, the question how such suggestions are obtained will be the central point of the discussion.

Passing over smaller discoveries, the principal result of my closer studies of it has been the very great part which an operation plays in it which throughout modern times has been taken for nothing better than a proper butt of ridicule. It is the operation of *abstraction*, in the proper sense of the term, which, for example, converts the proposition "Opium puts people to sleep" into "Opium has a dormitive virtue." This turns out to be so essential to the greater strides of mathematical demonstration that it is proper to divide all Theorematic reasoning into the Non-abstractional and the abstractional. I am able to prove that the most practically important results of mathematics could not in any way be attained without this operation of abstraction. It is therefore necessary for logic to distinguish sharply between good abstraction and bad abstraction.

It was not until I had been giving a large part of my time for several years to tracing out the ways in which mathematical demonstration makes use of abstraction that I came across a fact which a mind which had not been scrutinizing the facts so closely might have seen long before, namely, that all collections are of the nature of abstractions. When we pass from saying, "Almost any American can speak English," to saying, "The American nation is composed of individuals of whom the greater part speak English," we perform a special kind of abstraction.

This can, I know, signify little to the person who is not acquainted with the properties of abstraction. It may, however, suggest to him that the popular contempt for "abstractions" does not aim very accurately at its mark.

When I published a paper about Number in 1882, I was already largely anticipated by Cantor, although I did not know it. I however anticipated Dedekind by about six years. Dedekind's work, although its form is admirable, has not influenced me. But ideas which I have derived from Cantor are so mixed up with ideas of my own that I could not safely undertake to say exactly where the line should be drawn between what is Cantor's and what my own. From my point of view, it is not of much consequence. Like Cantor and unlike Dedekind, I begin with multitude, or as Cantor erroneously calls it, Cardinal Number. But it would be equally correct, perhaps preferable, to begin with Ordinal Number, as Dedekind does. But I pursue the method of considering Multitude to the very end; while Cantor switches off to ordinal number. For that reason, it is difficult to make sure that my higher multitudes are the same as his. But I have little doubt that they are. I prove that there is an infinite series of infinite multitudes, apparently the same as Cantor's *alephs*. I call the first the denumeral multitude, the others the abnumerable multitudes, the first and least of which is the multitude of all the irrational numbers of analysis. There is nothing greater than these but true continua, which are not multitudes. I cannot see that Cantor has ever got the conception of a true continuum, such that in any lapse of time there is room for any multitude of instants however great.

I show that every multitude is distinguished from all greater multitudes by there being a way of reasoning about collections of that multitude which does not hold good for greater multitudes. Consequently, there is an infinite series of forms of reasoning concerning the calculus which deals only with a collection of numbers of the first abnumerable multitude which are not applicable to true continua. This, it would seem, was a sufficient explanation of the circumstances that mathematicians have never discovered any method of reasoning about topical geometry, which deals with true continua. They have not really proved a single proposition in that branch of mathematics.

Cayley, while I was still a boy, proved that metrical geometry, the geometry of the elements, is nothing but a special problem of projective geometry, or perspective. It is easy to see that projective geometry is nothing but a special problem of topical geometry. On the other hand, since every relation can be reduced to a relation of serial order, something similar to a scale of values may be applied to every kind of mathematics. Probably, if the appropriate scale were found, it would afford the best general method for the treatment of any branch. We see, for example, the power of the barycentric calculus in projective geometry. It is essentially the method of modern analytic geometry. Yet it is evident that it is not altogether an appropriate scale. I can already see some of the characters of an appropriate scale of values for topical geometry.

My logical studies have already enabled me to prove some propositions which had arrested mathematicians of power. Yet I distinctly disclaim, for the present, all pretension to having been remarkably successful in dealing with the heuretic department of mathematics. My attention has been concentrated upon the study of its procedure in demonstration, not upon its procedure in discovering demonstrations. This must come later; and it may very well be that I am not so near to a thorough understanding of it as I may hope.

I am quite sure that the value of what I have ascertained will be acknowledged by mathematicians. I shall make one more effort to increase it, before writing this second memoir.

No. 3. *On the Categories.* My aim in this paper, upon which I have bestowed more labor than upon any other, beginning two years before my first publication on the subject in May 1867, is far more ambitious than that of Kant, or even that of Aristotle, or even in the more extended work of Hegel. All these philosophers contented themselves mainly with arranging conceptions which were already current. I, on the contrary, undertake to look directly upon the universal phenomenon, that is upon all that in any way appears, whether as fact or as fiction; to pick out the different kinds of elements which I detect in it, aided by a special art developed for the purpose; and to form clear conceptions of those kinds, of which I find that there are only three, aided by another special art developed for that purpose.

In my present limited space, I cannot make myself clear, still less convincing. Yet I will give such hint as I can of the three kinds of elements. I might name them Qualities, Occurrences, and Meanings. In order to get an idea of what I mean by a Quality, imagine a being whose consciousness should be nothing but the perfume of a damask rose, without any sense of change, of duration, of self or anything else. Put yourself in that being's shoes, and what of the universal phenomenon remains is what I call a Quality. It may be defined as that whose mode of being consists simply in its being what it is. It is self-essense. Suppose next that the consciousness we have imagined should undergo the simplest possible experience; that, for example, the rose-odor should suddenly change to violet odor. If it is to remain the same consciousness, there must be a moment in which it is conscious of both odors. It cannot in this moment be conscious of the flow of time; but the former rose-odor will appear as its *ego*, as its consciousness, while the new violet-odor will at that moment be its *non-ego*, the object of its consciousness. We have this sort of consciousness whenever we experience an event. The old, which has just come to an end, appears as an *ego*, with the new, which is just about to begin, over against it as a *non-ego* instantly passing into the *ego*. The sense of actuality, of present fact, is thus essentially a consciousness of duplicity, of opposition. When we have thus got the idea of an Inner and an Outer, we can review our experience and place ourselves back to a moment when both the former and the latter states were non-egos, and thus we get the idea of a force acting between Outward objects. I do not mean to say that historically we actually do so reflect: probably not. But I mean that that would be a logical reflection. Thus we might logically derive the notion of a thing, as something whose mode of being consists in a reaction against something else. This is my second category. The occurrence is essentially present. When it is not present its peculiar mode of being is gone. There is no time-constituent in it; for the flow of time involves a very different element. There is always a certain resistance to the Unexpected. It is usually broken down so instantly that it can only be detected in cases in which peculiar circumstances cause its continuance. But that the new experience always has to overcome a resistance on the part of the old is proved by the fact that we feel it to be irresistible. We feel its force. Now, there can be no force where there is no resistance. The two are but reverse aspects of the same phenomenon. This resistance is a counter-force. Hence the sense of actual fact is a sense of reacting efforts.

So far, we have left out of account the staple element of the universal phenomenon. Since we have been considering things as temporal, we may as well continue to take the same point of view. The future grows into accomplished fact by a gradual unrolling. The new becomes gradually old. Its effects remain, but they dwindle in importance toward utter oblivion. According to legitimate physical presumption, the evidence certainly now is (although we may not think it likely that it is quite true) that all physical forces are at bottom conservative. Now conservative forces necessarily produce cyclical effects. It is true, that if two particles are attracted precisely inversely as the cube of their distance, or by any law equivalent to that, the one will move in a spiral nearer and nearer to the other forever. This is an interesting point; and I have never seen it stated with precision. Formulae given on p. 378 of my father's *Analytic Mechanics* show that if P is the rate of description of area of the Boscovichian point moving round a fixed attracting centre, then if we use a system of rectangular coordinates in which x shall be equal to the square of the reciprocal of the radius vector, and y equal to the square of the velocity, then the straight line whose equation is $y = 4P^2x$ will determine the condition of the moving particle reaching an apse; that is, a maximum or minimum distance. Another curve, dependent on the law of the variation of the attraction with the distance will determine how v^2 will vary with $\frac{1}{r^2}$ [Fig. 3]. If the attraction

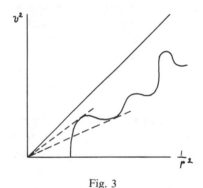

Fig. 3

varies less rapidly than the inverse cube of the distance this second curve will be concave downwards; if more rapidly, concave upwards. But if it ever crosses the straight line $y = 4P^2x$ the body will have that distance as a maximum or minimum distance. If it is tangent to that straight line, it may describe the circle at that distance. When it is below the straight line its velocity will be insufficient and the distance will diminish; so that x will increase. . . . [2]

---

[2] Another draft expands the reference to "page 378 of my father's *Analytic Mechanics*" as follows: "I detest vagueness; and will indulge this passion by stating the matter exactly, according to a formula on page 378 of my father's *Analytic Mechanics*. Ω being the potential, Ω_0 being the potential at the distance at which the velocity would be zero, r is the distance and p_1 is a constant depending on the size, etc. of the system (being in fact the length of the perpendicular upon the path when $\Omega - \Omega_0 - 1$), then the orbit will become perpendicular to the radius vector when

$$\Omega - \Omega_0 = \frac{2p_1^4}{r^2} \text{ "}$$

We cannot safely employ in logic any kind of reasoning which is subject to doubts which a science of logic is needed to remove. We are, therefore, restricted to mathematical reasoning. Now mathematical reasoning requires a diagrammatic or pure constructive notion of the thing reasoned about. But the ordinary logicians talk of acts of the mind, concepts, judgments, acts of concluding, which are mixed ideas into which enter all sorts of elements in a manner which prevents any strict mathematical reasoning about them. All these ideas of the mind are, however, representations, or signs. We must begin by getting diagrammatic notions of signs from which we strip away, at first, all reference to the mind; and after we have made those ideas just as distinct as our notion of a prime number or of an oval line, we may then consider, if need be, what are the peculiar characteristics of a mental sign, and in fact may give a mathematical definition of a mind, in the same sense in which we can give a mathematical definition of a straight line. We cannot by any purely mathematical definition build up the peculiar idea of straightness, since that is nothing but a feeling. We can only define a straight line as one of a continuous family of lines having certain relations to one another. But there might be just such a family composed of lines none of which would appear straight to us. In like manner, we can define the formal character of mind in a manner perfectly adequate to all the purposes of logic. But there is nothing to compel the object of such a formal definition to have the peculiar feeling of consciousness. That peculiar feeling has nothing to do with the logicality of reasoning, however; and it is far better to leave it out of account.

No. 12. *On the Definition of Logic.*   Logic is *formal semiotic.* A sign is something, *A*, which brings something, *B*, its *interpretant* sign, determined or created by it, into the same sort of correspondence (or a lower implied sort) with something, *C*, its *object*, as that in which itself stands to *C*. This definition no more involves any reference to human thought than does the definition of a line as the place within which a particle lies during a lapse of time. It is from this definition that I deduce the principles of logic by mathematical reasoning, and by mathematical reasoning that, I aver, will support criticism of Weierstrassian severity, and that is perfectly evident. The word "formal" in the definition is also defined.

No. 22. *Of the Logic of Chance.* The origin of probability, and its nature. The connection between objective probability and doubt. Nature of a "long run." In what senses there can be any probability in mathematics. Problems in probability as affording mathematical methods. It is not necessary that there should be any definite probability that a generic event should have a specific accompaniment. Necessity of considering in each case whether this is so or not. There is no definite probability, for example, that a witness will tell the truth. To suppose that events are subject to law but that those laws are unknown to us deprives the doctrine of chances of all validity. It is necessary that it should be known that there is *no* law of a kind which could interfere with the security of the party experiencing the long run. This becomes especially necessary when we assume events to be independent. The calculus of probability is applicable to scientific problems. Nature of the assumption that it is applicable to insurance companies. Rothschild at Monte Carlo determines to bet a franc at each bet and not come away until he has netted a gain of a franc. Probability for a selfish individual, what?

The rules of probability developed. Application to high numbers. Rationale of least squares. Pearson's developments.

Inverse probabilities. The fallacies of the books. Examination of attempts to defend the Laplacian doctrine.

This memoir is intended to be a complete *vade mecum* of the Doctrine of Chances, discussing every important or interesting question yet suggested with many new ones.

---

3. *Analysis of the Conceptions of Mathematics.* My work in this direction is already somewhat known, although very imperfectly. One of the Learned Academics of Europe has crowned a demonstration that my definition of a finite multitude agrees with Dedekind's definition of an infinite multitude. It appears to me that the one is hardly more than a verbal modification of the other. I am usually represented as having put forth my definition as a substitute for Dedekind's. In point of fact, mine was published 6 years before his; and my paper contains in a very brief and crabbed form all the essentials of his beautiful exposition (still more perfect as modified by Schröder). Many animadversions have been made by eminent men upon my remark, in the *Century Dictionary*, that the method of infinitesimals is more consonant with [the] then (in 1883) recent studies of mathematical logic. In this memoir, I should show precisely how the calculus may be, to the advantage of simplicity,

based upon the doctrine of infinitesimals. Many futile attempts have been made to define continuity. In the sense in the calculus, no difficulty remains. But the whole of topical geometry remains in an exceedingly backward state and destitute of any method of proof simply because true continuity has not been mathematically defined. By a careful analysis of the conception of a *collection*, of which no mathematical definition has been yet published, I have succeeded in giving a demonstration of an important proposition which Cantor had missed, from which the required definition of a continuum results; and a foundation is afforded for topical geometry, which branch of geometry really embraces the whole of geometry. I have made several other advances in defining the conceptions of mathematics which illuminate the subject.

4. *Analysis of the Methods of Mathematical Demonstration.* I have hitherto only published some slight hints of my discoveries in regard to the logical processes used in mathematics. I find that two different kinds of reasoning are used, which I distinguish as the *corollarial* and the *theorematic*. This is a matter of extreme importance for the theory of cognition. It remains unpublished. I also find that the most effective kind of theorematic demonstration always involves the long despised operation of *abstraction*, which has been a common topic of ridicule. This is the operation by which we transform the proposition that "Opium puts people to sleep" into the proposition that "Opium has a soporific virtue." Like every other logical transformation, it can be applied in a futile manner. But I show that, without it, the mathematician would be shut off from operations upon lines, surfaces, differentials, functions, operations, — and even from the consideration of cardinal numbers. I go on to define precisely what it is that this operation effects. I endeavor in this paper to enumerate, classify, and define the precise mode of effectiveness, of every method employed in mathematics.

---

14. *On the Methods of Establishing and Discovering the Doctrines of Logic.* It need not be said that a science whose methods are all at sixes and sevens is in poor case. I shall show that there are at present actually in use six *plus* seven, or thirteen methods in use for establishing logical truth, without counting the method of authority which is really operative, although unavowed. While there are some logicians who are more or less scrupulous in their choice of methods, most of them resort indifferently to any one of twelve, the only one they scrupulously avoid being the only one that is generally valid. For I shall prove conclusively

that the majority of the methods are absolutely worthless, and that of the others only one is properly applicable in all parts of logic. That one method consists in proceeding from universally observed facts, formulated abstractly, and deducing their consequences by mathematical reasoning. We are here with certain objections which weigh with almost all logicians but which I shall undertake to show are merely due to feeble grasp of the conceptions of logic. The first of these objections, which lies behind them all, is that, logic being the science which establishes the validity of reasoning, it begs the question to employ reasoning to establish the principles of logic. To this I reply that as long as all doubt is removed by a method, nothing better can be demanded. But owing to the confused state of mind of logicians, they make various attempts to answer this, such as that doubt is not removed if we question the validity of the reasoning. The rejoinder is obvious enough. Of course, it follows that pure mathematics does not stand in need of any science of logic to determine whether a reasoning is good or not; and by a review of the different disputes which have arisen between mathematicians, I show that this is the case; and I contrast this with a number of instances in the history of other sciences, where logical doctrines were sadly needed.

In regard to methods of discovering logical truth, there are few logicians who show any vestige of any definite method except that of reading what others have written. There are, however, a few methods which have been employed, which I consider. I show that the most successful of these really consist in an unconscious and ill-defined application of one method which I describe.

---

21. *Of First Premisses.* (Since somebody may think that I write *premiss* instead of *premise* from negligence, may I be permitted to say that desperately negligent as I am of non-logical matters, I endeavor to attend to all the minutiae of logic. The word *praemissa* is a substantive meaning a premiss, came into Latin very late and was never very common. Consequently, the English word was for a long time little used; but when it was used, it was always spelled *premiss*. But when it became more common so as to be written by persons of insufficient learning, it was confused with another word, the legal word, generally used in the plural, *premises*. This word is simply a French legal adjective meaning "aforesaid," and commonly used in the phrase "les choses prémises." It thus passed into English in its plural form; and this plural form masked its adjectival nature, so that the unlearned did not know what it meant. Probably not one uneducated man today who talks glibly

about the *premises* could tell what the word *premises* means. So when the logical word came in (the word *premises* having presumably already been pronounced with its first s hard), the vulgar thought it somewhat mysterious and doubtless the same mysterious thing which the lawyers spoke of in the plural. Hence, the word by the vulgar and eventually by the refined, though *not by logicians*, was spelled with a single s, *premise*. It was not until much later that logicians, to give themselves a mundane air, took up this false spelling.) Kant divided propositions into Analytic, or Explicatory, and Synthetic, or Ampliative. He defined an analytic proposition as one whose predicate was implied in its subject. This was an objectionable definition due to Kant's total ignorance of the logic of relatives. The distinction is generally condemned by modern writers; and what they have in mind (almost always most confusedly) is just. The only fault that Kant's distinction has is that it is ambiguous, owing to his ignorance of the logic of relatives and consequently of the real nature of mathematical proof. He had his choice of making either one of two distinctions. Let definitions everywhere be substituted for *definita* in the proposition. Then it was open to him to say that if the proposition could be reduced to an identical one by merely attaching aggregates to its subjects and components to its predicate it was an analytic proposition; but otherwise was synthetic. Or he might have said that if the proposition could be proved to be true by logical necessity without further hypothesis it was an analytic one; but otherwise, was synthetic. These two statements Kant would have supposed to be equivalent. But they are not so. Since his abstract definition is ambiguous, we naturally look to his examples, in order to determine what he means. Now turning to Rosenkranz and Schubert's edition of his works, Vol. II (the Critik d.r.V.) p. 702 we read, "Mathematische Urtheile sind ingesammt synthetisch." That certainly indicates the former of the two meanings, which in my opinion gives, too, the more important division. The statement, however, is unusually extravagant, to come from Kant. Thus, the "Urtheile" of Euclid's Elements must be regarded as mathematical; and no less than 132 of them are definitions, which are certainly analytical. Kant maintains, too, that $7 + 5 = 12$ is a synthetical judgment, which he could not have done if he had been acquainted with the logic of relatives. For if we write G for "next greater than," the definition of 7 is $7 = G6$ and that of 12 is $12 = G11$. Now it is part of the definition of *plus*, that $Gx + y = G(x + y)$. That is, that $G6 + 5 = G11$ is implied in $6 + 5 = 11$. But the definition of 6 is $6 = G5$, and that of 11 is $11 = G10$; so that $G5 + 5 = G10$ is implied in $5 + 5 = 10$, and so on down to $0 + 5 = 5$. But further it is a part of the definition of *plus* that $x + Gy = G(x + y)$ and the definition of 5 is $5 = G4$, so that $0 + G4 = G4$

is implied in $0 + 4 = 4$, and so on down to $0 + 0 = 0$. But this last is part of the definition of *plus*. There is, in short, no theorematic reasoning required to prove from the definitions that $7 + 5 = 12$. It is not even necessary to take account of the general definition of an integer number. But Kant was quite unaware that there was such a thing as theorematic reasoning, because he had not studied the logic of relatives. Consequently, not being able to account for the richness of mathematics and the mysterious or occult character of its principal theorems by corollarial reasoning, he was led to believe that all mathematical propositions are synthetic.

---

22. *Logic of Chance.*   Deduction, as such, is not amenable to critic; for it is necessary reasoning, and as such renders its conclusion evident. Now it is idle to seek any justification of what is evident. It cannot be rendered more than evident. Fallacies, it is true, may be criticized; but this subject will be postponed until all the legitimate modes of argument have been considered.

But when deduction relates to probability, it becomes open to criticism, not insofar as it is deductive, but insofar as it relates to a logical conception which in a sense deprives the reasoning of its necessary character. I therefore in this memoir examine the nature of probability, and the processes of the doctrine of chances. I flatter myself that I shall put the whole matter, both of the origin of probability and of the application of the calculus, in a much clearer light than has hitherto been done. Objective probability is simply a statistical ratio. But, besides that, doubt has degrees of intensity, and although these have no necessary signification, it might be useful for us to believe more intensely in propositions which would less often deceive us than in such as would oftener deceive us. In point of fact, we naturally "weigh" or "balance reasons," as if the degree of our trust in them were significant of fact. This is a matter requiring minute examination.

In the first place, regarding probabilities as statistical ratios, probability is exclusively confined to cases where there is a "long run" of experience, that is an endless series of events of a general character, of which some definite ratio have a special character, which shall not occur at any regular law of intervals. It is not necessary that this ratio should remain constant throughout the experience. But it is requisite that there should be such a ratio. It is easy to imagine cases in which there should be no such ratio; perhaps even a universe in which there should be no such thing as probability. (I will endeavor to determine this with certainty before drawing up the memoir.) It is commonly said that there is a law of the occurrence of the event, only it is unknown to us. But it is easy to show that the utility of the calculus *depends on there being no law* of the kind which would concern the application. Ignorance is not sufficient.

The rules of probability are easily deduced, involving the conception of independent events, that is events such that the product of the number of occurrences of both into the number of non-occurrences of both equals the product of the number of occurrences of the first only into the number of occurrences of the second only. From this follows the probability law.

Now as concerns the connection between probability and doubt, we find the books stuffed with errors. It is, for example, generally said that probability 1 represents absolute certainty. But on the contrary, probability 1 is that of an event which in the entire long run fails to occur only a finite number of times. In the next place, the majority of the books give formulae from which it would follow that the probability of a wholly unknown event is $\frac{1}{2}$. It is evident that probability, in this crude form, is quite unadapted to expressing the state of knowledge generally. The relation of real evidence to a positive conclusion is not a mathematical function. From a bag of beans, I take out a handful, *in order to test a theory which I have some other reason for entertaining*, that two thirds of the beans in the bag are black. I find this to be nearly so in the handful, and my theory is confirmed, and I now have strong reason for believing it approximately true. But it is not true that there is any definite probability that it is true. For what would such a ratio mean? Would it mean that once in so often my conclusion is true? That depends on the general commonness of different distributions of beans in a bag, which is a positive fact, not a mathematical function. Mathematical calculation is deductive reasoning, applicable solely to hypotheses; and whenever it is applied to do the work of induction or abduction it is utterly fallacious. This is an important general maxim.

This consideration affects the method of least squares, if this method is looked upon in an extravagant theoretical manner; but not if it is regarded as a way of formulating roughly an inductive inference. Mr. Pearson's extensions, though they are excessively complicated, and thereby violate the very idea of least squares, are not without value. But other somewhat similar modifications of probability are called for; and I shall endeavor to work out one or two of them.

I give in this memoir a summary of all the ordinary scientific man needs to know about probability, in a brief intelligent manner.

26. *Of Fallacies*.   This is a subject which has very little attracted the attention of the stronger logicians and is consequently in the most deplorable condition. I divide them into three classes, as follows: 1st, those fallacies which are mere slips, such as one may fall into in adding a column of figures, which is, indeed, a fallacy; 2nd, those which arise from misunderstandings, such as the *ignoratio elenchi* and *petitio principii*; 3rd, those which have their origin in loose *logica utens*, or more frequently, in inexact *logica docens*. To these may be added, 4th, sophisms which really deceive nobody, but which present problems in logic often highly instructive. I make an attempt to enumerate all varieties. Those of the first class are hardly worth notice; yet still not utterly useless, any more than it would be to call attention to the ways in which there is danger of error in performing an algebraical computation. My remarks about the *petitio principii* I hope will be useful. In the third class, I call attention to a number of fallacies that are not mentioned in any of the books. Such, for example, is the extension of the doctrine of the *burden of proof* to cases where it has no meaning, but where formalistic reasoners appeal to it as a source of knowledge, as if it were a law of nature. Another class of examples of fallacies, to which logicians are especially liable (and logicians are the most fallacious reasoners in the world), are objections to arguments as being fallacious which are, in reality, sound, but are merely misunderstood by the objector to be arguments of a different kind from what they profess to be. The books are full of pretended refutations of fallacies, where the reasoning criticized is really sound. Indeed, my observation leads me to conclude that persons of good sense whose minds are not vitiated by logical notions rarely fall into fallacies, unless they be mere slips. On the other hand, I know no class of books in which fallacies so abound as works on logic and philosophy. I have carefully read a large number of German treatises on logic of a somewhat original and superior kind, — certainly at the least estimate over fifty of them. But I do not think I ever met with a single one, — not even that of Schröder, — which does not somewhere fall into an unquestionable and utterly indefensible logical fallacy. This is not true of English books; but there are few English logics of any strength. The Germans, I think, are naturally stupid about logic; although some of the most magnificent reasoners have been Germans. Keppler is quite incomparable in inductive logic; Weierstrass and Georg Cantor superb in mathematical subtlety, for all the latter's being one of the "Baconians" in Shakespeare-ology. Among logicians, Leibniz, Lambert, Kant, Herbart, are men of distinguished power. But there is a vicious tendency to Subjectivism in Germans wherever they deal with any subject that tempts that disposition. I do not wish to be supposed

not to admire the Germans; but when I see so many young Americans copying all their faults and generally worshipping them, I am moved to say that they are not gods.

27. *Of Methodeutic.*   The first business of this memoir is to develope a precise conception of the nature of methodeutical logic. In methodeutic, it is assumed that the signs considered will conform to the conditions of critic, and be true. But just as critical logic inquires whether and how a sign corresponds to its intended *ultimate* object, the reality; so methodeutic looks to the purposed *ultimate* interpretant and inquires what conditions a sign must conform to, in order to be pertinent to the purpose. Methodeutic has a special interest in Abduction, or the inference which starts a scientific hypothesis. For it is not sufficient that a hypothesis should be a justifiable one. Any hypothesis which explains the facts is justified critically. But among justifiable hypotheses we have to select that one which is suitable for being tested by experiment. There is no such need of a subsequent choice after drawing deductive and inductive conclusions. Yet although methodeutic has not the same special concern with them, it has to develop the principles which are to guide us in the invention of proofs, those which are to govern the general course of an investigation, and those which determine what problems shall engage our energies. It is, therefore, throughout of an economic character. Two other problems of methodeutic which the old logics usually made almost its only business are, first, the principles of definition, and of rendering ideas clear; and second, the principles of classification.

28. *Of the Economics of Research.*   Political economy, in its general analysis by Ricardo and others, is a fine example of logical method. Its chief fault is that no coëfficient of average stupidity is introduced and no coëfficient of average sentimentality, which could have been introduced into the formulae. Of course, their values would have to be determined for each class of society. Political economy now goes by the name of economics, a change of title which obscures an important feature of the science, that it relates to very large collections of individuals whose average character must be much more fixed than those of the single individuals. The chief factors to be considered are the demand at different prices and the cost of different amounts supplied. In the case of research we have something analogous although measures cannot be made with any precision. The amount of the commodity is to be represented by the amount of knowledge of a given subject. The price is represented by the utility of an addition to knowledge, especially

the scientific utility. The cost is the amount of energy, time, money, etc. required to produce a given increase of knowledge. The irregularities are excessive. The peculiarities of the individual case must always be considered. Nevertheless, there are certain general rules, subject to frequent exceptions, the consideration of which is far from being entirely useless. Two such rules are, that the more we already know of a subject, the less is likely to be the utility of a given increase of knowledge, and that the more we already know, the greater is likely to be the cost of a given increase of knowledge. But if for the amount of knowledge we substitute the number of persons informed, both rules will be reversed. Hence, by far the most valuable knowledge is that which is common experience. This does not, in itself, decide the question between the respective utility of diffusing and advancing knowledge; yet I think it is evident that until people generally know enough to conduct affairs with reasonable economy, it is bad economy to spend much on the advancement of science. Ten millions is a small sum when we are think- ing of seventy millions of people. But if a hundred million were expended in teaching the people of the United States some things that are known respecting our protective tariff, it would produce a larger amount to be applied to the advancement of science. I do not begrudge the money spent upon churches, because what is taught in churches is, *in itself considered*, the most valuable of all truth. But I wish one tenth of that amount could be appropriated to diffusing economic knowledge, be- cause that knowledge would produce the wealth requisite for the advancement and diffusion of all other knowledge. A great capitalist who is generous is a strange and wonderful phenomenon, while the people are naturally generous to the point of extravagance. In the light of these considerations, it becomes a maxim of the economy of research that great encouragement should be given to applications of science. For although steam and electricity are things of trifling value in them- selves, since people were nearly as good and happy before the days of steam and electricity, yet they become of extreme utility in causing great expenditures to be made for the advancement of pure science.

  Now coming to pure science, the economy of research demands the opening up of new branches of knowledge as soon as the study of them can be conducted scientifically, rather than in carrying to extreme per- fection sciences from which the richest juice has already been pressed. Carry forward the research that is promising: neglect the one whose outlook is dismal. If for one inquiry several hypotheses are equally attractive, and in another but one, prefer the latter. In any given inquiry, other things being fairly equal or even considerably against equality, prefer the hypothesis which if false can easily be proved to be so; if it can very easily be dispatched, adopt it at once, and have done with it. But while you hold it, hold it in good faith, so as to do it full justice. Among hypotheses choose one whose elements are well understood, so that unknown complications, and consequent expense of energy cannot arise. Prefer general hypotheses to special ones, provided the more general are so by being simpler; if they are so by being complex, it is necessary to consider the economics of testing them more particularly. For instance, instead of supposing $y = a + bx + cx^2 + dx^3 +$ etc. and determining the coefficients, ask whether y has a constant term, next whether it tends to infinity with x, next whether its increments are approximately proportional to those of x, etc.

There are many economic reasons for preferring hypotheses which seem simple. I do not here mean by simple, having only one indeterminate element, although that is a manifest ground of preference; but I mean simple to human apprehension. Especially, in using abduction you already commit yourself to the hypothesis that the truth is comprehensible to you, and therefore that what is akin to your mind is likely to be true. Being committed to this, you scarcely make an additional hypothesis in assuming that that which is more akin to your natural way of thinking is more likely to be true.

Nothing unknown can ever become known except through its analogy with other things known. Therefore, do not attempt to explain phenomena, isolated and disconnected with common experience. It is a waste of energy, besides being extremely compromising. Turn a deaf ear to people who say, "Scientific men ought to investigate this because it is so strange." That is the very reason why the study should wait. It will not be ripe until it ceases to be so strange.

Do not waste your time over questions concerning which facts are scanty and not to be gathered.

All these maxims are so many theorems of logic which I shall endeavor in my memoir to present in systematic form.

30. *On Classification.* In 1867 I worked out a theory of natural classification which I never published, because the naturalists did not seem to take to it. I had been a special student under Louis Agassiz for about six months, with a view to studying his method of classification, that subject being a branch of logic. Since then, I have endeavored to penetrate further into the matter, and I think with some success. I continue to think that the definition I, then, gave of an *important* character is just. Namely, if one asks a naturalist why he considers a character "important," he surely must give some reason: he cannot be content with saying that it impresses him as such. Now his reason will either be

that this character involves certain others, as for example a particular likelihood to taking certain forms, or it will be that this character is of an order of characters, such, for example, as its relating to the skeleton of the animal, which are generally important. This importance must ultimately resolve itself into an importance of the first kind; so that *importance* consists in a character's universally carrying with it certain others, be those others no more than tendencies. The objection made by the naturalists was that above *families*, or some said above *species*, taxinomic characters do not generally carry others with them. But in saying this they were evidently limiting too much their conception of a character. For there must be some reason for regarding a character as important, and it is obvious that, in the last analysis, that means that the character *imports* some other. In fact, the true objection to the definition is not, as the naturalists said to me at that time, that so few characters are important, but, on the contrary, that all characters, even quite trivial ones, appear as important under that definition. This consideration leads, at once, to the needed correction of the conception of importance; and a very fundamental correction it is. Namely, it is that an important character must not only entrain others, but it must entrain another which has relation to the purpose in view. That brings us back to Agassiz's conception of natural classification, which all my study confirms me in holding to be correct. Namely, every classification whatsoever, be it merely arranging words in alphabetical order, has reference to some purpose, or some tendency to an end. By a tendency to an end, I mean that a certain result will be brought about, or approached, and in such a way that if, within limits, its being brought about by one line of mechanical causation be prevented, it will be brought about, or approached, by an independent line of mechanical causation. This definition is the one always virtually used by physiologists in determining whether there is a tendency to an end. Every classification has reference to a tendency toward an end. If this tendency is the tendency which has determined the class characters of the objects, it is a natural classification. Nobody in his senses would classify human creations otherwise. An end is an intellectual idea; for by the definition of an end, it is not brought about by mechanical force, and it is something of which there is a unitary conception. Persons whose conceptions are in need of logical training may misunderstand the statement that the end is not brought about by mechanical force. This is because crude and incomplete notions of "energy" and mechanical force have so taken possession of empty heads, that they do not perceive that according to the general equation of motion no state of things is due exclusively to the action of forces, because the equation of motion is merely a differential equation of the second order; so that there are six circumstances for each particle that are not due to force. Now in case these trillions of circumstances present any general character, as they always must, or the problem would not attract any attention, a general character of the result is due to other factors than force; and it very generally happens that the most important characters are due to other factors. Take, for example, the phenomenon of the diffusion of gases. Force has very little to do with it, the molecules not being appreciably under the influence of forces. The result is due to the statistics of the equal masses, the positions, and the motions of the molecules, and to a slight degree only upon force, and that only insofar as there is a force, almost regardless of its character, except that it becomes sensible only at small distances. These features of a gas, that it is composed of equal molecules distributed according to a statistical law, and with velocities also distributed according to a statistical law, is an intellectual character. Accordingly, the phenomenon of diffusion is a tendency toward an end; it works one way, and not the opposite way, and if hindered, within certain limits, it will, when freed, recommence in such way as it can. Not only is an end an intellectual idea, but every intellectual idea governing a phenomenon produces a tendency toward an end. It is very easy to see by a general survey of nature, that force is a subsidiary agency in nature. There ought, therefore, to be discoverable natural classifications in nature; and Agassiz was right in saying that such a classification must have reference to an intellectual idea. I need not say that the idea itself, like almost every profound and important idea of philosophy, was very old. It was Aristotle's, if not older. The theory of natural development is in nowise opposed to this, whatever flavor it may take, least of all in the Darwinian flavor. For natural development takes place in one way, not in the opposite way; and the Darwinian machinery for it is reproduction which is manifestly a tendency to an end. The neo-Darwinians seem to wish to make reproduction and variation as mechanical as they can. This is a praiseworthy effort, because it must inevitably eventuate in making the truth more plain that they are not mechanical, in the sense of being governed mainly by force. I do not know enough about biology to entertain a definite opinion that the work of classification is now conducted by a wrong method. I only note that naturalists certainly entertain a number of opinions about classification which are not true of classification generally. But how far these errors affect their work, I do not know. I fancy that the study of nature must largely force the right ideas upon them largely.

As a specimen of what I refer to as erroneous notions entertained by naturalists about classification, I may mention the idea that if two classes merge into one another they cannot be natural classes. If we turn to classifications of human works, where the true principles of natural classification are beyond question, we soon find this idea refuted. In order to illustrate this, I shall, in the memoir, discuss the weights found by Prof. Petrie at Naucratis, and admirably worked over by him. I show, by an application of the principles of probability, beyond all reasonable doubt and so clearly that every naturalist must see the force of the argument, that in certain cases, where weights were intended to conform to two different standards, a weight intended to conform to the lighter standard was heavier than another weight intended to conform to the heavier standard. We can even say, roughly, how often this occurs. As a consequence of this, it is impossible to say which standard certain individual weights were intended to conform to. The two classes of weights merge, and as far as individual weights are concerned, merge inextricably, although they can be separated statistically. Therefore, a naturalist does not prove that two species are not natural classes by merely showing that they blend. I will give another example to show that the general principle which seems to underlie the naturalists' notion, namely, that an object has not distinct parts unless those parts have definite limits, is false. Namely, a lake with two islands in it certainly consists of two simple parts, if by a simple part we are to understand a part not enclosing an island. But the boundaries may be drawn as in Fig. 4, or almost any way.

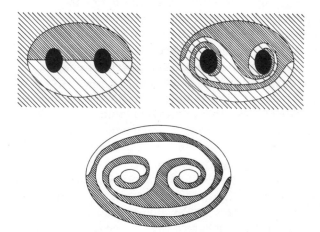

Fig. 4

I do not pretend to have had any signal success in my studies of classification. Yet what I have found out seems worth giving. I have made classifications of artificial contrivances whose genesis we can indubitably understand. In these cases, we find a course of experience in which my three categories are repeated in order over and over again. First, there is a form with its peculiar characteristic or *flavor*. The *reaction* of experience develops manifest inconvenience whence comes *thought*, resulting in one or more *new* forms (all novelty involves the first category) which in process of time has to contend with new *difficulties*, a new *analysis* is made, resulting in new improvements.

I have not, as yet, discovered any particular law of the succession of problems, — at least, none that I should care to put forward.

31. *On the Course of Research.*   One must suspect that a close relation exists between this problem and that of classification; and since this one ought, one would think, to be connected with some law exhibiting itself in the history of science, we should expect a deep, sympathetic study of the history of science to throw a light on the secret of the categories of the classificatory hierarchy. It was owing to a hope that this might turn out to be the case, and that those hierarchical categories might have other useful applications, that I have bestowed great study on the history of science.[3]

The general course of the history of science has been something like this. The first scientific problems to be taken up were medicine, pneumatology, cosmogony, etc. which mostly seem hopeless today. The result was that some successes began to be attained in arithmetic and in the simplest parts of astronomy, and shortly there was some development of geometry. We find in Pythagoras the beginnings of a true science of the categories. His numbers were categories; that is, elements of the phenomena; and they bear a certain general resemblance to my categories. The duality on which he so much insisted was my second category, that of reaction. His examples show this. He looked too much on the formal side of it, but this was a good fault. We next find the Greeks developing a most extraordinary understanding of esthetic truths. A little later, in Socrates, we meet with a lofty ethical science. Logic follows in Plato, thoroughly worked out in Aristotle. Metaphysics also takes important steps; and that of Aristotle (a mere reworking of Plato) is in some respects better than what is current today. We also find in Aristotle decided success in Psychology, the doctrine of association being well stated. His mechanics was excessively bad. His biology very rudimentary. Then came further successes in the simpler parts of astron-

---

[3]   The editor of these Peirce papers in mathematics has been working for a number of years on a volume of Peirce's studies in the history of science.

omy. Statics was established. Grammar became worked out. Thus the order of development was substantially, and quite minutely, that of my table of the classification of the sciences, which I drew up exclusively to express the *present* state of the sciences as living today. The only exception is that the beginnings of several descriptive sciences were made although I place them at the bottom. Omitting them, and also geometry in which additions were continually [in the] making, the order was

> Arithmetic – The Categories – Esthetics – Ethics – Logic – Metaphysics – Psychology – Statics – Grammar

Modern science is too complex to permit any such arrangement. The general law is that of progress from the more abstract to the more concrete. The history of any well-developed science exhibits the same law. In optics, the doctrine of rays and perspective came first. The law of reflexion was early discovered. The law of refraction was the first modern discovery early in the 17th century. The velocity of light was ascertained in 1676. Polarization, diffraction, and dispersion were discovered about the same time, as well as phenomena which were really those of interference. Thus, the main phenomena were already known. The general theory of undulations was suggested by Huygens, and Hooke showed that it would explain the colors of thin plates. It was approved by Euler. But it was not until 1817 that Young saw that the vibrations were transverse. The electrical theory of light dates from 1873.

Here, therefore, a purely geometrical account of the phenomena of ordinary experience was worked out. Then the main phenomena were discovered and mathematically formulated. Then the formal theory of the constitution of light was lit upon and worked out mathematically and finally the material theory of its constitution arose from a mathematical analysis of another branch of physics.

I have accumulated a considerable store of truth concerning the course of scientific discovery of almost all branches; but I have not yet brought it into the form of a system, as I propose to do in this memoir.

---

22. *Logic of Chance.*   Although deduction is not directly and as such amenable to critic, yet it becomes such when it deals with probability and certain allied conceptions. The criticism is not properly of the deductive process but of those conceptions. I here examine the philosophy of probability and show, among other things, that it is by no means true that every contingent event has any definite probability. I

describe the construction of an urn of black and white balls such that there is no definite probability that a ball drawn will be white. By way of illustration, I show that there is no definite probability that a witness will tell the truth. Another point I make clear is the distinction between probability unity and absolute certainty. This is illustrated by the case where a large number of players, playing against a banker, at a perfectly even game, each bet one franc each time until he nets a gain, when he retires from the table and gives place to a fresh player. The probability is 1 that any given player will ultimately net a gain, and therefore that all will do so; and yet the probability is 1 that in the long run the bank will not lose, or at any rate, there is an even chance that if the banker does not come out precisely even he will win too. I show that the "moral value" of a player's chances is quite irrelevant to the Petersburg paradox; and I correct various other errors current about probability. Hume's argument about Miracles will be analyzed.

Rules by which all errors in the use of the doctrine of chances will be plainly laid down, and their use exemplified.

No. 26. *On Fallacies.*   Few logicians of theoretical strength have manifested any taste for the general doctrine of fallacies. It is no part of pure logic, but is an application of logic in attacking which one comes into direct competition with Aristotle in one of the directions in which he was the most incomparable. Most of the best modern treatises altogether ignore this subject. Mill attempted to say something fresh about it with little success. But though disagreeable, the topic is a useful one to treat, and I feel it my duty not to shirk it, but to treat it, not from a strict logical point of view but in a human, mundane manner, as well as my poor talent in that direction will.

In that quite non-logical way, it seems to me useful to divide fallacies into five classes, according to their causes. Class I consists of fallacies due to mere slips, such errors as one may commit in footing up a long column of figures. It is useful to point out, first, in what sort of operations [they] are most likely to occur, and secondly, what can best be done to avoid them. Every slip by which one inadvertently substitutes one thing for another is essentially of the same kind as a slip in logic, even if it be not strictly such. There is no use in drawing any sharp boundary. We may as well consider putting salt in one's coffee and sugar on one's potatoes as fallacies of this description. All such inadvertencies are confined to intellectual operations of such extreme simplicity as to be almost relegated to the government of physiological habit, but yet not quite so. In algebraical calculations errors in regard to + and − are more frequent than all others put together. This is because the reversal of a sign is so simple a matter that one performs it mentally without setting down anything; and thus one is tempted to perform half a dozen such reversals before setting down a result. But to do this one has to carry five signs of one kind and six of the other in one's head, with great risk of confusion. One may even acquire a *habit* of making a particular kind of error. It is found that a computer in going over his own work is more liable to repeat the same errors than to commit others. I have an annoying habit of writing perhaps in one case out of every fifty the word "than" in place of "that." Perhaps this may be due to my often thinking in French and writing in English. We may also commit slips owing to our heedlessly extending a habit from its proper occasions to others that have some resemblance to them, as in the sugar-salt inadvertency. When we find we have a habit of committing a particular kind of slip, so that we do so in a certain proportion of cases, we shall carefully form a contrary habit, a business which will often involve no small expenditure of energy. It is bad economy to employ the brain in doing what can be accomplished mechanically, just as it would have been bad economy for Napoléon to write his own dispatches. But where it must be done, it will be best to work out each result in two different ways. Class II consists of misunderstandings, particularly the *petitio principii* and *ignoratio elenchi*. We have to consider, first, how to avoid these ourselves, and second, how to treat them when made by opponents. If a man follows the directions of my methodeutic, a *petitio principii*, in his original reasoning, will be impossible. But in attempting to refute the reasoning of others, one has to take pains to understand just how they have reached their opinions. The common mistake is that one thinks of a line of thought by which an opponent *might* have reached his opinion, and then hastily assumes that it is that process that he has followed, when in reality it may be a different and far more profound course of reasoning that he has pursued. Many men tire one to death with their continual misapprehensions of this sort. Some writers take so much pains to write an easy, agreeable style that they lead, not only superficial persons, but even men of immense mental power, to suppose that their thought is equally of an easy, obvious kind, when in reality it may be quite the reverse. It is certainly the very worst style that one can use in philosophical discussions. Hume and J. S. Mill are great sinners in this respect. Thus, Prof. Langley, though I need not say how good a thinker he is, seems to have the idea that the kernel of Hume's opinion about miracles is that a miracle is a violation of a law of nature, though that remark was merely a concession by Hume to a mode of definition in vogue at his time, and his real argument is only strengthened if this be denied; and Kant, whose first business it was to understand Hume, equally mistakes his position about causality.[4] Of course, the very thing which makes these writers atrociously hard reading for careful readers, causes them to be applauded for their perspicuity by those who skim their pages. On the other hand, the writers who are really the easiest to read, if by reading we mean getting at their thought, are popularly considered dreadfully abstruse. Such, in the first line, are the mathematicians; and next Ricardo, Duns Scotus, Kant, etc. In order to understand a book, it is necessary to learn the whole history of the author's thought. A good "reading-book," in the sense of a book to exercise one's power of penetrating an author's real thought, is the first book of Euclid's *Elements*. To be sure, it is a very badly written book, an enigmatical book. That was Greek taste. A student has not really read it until he can answer these questions, and many others: Why, in the definition of a circle, does Euclid introduce the clause τῶν ἐντὸς τοῦ σχήματος κειμένων? Why does Euclid make it a postulate rather than an axiom that all right angles are equal, and an axiom rather than a postulate that τὸ ὅλον τοῦ μέρους μεῖζον? Why did he consider the postulate about right angles necessary? Why did he attain his purpose in this particular way? How could he assert such a monstrous proposition as that τὸ ὅλον τοῦ μέρους μεῖζον, when he well knew that a line unlimited at one end became no shorter by having a piece cut off from it? Why did he give the fifth postulate its extraordinary form? Why does he ignore projective and topical geometry? When a student has once *read* the first book of the *Elements*, he will have an idea of how philosophy has to be read. Coming to the consideration of cases in which one is disposed to accuse an opponent of committing a *petitio principii* or an *ignoratio elenchi*, one must not forget that such a plea involves the admission that the argument objected to logically deduces its conclusion from its premisses. Hence, those who say that a syllogism is a *petitio principii* themselves, at best, commit an *ignoratio elenchi*; for their plea admits that the conclusion of a syllogism follows from its premisses; and this is all that the logics affirm. This remark applies more forcibly to the *petitio principii* than it does to the *ignoratio elenchi*. For if you accuse an opponent of making an assumption which he had no right to make, you have not only to show that he had no right to make the assumption,

[4] Samuel P. Langley, Secretary of the Smithsonian Institution, often employed Peirce's talents in the translations of foreign scientific papers for publications of the Institution. The story of Peirce's controversy with Langley on "The Laws of Nature and Hume's Arguments against Miracles" may be found in "The Scientist-Philosopher C. S. Peirce at the Smithsonian," by Carolyn Eisele in *Journal of the History of Ideas* 18:4, and in "The Peirce-Langley Correspondence and Peirce's Manuscripts on Hume and the Laws of Nature," by Philip Wiener in *Proceedings of the American Philosophical Society* 92 (1947).

but you have further to show that this was not a mere side remark, but that it is essential to his argument. Now if the argument is logically bad, one cannot say that anything is essential to it as a good argument. In the case of the *ignoratio elenchi*, you have to show that what he proves does not really conflict with what you maintain. Then your duty is not to quibble about the possibly inaccurate way in which he may have expressed his conclusion, but must examine what it really is that his argument does prove. If it does not prove anything at all, you may still say "If this proves anything it is nothing that I have denied," and may loosely call it an *ignoratio elenchi* with no particular inconvenience, although a true *ignoratio elenchi* does prove something. Class III consists of fallacies having their origin in a loose *logica utens* or in a faulty *logica docens*. This is a large class of very common fallacies, perhaps the commonest of all. A large part of the memoir must be devoted to them. One common fallacy consists in urging that the "burden of proof" lies upon the opponent to a purely theoretical proposition in order to establish it without any knowledge of the matter at all. The burden of proof is strictly an affair of legal procedure, where, owing to the necessity of deciding each case one way or the other, certain rules of presumption are adopted by courts. There is something analogous in other cases in which questions must be decided, and in which there are some recognized rules for deciding them in the absence of data. But a purely theoretical question need not be decided at all, and therefore, in such a case, there is no "burden of proof." The person who talks of it may mean to say that there is some vague improbability in the proposition he opposes, which may be true. But then he should state his argument just as it really is, so that its true force or weakness may appear. For example, a noneuclidean geometer might say that the burden of proof is upon whoever says that the sum of the angles of a triangle is that of two right angles; to which his opponent will answer that it is certainly extremely near that, and that the burden of proof is upon whoever says it is not exactly so.

3

FERMATIAN INFERENCE AND DEMORGAN'S SYLLOGISM OF TRANSPOSED QUANTITY

A. THE CONCEPTION OF INFINITY (819)

The *syllogism of transposed quantity* is a kind of reasoning, so named by De Morgan, who first called attention to it, of which the following is an example.

Every dollar which an insurance company pays for losses must have been paid in for premiums (including accrued interest on them) or else the capital (including accrued interest) is impaired. Suppose, then, that, considering all the new property insured in one year, if the total losses which ever happens on this property exceed the total premiums paid on it, and suppose this is true for the new property insured every year. It follows that the capital must be impaired.

De Morgan considered this reasoning as absolutely valid; and, because he was an actuary, I have selected an example from his profession which the actuaries of his day would have considered sound reasoning, but which the actuaries of our day would laugh at. I give another example which shows more clearly the essence of the reasoning. It is given, with a satirical purpose, by Balzac, in the preface or introduction to the *Physiologie du mariage*. It runs something like this:

Every young Frenchman (to believe what he says) seduces a Frenchwoman. But no Frenchwoman is seduced but once, and for each Frenchwoman, you can name a distinct Frenchman (i.e. there are as many Frenchmen as Frenchwomen). Hence, every Frenchwoman must get seduced.

There are some collections for which such reasoning holds good, and others for which it does not hold good. The reasoning about insurance takes no account of the continual increase of business, in consequence of which it is not the question whether the property destroyed paid fully its losses, but whether the *new* premiums are continually sufficient to pay losses on the property formerly insured. (Whether this will be a sound principle in the end, is a question of whether this will always be kept up. The insurance men say it will work till there is such a universal crash that they will fail when everybody fails). The argument of Balzac overlooks the fact that the seduced are on the average younger than the seducers, and this vitiates the argument to the extent of the increase of population in that time.

Here is another example of this reasoning. In every count of a certain collection of things every number used in the counting is immediately followed by a number used in the counting, or else there is a maximum number used. But no number is immediately preceded by two different numbers; and there is a number used, namely *one*, which is not preceded by any number. Hence, there is a maximum number used.

To say there is a maximum number used in counting a collection, is to say that that collection is finite, or to avoid a much abused word, let us say *enumerable*. We thus see that the general character of an enumerable collection is that, to it, syllogisms of transposed quantity are truly applicable. We also see that there are collections of which this is not true, such as the whole collection of positive integer numbers; for there is no highest number. There is thus no difficulty in conceiving that a collection is inenumerable.

Fermat introduced a new method of reasoning about numbers. Suppose a certain property, P, is proved to belong, if to any number N, then to the next higher $N + 1$. Then, this property if it belongs to the first number, 1, belongs to all numbers. This kind of reasoning I call the *Fermatian inference*.[1]

It holds good of all enumerable collections, no matter in what order their members be arranged. It holds good of *some* inenumerable collections, provided they are properly arranged. It would not do to arrange all the odd numbers first with all the even numbers following them. It is necessary to arrange the objects in such a linear order that each one is preceded by an enumerable collection of objects, or by none.

It is susceptible of demonstration that there are collections with which we frequently deal which cannot be so arranged as to render the Fermatian inference truly applicable. In other words, the positive integer numbers cannot be assigned each to a different member of such a collection.

---

[1]   Cassius J. Keyser of Columbia University sent Peirce a copy of his paper "Concerning the Axiom of Infinity and Mathematical Induction" which had been read before the American Mathematical Society 29 Dec. 1902 and was later printed in the *Bulletin*, May 1903, in which he erroneously adopted Peirce's "Fermatian inference" as "Fermatian induction." Peirce wrote on the copy "I never called it 'induction.'" Keyser incidentally had great respect for Peirce and was responsible in part for Peirce's Hibbert Journal publications.

B. FERMATIAN INFERENCE (820)

What we wish to prove is that if the Fermatian inference holds and if nothing has two Qs then there are not two things Q to themselves.

Suppose this not true. Then

(1) $J \, \bar{\pi} \, \beta \, \sqsubset \, IN \,..\, KQJ. \left[\overline{Uh\beta} + Hh\beta \, (\overline{IQH} + \overline{Ih\beta}) + Nh\beta \right]$

(2) $XYZ \, \epsilon, \overline{XQZ} + \overline{YQZ} + \overline{Xh\epsilon} + \overline{Yh\epsilon}$

(3) $V \lesssim \sim \, VQV \cdot WQW \cdot Vh\zeta \cdot \overline{Wh\zeta}$

To prove the absurdity of this I first [deduce] their incompatibility with

(4) $\geq \langle \boxtimes \diamond \, \Phi \, \text{ᦞ} \dashv LM$

$\dfrac{(\overline{Ah\theta} + Sh\theta \cdot TQS \cdot \overline{Th\theta} + QVVh\theta)BQA \cdot Bh\vartheta}{(Lh\vartheta + \overline{MQL} + Mh\vartheta)(\overline{VQV} + \overline{Vh\vartheta})}$

For this purpose, I define σ

(5) $D \,..\, Dh\sigma = Dh\vartheta + DiA$ (i means identical with)

Identifying D with A we get

(6) $Ah\sigma$

In (4) $LM \,..\, \overline{Ih\vartheta} + \overline{MQL} + Mh\vartheta$

In (2) identify X with B, Z with A, ϵ with ϑ and multiplying $Bh\vartheta \cdot BQA$

gives $Y \,..\, (\overline{YQA} + \overline{Yh\vartheta})$

(7) $LM \,..\, Lh\sigma + \overline{MQL} + Mh\sigma$

In (4) identify θ with σ and we have

(8) $\langle \,..\, VQV \cdot Vh\vartheta$

But we prove that $\langle iA$

for in (4) identify θ with ϑ

and we get $\overline{Ah\vartheta} \cdot BQA \cdot Bh\vartheta$

Then in (2) identify ϵ with ϑ, X with B, Y and Z with A and multiplying we get $\overline{AQA}$ while CQC.

Then by (8) and (5) $V \,..\, VQV \cdot Vh\vartheta$

But this muliplied into (4) reduces it to zero.

Thus, if K be defined

(12) $D\theta \, \text{ᦞ} \dashv \,..\, \overline{Dh\theta} + Sh\theta \cdot TQS \cdot \overline{Th\theta} + VQV \cdot Vh\theta = Dhk$

V taken individually we have shown that

(9) $DE \,..\, \overline{DhK} + \overline{EQD} + EhK$

But by (1) and (9) UhK (10)
By (1) (9) and (10) N .. NhK (11)
From (11) WhK (13)
Then by (12) putting for $Dh\theta = DiW \cdot \overline{DiV} + WiV$ (14)
we have

$$\text{S}\dashv \ldots + (SiW \cdot \overline{SiV} + WiV)\,TQS\,(\overline{TiW} + TiV)\overline{WiV}$$
$$+ VQV(WiV)$$

whence WiV.

Theorem. If the Fermatian inference holds let Q be a Fermatian relation such that nothing has two Qs. Then I say no two different things are Q to themselves.

Proof.   Let V be something Q to itself
            and W be something Q to itself
I have to prove V and W identical.

Let K be such a character that to say anything A has it is the same as to say that any character whatever which is true if of anything then of every Q of that thing and which is true of A is also true of V.

Then I say if anything A has the character K, everything Q of A has the same character.

For if not, let B be something Q of A which has not the character K. That is to say, there is some character ϑ that belongs to B, and which belongs to everything which is Q to a thing it belongs to, and yet does not belong to V.

Now let the character σ consist in either possessing ϑ or being A. Of course A has the character σ.

A thing Q to a thing having the character σ is either Q to a thing having the character ϑ or is Q of A. But B, the only thing Q of A has the character ϑ and everything Q to a thing having the character ϑ has the character ϑ. Hence, everything Q to a thing having the character σ has the character ϑ and consequently the character σ. Since therefore A has the character σ which also belongs to everything the Q of anything to which it belongs. Then if A has the character K, σ is true of V. K is also true of V; hence if K is not true of B, V and B are not identical. Since, then, B is Q of A, V is not Q of A since nothing has two Qs and consequently A and V are not the same, for V is Q of V. Since then V has the character σ but is not A it has the character ϑ. Thus the supposition that there is a character that belongs to B (or the Q of A) and belongs to everything Q of a thing to which it belongs and yet does not belong to V is reduced to absurdity. In other words, if anything A has the character K, the Q of A has the same character.

Now if U has any character which belongs to the Q of whatever it belongs to, everything in the universe (including V) considered has that character. Hence, by definition, V has the character K. Since then K is a character which belongs to the Q of whatever it belongs to, by the Fermatian inference everything in the universe has the character K. Therefore W has the character K.

Consider then the character which consists in being W but not V. This is a character which belongs to W, unless W and V are identical. And if it belongs to anything (which can only be W) it belongs to the Q of that thing (for the Q of W is W). Hence, since the character K belongs to W, unless W and V are identical the character of being W but not V belongs to V, which is absurd. Hence, W and V must be identical. Q.E.D.

C. THE CRITIC OF ARGUMENTS (589)

III. SYNTHETICAL PROPOSITIONS À PRIORI

That geometry contains propositions which may be understood to be synthetical judgments *a priori*, I will not dispute. Such are the propositions that the sum of the three angles of a triangle is invariably and unconditionally equal to 180°, that there are only 5 regular solids, etc. The theory of functions has had in its time a good many such propositions. But the difficulty is that, considered as applicable to the real world, they are *false*. *Possibly* the three angles of every triangle make exactly 180°; but nothing more unlikely can be conceived. It is false to say it is necessarily so. Considered, on the other hand, as purely formal, these propositions are merely ideal. Some of them define an ideal hypothesis (in the mathematical sense), and the rest are deductions from those definitions. Nothing but ignorance of the logic of relatives has made another opinion possible.

To illustrate this, I will prove from definitions that $7 + 5 = 12$. In doing so, I shall in the first instance, assume the "fundamental proposition of arithmetic," which is that if the count of a lot of things stops by the exhaustion of those things, every count of them will stop at the same number. I assume this, because it will scarcely be maintained by anybody that the only reason $7 + 5 = 12$ cannot be proved from definitions is that that general proposition has to be assumed. Still, lest such a notion should arise, I will append a proof that this proposition is not synthetic. Assuming it for the present, I note that it implies that every finite lot of things has a number uniquely determined.* (* I do not write *eindeutig bestimmt* for two reasons; first, I prefer while I am about it to write English; and second, if I did not like the English phrase I would resort to the French one of which the English and German alike are translations.) Now, I define the sum of two numbers, X and Y, as the number of a class, the logical aggregate of two mutually exclusive classes whose numbers are X and Y. Consider, then, any three numbers, A, B, C, which belong to three mutually exclusive classes, a, b, c. The sum

$B + C$ is, by the definition, the number of the logical aggregate of b and c; and the sum $A + (B + C)$ is the number of the logical aggregate of a and of the aggregate of b and c. The sum $A + C$ is the number of the logical aggregate of a and c; and the sum $(A + C) + B$ is the number of the aggregate of the aggregate of a and c and of b. But it is easy to show the two threefold logical aggregates are identical. For the logical aggregate of two classes, x and y, includes all that x includes and all that y includes, and nothing else. It follows that both the aggregates in question include all that a includes, all that b includes, and all that c includes, and nothing else. Consequently, being one and the same class, it has but one number, and

$$A + (B + C) = (A + C) + B.$$

Now, I take the definitions of the numbers from 2 to 12 to be $1 + 1 = 2$, $2 + 1 = 3$, $3 + 1 = 4$, etc. Hence $7 + 5 = 7 + (4 + 1)$. Then, by the formula just obtained, $7 + (4 + 1) = (7 + 1) + 4 = 8 + 4$. In like manner, we could evidently show successively $7 + 5 = 8 + 4 = 9 + 3 = 10 + 2 = 11 + 1 = 12$.

In regard to the socalled "fundamental proposition of arithmetic," I remark that the only part of it whose proof presents the slightest difficulty is its implication that if a lot of things counted in one order is exhausted, then it is exhausted in every other order; for that the count, if it stops at all, must stop at the same number is all but an evident corollary from the definitions. Yet I remark that several proposed proofs of the proposition quite overlook the part of it which is alone a little refractory to proof, and tacitly take it for granted. My proof requires "hard thinking." So I shall put it in small type, which who will may skip.

I shall use the language of the logic of relatives. Namely, supposing λ signifies a class of ordered pairs of which PQ is one (QP may, or may not, belong to the class), then I shall say that P is a λ of Q and that Q is λ'd by P.

I next define a *finite* class. Suppose a lot of things, say the As, is such that whatever class of ordered pairs λ may signify, the following conclusion shall hold. Namely, if every A is a λ of an A, and if no A is λ'd by more than one A, then every A is λ'd by an A. If that necessarily follows, I term the collection of As a *finite* class.

I now proceed to prove the difficult part of the proposition, namely, that every collection of things the count of which can be completed by counting them in a suitable order of succession is finite. For suppose there be a collection of which this is not true, and call it the As. Then, by the definition of a finite class, there must be some relative, or class of ordered pairs, λ, such that while every A is a λ of an A, and no two As λs of the same A, there is some A not λ'd by any A. Then, I say that if this A, not λ'd by any A be removed from the class of As, the same thing will remain true. Namely, 1st, every A is λ of an

A, for so it was before the removal, and no A λ'd by an A has been removed; 2nd, no two As are λ of the same A, for the removal could not increase the number fulfilling any positive condition; and 3rd, there is still an A not λ'd by any A, namely, that A which was λ'd by the removed A, and by no other A. Now, the class of As is said to have been counted, and by the definition of counting, some number must have been called out in counting the A that was afterward removed. Let every number higher than that be lowered by unity, and a count of the class after A is removed results. It follows, then, that if there be a collection not finite the count of which can, by a suitable arrangement, be terminated by any number N, then the same is true of some collection the count of which can be terminated by a lower number. This implies there is no lowest number; but by the definition of number, there is a lowest number, namely, *one*. Thus, the hypothesis that a class whose count in any order can be completed is not finite is reduced to absurdity.

Now, suppose a finite class to be counted twice. By the definition of a finite class, each count must stop. For make λ mean "next followed in the counting by" and the definition states that if the counting does not stop, then there is no A at which it begins, which is contrary to the definition of counting. If the two counts do not stop at the same number, call that the superior which stops at the higher number. Let the cardinal numbers used in this "superior" count be called the Ss. Let a number of this count be said to be "successor" of the number which in the inferior count was called out against the same thing. Then, every S is successor to an S, but no two Ss are successors of the same S (since, by the definition of counting, no number was used twice in the inferior count). Consequently, the number of Ss being finite, by the definition of a finite class, every S is succeeded by an S, or, in other words, every S, including the greatest, was used in the inferior count. Hence, the two counts end with the same number.

Now, for the benefit of those who have skipped all that, let me give an abridged edition. The sum of 7 and 5, by definition, consists in counting the 7 and then continuing on with "eight," "nine," etc. pronounced against the things which make up the five. It is presupposed the five things count 5 in all orders. Then, whatever one is counted as eighth, was in some order of counting the five, the fifth. Consequently, the $7 + 5 = 8 + 4$.

Thus, it seems clear to me that the proposition in question is analytical or explicatory; but, no doubt, Kant had a very narrow conception of explicatory propositions, owing to his knowing nothing of the logic of relatives. Some have been of the opinion that while arithmetical propositions are analytic, geometrical ones are synthetic. But I am certain they are all of the same character. Unquestionably, it was Riemann's opinion that all the synthetical propositions of geometry are "matters of fact," and "like all matters of fact not necessary, but only empirically certain; they are hypotheses." This I substantially agree with. Considered as *pure* mathematics, they define an ideal space, with which the real space approximately agrees. As Riemann also says: "Geometry assumes, as things given, both the notion of space, and the first principles of spatial constructions. She gives definitions of them that are merely nominal, while the true determinations appear in the form of axioms." (He should have said *postulates*; but he knew only a corrupt text of Euclid.) Bernhard Riemann is recognized by all mathematicians as *the* highest authority upon the philosophy of geometry, and these quotations are from his greatest memoir.

I add the exposure of the fallacy of Euclid in his proof that the three angles of a triangle sum up to two right angles. In the first place, his fifth postulate* (*By a postulate, he meant a proposition not evident but assumed to be true.) takes for granted that the sum of the three angles is not less than two right angles. It reads: "If two straight lines in a plane cross a third so as to make the sum of the internal angles on one side less than two right angles, then those lines will meet if sufficiently produced, on the side on which the sum of the internal angles is less than two right angles." The 16th proposition of the 1st book of Euclid's Elements is: In any triangle, if one side is produced the exterior angle exceeds either of the opposite angles of the triangle. This once granted, it is easily shown that the sum of the three angles is not greater than two right angles. But the proof of the 16th proposition is fallacious. Namely, Euclid bisects the side AC in E, draws BE and produces it to Z, making $EZ = BE$. Then he joins CZ. Thereupon, he says that the angle ACZ, which equals the angle A, is less than the angle ACD. [Fig. 1]. But this is a monstrous assumption, without a scintilla of proof. It is due to the triangle being so drawn that the proposition is very

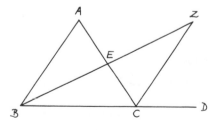

Fig. 1

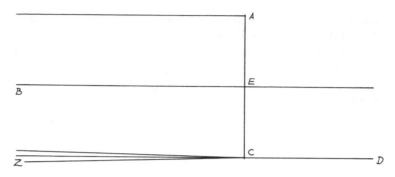

Fig. 2

manifestly true *of it*. Suppose the triangle were very long [Fig. 2]. Then, it may be, that Z would be so situated that CZ would not lie in the angle ACD; and the appearance of reasoning would disappear. It is astonishing that this fearfully bad reasoning should have been repeated by one text-book writer after another, and should appear in that of the "Society for the Improvement of Geometrical Teaching."

We begin, then, by defining the ideal relations of the positive integer numbers. Whether or not it be in our power to say how the series,

> *one, two, three, four, five, six, seven*, etc.

differs from every other row that has a beginning but no end, as from

> *first, second, third, fourth, fifth, sixth*, etc.,

and from

> *one, three, five, seven, nine, eleven*, etc.,

certain it is that mathematics has no interest in searching out such difference. To every system exactly analogous to number all the theorems of arithmetic, *mutatis mutandis*, apply; and in the pure ideal of mathematics, it is the same system.

What then are the mathematical characters of the system of numbers? Two only. First, the syllogism of transposed quantity does not apply to numbers. In other words we cannot reason about them as we do about Hottentots in the following:

> Every Hottentot kills a Hottentot;
> No Hottentot is killed by more than one Hottentot:
> ∴ Every Hottentot is killed by a Hottentot.

We cannot reason about number in this way; since we should thus be led into such absurdities as the following:

> Every integer is the half of an integer;
> No integer is the double of two different integers:
> ∴ Every integer is the double of an integer.

To say that such reasoning is bad is to say that we can find a relation in which no whole number stands to more than one number, and some number to none, while yet to every number some number stands in that relation. This can be written algebraically, as follows:

$$\text{⅋} \, ABC\alpha D \text{⊓} \text{⊓} G .. (\overline{ARC} + \overline{BRC} + \overline{Ah\alpha} + Bh\alpha) \, DRE.\overline{GRF}.$$

The second property of positive whole numbers is that the Fermatian inference is positively applicable to the whole collection of them. Pierre de Fermat, born 1601, died 1665 (some say, 1655), was a learned and hardworked lawyer of Toulouse. In his spare moments, he studied mathematics; and his is the singular glory of having invented a form of inference utterly unknown before. If it had ever been used, it was but rarely, casually, loosely, and without recognition of its peculiarity. I am inclined to pronounce Fermat's invention the greatest feat of pure intellect ever performed. I do not say it was the most difficult; still less that the greatest of feats was necessarily the work of the greatest of intellects. Nor do I mean, for an instant, to rank any purely intellectual achievement along with those of sagacity, that faculty which makes evident to the philosopher the secrets of things not seen. Fermat himself felt this: so, at least, I interpret his reply to some arrogances of Descartes, "M. Descartes ne sauroit m'estimer si peu que je ne m'estime encore moins." Nevertheless, he who could open a way of reasoning so untravelled that the arithmetical theorems to which it had conducted him, copied after his demise from his loose papers and from the margins of books upon which he had jotted them down, and published without the directions which his lot forbade his writing out, remained a mystery or a perplexity even unto seven generations of Pascals, of Huygenses, of Wallises, of Eulers, of Legendres, of Dirichlets, and of Lebesgues, can surely not be denied an exalted rank in the roster of the sons of Man.

Fermat's method of proving that all positive whole numbers possess a given property, β, is to show that if there be any number, N, which does not possess this property, then there must be another number smaller than N, which equally wants that property. Now this, by the syllogism of transposed quantity (which is applicable to any given number, though not to the entire collection of numbers) is absurd. The critical reader (for whose sake I write) will be disposed to challenge the assertion that it is the syllogism of transposed quantity which shows the absurdity of every β number being larger than a β number. For, he will say, in the syllogism of transposed quantity, it is the fact that there are not two different things in the given relation to any one thing which makes it imperative that every individual of the collection should play the part of correlate in that relation; while here, it is the fact that what is smaller than something smaller is itself smaller is that which leads not to the result that every number would be smaller than a number but to the absurd result that two numbers would be each smaller than the other. By way of reply, let me ask whether the following inference is of the nature of a syllogism of transposed quantity. Imagine a number of dots scattered upon a sheet of paper, and suppose that the erasure of any one inevitably leads to the erasure of a fresh one. Then, it follows that if one is erased all must be erased. You might say that the reasoning is that what is later than later is itself later; and that all must be erased because the erasure of a late one cannot cause the erasure of one already earlier erased. In that point of view, the inference is seen to be the same as the exhaustion by continual decrement. Or, you may say, that the necessity of the conclusion arises from the impossibility of the erasure of any one being the immediate consequence of two different erasures, owing to the erasures being successive. From this point of view, we see that the inference is a syllogism of transposed quantity. In the inference about numbers, we have a series of numbers each smaller than the last, and each necessitating the non-β-ness of the next. Now, since as compared with any number, N, of this series, all the rest are either larger or smaller, and of the larger, only one *immediately* necessitates the non-β-ness of N, while of the smaller ones none can do so, without being both greater and smaller than the same number, it follows that N can be immediately necessitated to be non-β by only one number, and so by the syllogism of transposed quantity every one of the series is so necessitated by another, that is, the series has no first, which is absurd. On the other hand, consider the reasoning about Hottentots. Here we have one or more series (or cycles) of Hottentots each killing another. Call the killer of another, an immediate successor of him; and call every successor of a successor a successor. Then, because every Hottentot is the immediate successor of a Hottentot, we infer that either a Hottentot can have two immediate successors, or else every Hottentot has a successor. The reasoning would not be essentially different if we assumed that there was a Hottentot without a successor and inferred that the conclusion that for any reason no Hottentot could have two immediate successors would land us in an absurdity. (I don't know why the phrase should be to "land" us in an absurdity: it seems more like sticking on a sunken reef.)

Having explored this *cul-de-sac*, we return to the proposition that the Fermatian inference concludes that something is true of the whole collection of numbers by using the syllogism of transposed quantity to reduce the opposite supposition to an absurdity. Yet we have seen that the syllogism of transposed quantity does not hold good of the entire collection of numbers. This illustrates one of the elementary principles of the logic of relatives. Let N be a number, C a stage of counting, R the fact that a number is reached in counting. Then,

$$N \bigcirc .. NRC$$

is true; that is, taking any number whatever, there is a stage of counting which reaches it. But

$$\bigcirc N .. NRC$$

is not true; that is, we cannot say there is a stage of counting that reaches whatever number we may name. Now an enumerable collection, or one the count of which can be completed, is finite, and the syllogism of transposed quantity holds good of it. The former of the above two propositions states, then, that this is true of every number. But the latter of the two propositions states that the count of the entire collection of numbers can be completed, so that the syllogism of transposed quantity holds good of it; which, of course is not true.[1]

---

[1]   From fragments in MS. 589:

"The two characters mentioned together constitute the definition of number, so far as number has a definition which concerns mathematics. They involve no universal synthetical judgment; that is, they assert nothing universally of a class whose limits are otherwise determined. Something conceivable is countless, or inenumerable, and yet dinumerable (or Fermatian); or, otherwise stated, there is a conceivable class of which the syllogism of transposed quantity does not hold good, while the Fermatian inference does. That fact of conceivability is synthetic, if you will; but it is only particular not universal. It relates to *some*, not to *all*. It says that a certain description of thing can be conceived, only.

"Before I leave the subject, I intend to make it clear that nothing can be asserted with certainty concerning numbers, which does not follow as a necessary consequence from numbers constituting a class with the two properties mentioned. Thus, the whole of arithmetic contains not one universal synthetical proposition.

"What is counting? It is an operation discovering that a given class, y, has the relation, P (counts up to) a certain number, N. We *count* the objects of the class; that is, every one has a number *told* against it, or is checked off against the number,—say, is in the relation, 0, to a number. Nothing in a counting operation is 0 to two different numbers; nor are two different objects 0 to one number. Further, every number whose next greater is told against a member of the class counted is itself told against a member of that class. If then, these conditions being fulfilled, there is a number, N, told against a member of the class, while its next greater is not so told, the class is P to that number. And conversely, if there be a number, N, to which the class is, P, then there is a relation, 0, fulfilling all the aforesaid conditions, and some member of the class is 0 to N, but none is 0 to the next greater than N."

Even collections to which the Fermatian inference is positively applicable are so only when considered in a certain order. (I shall express this more accurately below.) Take, for example, the collection of irreducible rational fractions, including $\frac{0}{1}$ and $\frac{1}{0}$. Arrange these, so that of any two $\frac{N}{D}$ and $\frac{N'}{D'}$, the one which has the smaller value of $N + D$ precedes the other, or if this has the same value for both, the one which has the smaller value of N precedes the other. Or, in place of $N + D$, we may use $N \times D$. In the first case, we get the series

$$\frac{0}{1}, \frac{1}{0}, \frac{1}{1}, \frac{1}{2}, \frac{2}{1}, \frac{1}{3}, \frac{3}{1}, \frac{1}{4}, \frac{2}{3}, \frac{3}{2}, \frac{4}{1}, \frac{1}{5}, \frac{5}{1}, \frac{1}{6}, \frac{2}{5}, \text{etc.}$$

In the second case, we get the series

$$\frac{0}{1}, \frac{1}{0}, \frac{1}{1}, \frac{1}{2}, \frac{2}{1}, \frac{1}{3}, \frac{3}{1}, \frac{1}{4}, \frac{4}{1}, \frac{1}{5}, \frac{5}{1}, \frac{1}{6}, \frac{2}{3}, \frac{3}{2}, \frac{6}{1}, \text{etc.}$$

In both cases every fraction is preceded in the series by a finite collection of fractions, for which the syllogism of transposed quantity holds good, so that in that order the Fermatian reasoning holds good of the entire series. But imagine the fractions arranged in the order of their values, beginning with $\frac{0}{1}$. Then, every other fraction is preceded by an infinite multitude of fractions, to which the syllogism of transposed quantity does not hold. Consequently, Fermatian reasoning, such as the following, is fallacious.

If any rational fraction is a square, there is a less fraction which is a square.

Hence, no such fraction is a square.

We thus see that to say that the Fermatian inference holds good of any collection is to say that this collection can be arranged in a series such that taking any member of it, the syllogism of transposed quantity holds good for the part of the series from the beginning up to that member.

If there is a character, α, such that, if it belongs to any number, it belongs to a smaller number, then there is another character, β (that of being a number below which there is none having the character α), such that if any number has the character, β, then the next higher number has this character. On the other hand, if there is a character, β, which if it belongs to any number, belongs to the number next higher, then there are characters connected with these such that, if they belong to any number, they belong to lower numbers. Accordingly, we may consider the Fermatian inference under the form,

The *first* of a certain series has a certain character;

But if any member of the series has that character, so has the next following member:

Hence, every member of the series has that character.

But we need not limit the inference to series, precisely. For a class to which the Fermatian inference applies, it is possible to find a relation, Q, and a member of the class, U, such that taking any character whatever, β, either U has not this character, or it is possible to find a member of the class, H, such that it has the character, β, and such that whatever member I of the class be taken, either I is not Q to H or I has not the character β, or else, finally, taking any member N of the class whatever, N has the character β.

But unless we specify what kind of relation Q is to be, this would be true of all classes, whether the Fermatian inference holds good of them or not. Namely, it is only necessary to take for the relation, Q, any one all whose correlates have the character, β. In this sense, the Fermatian inference is *negatively* applicable to all classes. But in order to confine the description to the class to which it is positively applicable, it is necessary and sufficient to add the proviso that Q is a relation in which every member of the class stands to some member. Taking this into account, the fact that the Fermatian is applicable to numbers is expressed in the following:

$$\bigcirc \sqsubset K J \pi \beta \sqsupset I N .. KQJ\left[\overline{Uh\beta} + Hh\beta(\overline{IQH} + \overline{Ih\beta}) + Nh\beta\right]$$

We here say that a relation Q, can be found such that from the premise if any number, H, has any property then *some* number Q to H has the same property, it follows that if U has that property every number has it. It might be doubted whether, in place of *some*, we ought not rather to say *all*, thus modifying the shorthand expression, so that, in place of

$$I .. (\overline{IQH} + \overline{Ih\beta}),$$

we should substitute

$$\dashv .. IQH \cdot \overline{Ih\beta}.$$

In most cases, we should not insist upon the conclusion following from the premise in *some*, because there is only one thing Q to (next after) any one H. But unless the principle is stated as here, the inference would hold good of all classes whatsoever, and would not serve to discriminate between two grades of infinity. Thus, the points of space, excluding those at infinity, form a collection so numerous that the Fermatian inference properly does not apply to them. Still if the following premises are true, "A certain point has the character β, and if any point has the character β, all points within an inch of it have the same character," then the conclusion will follow that all points not at infinity have that character.

That there are possible manifolds to which the Fermatian inference does not apply is evident as soon as we observe what this proposition involves. We have only to negative the shorthand statement that the inference in question does apply, by turning over every letter in the row of indicators, changing all multiplication into addition, and *vice versa*, and adding or removing a line over every leash. We thus have

$$QU \smile K \triangledown H \smile \gtrless \cdot \cdot \overline{KQJ} + Uh\beta(\overline{Hh\beta} + IQH \cdot Ih\beta)\overline{Nh\beta}$$

That is to say, taking any relation Q whatever, provided everything has something Q to it, then no matter what individual, U, be taken of the class, a character β can be found which U possesses while something else wants it, although whatever possesses that character β has something which is Q to it and which possesses the character.

1st. We say that a relation, Q, making the inference valid *can* be found. Not every arrangement would answer the purpose. We could not, for instance, range all the odd numbers first and all the even ones after them, and then reason thus:

> The first is odd;
> The number next following any odd number is odd:
> ∴ All the numbers are odd.

But many arrangements make the inference valid. Thus Q may mean "next but one following a number not divisible by 5 or next but 2 preceding a number divisible by 5." This gives the arrangement

1 3 5 2 4 6 8 10 7 9 11 13 15 etc.

Every theorem of arithmetic will have its analogue in reference to this arrangement.

2nd. It might seem to follow from the property as stated that every number has a number next following it. But all that really follows is that the numbers *can* be so arranged that the Fermatian inference holds, while at the same time every number has a number in some sense next following it. This may be in a cyclical arrangement; so that it is not implied that there is an infinite series, though this follows from the inapplicability of the syllogism of transposed quantity.

3rd. The second premise of the Fermatian inference is that if any number, H, has the character β, then *some* number Q to H has this character. We must say *some*, and not *every*; for if we did, the whole statement would become utterly empty. For the relation, Q, might then, for example, mean *coexistent with*. For if it be true of a character, β, that if any one thing has it all things coexistent with that thing likewise have it; then, of course, if one thing has this character, all things have it.[2]

---

[2]   The following statements are found in MS. 35.

   "In a paper on the logic of number, published in [1881], I remarked that multitudes could be separated into three classes, according to the applicability to them of the Fermatian inference and the syllogism of transposed quantity. To understand the

It is compatible with the validity of the Fermatian inference that there should be several things, Q to the same thing, as well as things to which there are different Qs, although in these cases the syllogism of transposed quantity would hold. . . .[3]

---

essential character of the method of reasoning invented by Fermat, let it be required to prove that every one of a certain multitude of objects, say the Ss, has a certain character, Q. We must first find one or more of the Ss which evidently have that character. We must then discover a relation among the Ss having the following properties. (To express them, I use the expressions "A is L of B," "B is L'd by A," to mean the same thing, namely that A stands in the relation to B expressed by L.)

"(1). Anything L of a second that is L of a third is itself L of that third.

"(2). If anything, A, is L'd by a second, B, but not L'd by anything else L'd by B, then B is the only thing so related to A, and there is nothing C of which it is true that A is L'd by C; unless A is L'd by something L'd by C.

"(3). If anything A is L'd by a second, B, and also by a third itself L'd by B, then there is something, C, L'd by B, such that A is L'd by C and is not L'd by anything L'd by C.

"(4). Every S not known to possess the character Q is L'd by a second S while not L'd by anything itself L'd by that second S.

"(5). Every S not known to possess the character Q is L'd by an S that does possess that character.

"(6). Every S that is L'd by a second S and not L'd by anything itself L'd by that second S, possesses the character Q provided that second S possesses it.

"From these six premises, the conclusion follows. . . .

"The demonstrations that various numbers are incommensurable are perfect. That the entire collection of such numbers is innumerable seems to admit of no doubt; since they would, in general, require the use of an endless series of numbers in order that they might all be distinguished from one another. Now, if they cannot all be exactly expressed, they cannot be ranged in an exact series.

"There are certainly as many points on a line or in an interval of time as there are of real numbers. These are, therefore, innumerable collections. Many writers have hastily assumed that the points on a surface or in a solid are greater than those on a line. But it is evident that this is not so, since to three or any other definite number of coordinates one distinct number corresponds. This is plain, since this number may be written by first taking the figures used in the writing of the coordinates in turn. For example, suppose we have

$$x = .123112233 \text{ etc.}$$
$$y = .455666445 \text{ etc.}$$
$$z = .777889078 \text{ etc.}$$

we can reduce these to one number by dealing out the figures thus
.1472573571681682692940347358 etc."

[3] Additional excerpts from fragments in MS. 389 are of interest here. Peirce proves a third corollary: "Where the Fermatian inference is positively applicable, it is possible so to select Q that no two members of the dinumerable class are Q to the same member." Peirce remarks that

"The discovery of this corollary requires an operation of the mind even less mechanical than that required for those which went before; because we have to imagine a relation different from any mentioned in the premise. Moreover, even when the idea is once seized the Fermatian inference is required to reach the conclusion. Thus, the assumption that this inference is applicable to the class considered is used twice over as a premise, in different ways. The suggestion of such an idea must take place by a mental process resembling synthetic probable inference. But the proof from the premises is perfectly demonstrative and analytical.

"The following proposition may be called a theorem rather than a corollary.* (*These words derive their meaning from their use by editors of Euclid. The *Elements*, in the edition of Heiberg, contains 465 propositions (theorems and problems), 132 definitions, 27 corollaries (of which several are spurious), 17 lemmas, 9 axioms, 5 postulates, and 2 scholia. The few genuine corollaries are called in the Greek, *porisms*, a word whose meaning has been more disputed than any other, not theological. But they are of the nature of obvious deductions, not requiring any fresh constructions. (The longest are to X.111 and XII.17.) All the editorial additional propositions were called corollaries. No more definite meaning has ever been attached to the word.) In case the Fermatian inference is positively applicable, and Q is so taken that no member of the class has two members Q to it, then there can be not more than one member that is Q to itself. . . .

"This reasoning requires some remark. In the first place, I remark that nothwithstanding that when stated with strict precision its complexity is terrible, yet as compared with the generality of mathematical reasonings, it is very uncomplicated. This will be believed when I point out that I have made it as simple as I possibly could, and that the proposition is no more than this that a line of dots with but one initial dot and but one dot next following any given dot, cannot have two terminal dots. Nothing simpler than that could well be dignified with the title of *theorem*. Now since this reasoning, relatively simple as it is, is hardly intelligible when stated with complete precision in ordinary language, the reader will readily see that the shorthand mode of expression, is quite indispensable in making a precise analysis of any mathematical reasoning such as mathematicians would deem complicated and difficult. I think, that even in this simple case, the reader would have more confidence that no fallacy had crept in, if he were to see the conclusion worked out in logical algebra. . . .

"The whole life of the reasoning lies in the use of these apparently empty premises. This is most characteristic of mathematical procedure. How could a machine be induced to originate ideas in this way? Yet if a machine cannot reason and the brain can, it is plain the brain is not a machine, and necessitarianism is at once upset. . . .

"We have, thus, proved that if there be a class to which the syllogism of transposed quantity is inapplicable, there is another resulting from excluding an individual from the first class to which that syllogism is equally inapplicable. This is a theorem which even a careful reasoner would be very likely to tacitly take for granted, without observing that it was not self-evident. The length of our proof measures the extreme severity of the mode of logical analysis here used. No unwarranted assumption, however small, can pass through this sieve. . . ."

4

LOGICO-MATHEMATICAL GLOSSES (812)

Voltaire laughed at Maupertuis offering to prove the existence of a God by $\frac{ab}{z}$; and his foolish chaff affects to this day the reputation of that profound philosopher. One of our newspapers whose function is to maintain certain economical theses, to a refutation I offered of one of them replied that my reasoning was "too much like the differential calculus." It was certainly like the differential calculus, being substantially the same as Ricardo's reasoning about rent; nor was the "too much" altogether without its truth. The same journal only the other day superciliously condemned the use of mathematical language in the expression of economical principles. The reason given was that political economy is of no use until the majority of the people embrace it. Its straits must be desperate then; and no wonder a language that leaves no openings for the shifts of ambiguity is eschewed. Professors of philosophy educated in theological seminaries are apt to suspect something sinister lurks beneath any mathematical treatment of logic, and down to this day continue to censure such treatment magisterially. The very last mail brought me the programme of a university where Sir William Hamilton's diatribe against the study of mathematics is still seriously discussed. Every mathematician understands that use of an algebraical system of expression implies nothing in particular about the subject-matter, — not even that it is of a quantitative character. But three things recommend such a system: first, it is free from ambiguities and unrecognized prejudices; second, it thoroughly analyzes meanings and expresses them with great precision; and third, it is readily handled by the mind and facilitates that mental experimentation in which all sound reasoning consists.

I desire from time to time when the editors of the *Bulletin* can find room for a little padding, to offer some remarks (less desultory, perhaps, than they may seem) about the logic of mathematics; and I will begin my considering the Boolian calculus of logic.

This algebra marks the merit of propositions upon a scale of two.

Good and bad, or true and false, are its only grades. It is a system of quantity having but two values; and each of its equations is, as it were, an arithmetical congruence having 2 for its modulus.

Most writers on the subject make the letters denote classes of things, primarily, and propositions only secondarily. They follow the old logicians in taking the categorical proposition to be more primitive than the conditional and disjunctive propositions. At first blush, this seems evident; for conditional and disjunctive propositions are built up out of other propositions. But a careful examination under the light of the logic of relatives shows that every categorical proposition of a general nature, such as alone can be fully expressed by the ordinary conventions of language, affirms or denies the dependence of one fact upon another. Thus, to say that all men are mortal, or whatsoever is a man is mortal, amounts to saying that any object, no matter how selected, if it be a man, then is mortal. The only distinction between the categorical, under that aspect and the conditional proposition, is that the former contains an "it" which is virtually a relative pronoun. The categorical proposition is, in short, simply a *relative* conditional. Let h signify "is a man," and m "is mortal." Then, the proposition may be written in a notation explained [in] *Am. Journ. Math.* Vol. VII.

$$\Pi_x(\bar{h}_x + m_x)$$

where Π_x means that x may be any object in the universe, $\bar{h}_x$ that x is not a man, m_x that x is mortal, $\bar{h}_x + m_x$ that either x is not a man or x is mortal. The part under the parenthesis alone belongs to the Boolian algebra. It is the sort of expression I call a "Boolian." The fact that h_x and m_x are not simple letters but letters with affixes has nothing to do with the procedure of the calculus. We thus see that every proposition which makes complete sense in ordinary language expresses a relation of dependence between propositions that do not severally make complete sense. But this incompleteness in no wise affects the application of the Boolian algebra. We, therefore, assume that every letter in this algebra signifies a fact, partial or complete.

Every mathematician understands that the use of an algebraical system of expression implies nothing in particular about the subject-matter,—certainly, not, for instance, that it is of a quantitative character. It is merely adopted as a language which has three merits, first, that of being free from ambiguities and unrecognized prejudices, second, that of analyzing the meaning and expressing it with great precision, and third and greatest, that of being readily handled by the mind and thus lending itself to that mental experimentation in which all sound reasoning consists.

I will begin these notes with some remarks about the logic of relatives, supposing the reader to have glanced at my paper in the seventh volume of the *American Journal of Mathematics.*

The Boolian algebra deals with a system of quantity (if one conceives it from the point of view of quantity) having but two values. Boole himself simply used ordinary algebra, supposing every quantity interpretable in logic, as x, to be subject to a quadratic equation with fixed roots. This may be written

$$(v - x)(x - f) = 0;$$

and it is not necessary to assign special numerical values to the roots. But Boole assumed $x^2 = x$, making the values 1 and 0. His object in making this choice was to bring the algebra into harmony with the ordinary calculus of probabilities; for he made $x = 1$ mean that the proposition signified by x was true, and $x = 0$ mean that it was false. (My v stands for *verum*, and my f for *falsum*.) But instead of measuring probability upon a scale running from $p = 0$ to $p = 1$, we might equally well measure it by Fp, where F is any univocal function, modifying the rules of the calculus accordingly; and in that case we should naturally take $v = F1$, and $f = F0$. Indeed, where we are considering the probabilities of arguments, it is convenient to consider instead of the "*probability*," the *odds*, or ratio of favorable to unfavorable cases, which

is $\dfrac{p}{1 - p}$, making the scale of probability range from ∞ to 0. Suppose, for example, there are two independent arguments each supporting a given conclusion. Let α be the ratio of the number of cases in the long run in which conclusions so related to facts like the premise of the first argument are true to the number of cases in which those conclusions are false; and let β be the corresponding ratio for the other argument. If, then, those two classes of arguments are quite independent, having no particular tendency to be true and false together, nor in opposition, then in the long run $\alpha\beta$ will be the ratio among cases in which two arguments, one of the one class, the other of the other, are both presented, of successes to failures. Suppose, on the other hand, that the conclusions of the two arguments are different, and incompatible, though possibly both false. Then the ratio of successes of one or other to failures of both is $\alpha + \beta$. These are much simpler than the formulae for the "probabilities." Namely, instead of $\alpha\beta$, we have with the ordinary

scale of measurement $\dfrac{pq}{1 - p - q + 2pq}$; and instead of $\alpha + \beta$, we have

$\dfrac{p + q - 2pq}{1 - pq}$. We often speak of "balancing reasons *pro* and *con*." The fact that this phrase is found natural is I think an argument in favor of Fechner's law.* (*The arguments usually given satisfy my logical tests; and the *best* of the objections to these arguments (the *worst* are mere rubbish) appear to me to arise from an untenable conception of the quantity of intensity of a sensation. On the other hand, the independent arguments which I have seen advanced against the law will serve me in my forthcoming book as good examples of logical futility. I ought to add the confession that were *Fechner's law* shown not to be the true one, the refutation of it would at the same time refute my synechistic hypothesis of the evolution of the universe.) Namely, the cases which in our experience have been favorable to a given mode of inference produce a feeling of their importance, proportional (other things being equal) to the logarithm of their number.

I do not think that the exact logician ought to hold that logic, or any branch of logic, however mathematically treated, is strictly speaking a branch of mathematics. It is like dynamics, which calls for, and has created, a special branch of mathematics for its study; yet the dynamics, the positive science, is one thing, and the mathematics, the purely hypothetical science, is another. The fact that from a given hypothetical premise a certain conclusion necessarily follows, can only be satisfactorily made out by mathematics; for mathematics is precisely the science of hypotheses. Yet the fact that every definite, selfconsistent hypothesis will lead to a definite, stable conclusion, instead of annulling and superseding itself, Hegelianally, is not a hypothetical fact, but an obstinate, positive fact.

On the other hand, I will not admit that the mathematician stands in any need of logic. The mathematician must reason, of course; but he needs no theory of reasoning, because no difficulties arise in mathematics which require a theory of reasoning for their resolution. The metaphysician *does* require a theory of reasoning; because in his science such difficulties *do* arise. All the special sciences (especially the nomological sciences) repose, more or less, on metaphysics, and therefore, at least indirectly, and some of them directly too, require a theory of logic. But pure mathematics can postpone such a theory.

In short, the doctrine of exact logic is, that as soon as the elementary positive facts on which logic reposes are made out, it is chiefly by mathematics, in its two functions of forming arbitrary hypotheses and of tracing their consequences, that logic is to be developed. But mathematics differs from logic in not directing its attention to those positive facts, as such, and in consequently not asking any of those questions which characterize logic.[1]

---

[1]   On 8 Sept., 1894, Peirce wrote to his friend Judge Francis Russell a letter (L 387) in which he described the then present state of his book entitled *The Principles of Strict Reasoning*. Chapter 10 was on the Boolian Calculus. Peirce wrote, "I show that the rules of the non-relative Algebra may be immensely simplified."

Among many other details the following plans for his *The Art of Reasoning* are of interest here.

"*Chapter 5. Of Inference in General.*   This is a new piece of work, largely occupied with discussions of terminology.

Simple Apprehension, Judgment, and Reasoning

Term, Proposition, Argumentation

Understanding, Verstand, *intellectus*.

"That most languages have no nouns, at least not so radically separated in the minds of these use them from verbs as are ours. The old Egyptian mode of conceiving the analysis of a sentence. . . . The Egyptian conception is the most convenient in Logic. . . .

"What good reasoning must be, Hegelians rate only that inference A1 which setting out with falsity lands us in the truth. But it is the Compulsion of *experience* that ought to be worshipped. Rules of inference. Tradition: what a wise respect for it is, what a childish clinging to it is. Descartes and his *je pense: donc je suis*. False reasoning in Elementary Geometry.

"The *reductio ad absurdum* and *Induction* compared. Stoical views of inductive reasoning. Necessary and probable inference. Diagrams and experiments with them. Examples of *Insolubilia*—mostly novel. Mathematics purely ideas.

"*Chapter 6. The Algebra of the Copula.*

"This difficult subject is here perfectly cleared up and systematized for the first time, and made to lead directly to the Boolian algebra.

"History of the principles of identity, contradiction, and excluded middle. . . .

"*Chapter 8. The Quantification of the Predicate.*

"Refutation of Hamilton.

"I then consider DeMorgan's propositional scheme from two points of view, which I call the *Diodoran* and the *Philonian*, with reference to a celebrated controversy between Diodorus and Philo. I myself advocate the Philonian view, DeMorgan the Diodoran. I first treat his scheme from his own point of view and show that putting his propositions at the vertices of a cube, every face and edge is significant. I then take the Philonian standpoint, and put his eight forms upon an anchor-ring; and I show the relation to Aristotelian syllogistic more clearly than before. I next treat *Spurious Propositions* and show that they are of extreme importance. In fact, all mathematical, or at least arithmetical propositions are spurious. I then take up a *Limited Universe of Marks*, and end with a *General Canon of Syllogism*."

"*Chapter 10. The Boolian Calculus.*   I show that the rules of the non-relative algebra may be immensely simplified by dropping all connection with arithmetic and I show that arithmetic itself dictates this if we regard the Boolian equation as an arithmetical congruence with modulus 2. I then pass to the *Logic of Relatives* which I now bring to a much more perfect and interesting form than before to greatly simplify the procedure.

The student will work it practically with facility and advantage. I show that it is necessary to think with it, and that is the chief use of it. . . .

"*Chapter 11. Graphs and Graphical diagrams.* Shows the value and limitations of the geometrical way of thinking.

"*Chapter 12. The Logic of Mathematics.* This is I think the strongest piece of logic I have ever done. It analyzes the reasoning of mathematics by means of the Calculus of Relatives. I don't see what loop-hole there is to escape my conclusion as to the nature of Mathematics, that it is merely tracing out the necessary consequences of hypotheses.

"*Chapter 13. The Doctrine of Chances.* The worst chapter in the book. A mere revision of my *Popular Science Monthly* article. This chapter demands amplification.

"*Chapter 14. The Theory of Probable Inference.* My old essay. Remarks on this I shall make presently.

"*Chapter 15. How to Make Our Ideas Clear.* Old article. I intended to add two more chapters. One on the Law of Continuity and the other on Logical Recreations and Exercises and the last I will throw together. Then there was intended to be a glossarial Index which would be made in ready proof sheet. . . ."

QUALITATIVE LOGIC (736)

PREFACE

This book is not intended to guide children in their first attempts to use their minds, nor does it address itself particularly to persons of great experience, who, while they may still correct their tendencies to be a little too credulous or too sceptical concerning any kind of evidence, have long since passed the time when any consideration of the theory of drawing inferences could influence their practice, — although such persons may find something to interest them in these pages. But there is an age, between boyhood and manhood, when there is a natural tendency to look at life in a rather theoretical way and when such speculation is of real use and service. The native force of intellect is at that age perhaps as great as it is ever destined to be, but want of skill in handling the reason and inexperience of the deceptions to which it is subject render the effective power of the mind very inferior to that which is later developed. If a young man at that time of life will only acquire a distinct preliminary conception of the methods of thought and of inquiry by which alone the truth can be ascertained, though this theoretical knowledge will not make him a powerful thinker, it will form an admirable foundation upon which to build habits of effective thinking.

To such a young man, I offer an outline sketch of the whole method of reason. If he is disposed to accept what I say with implicit belief, I shall at least try not to abuse his confidence, so that he may not some day wake up to find that it was all an idle and delusive dream. If he feels strong enough to subject what I say to a critical examination, so much the better. I shall endeavor to make my statements and to give my reasons in such a way as to facilitate his investigation.

The name of the study which forms the subject of this volume is Logic. But logic as it is set forth in most books upon the subject, is a study far worse than useless. It tends to make a man captious about trifles, and neglectful of weightier matters. It condemns every inference which is really sound, and admits only those which [are] really childish. The reason is that this logic has been handed down from the Middle Ages, the ages of Faith, i.e. of Unreason, before modern science, physical and moral, had begun. The lore of those days was laborious nonsense, their law was regulated warfare, their medicine fetichism, their history fables, their astronomy astrology, their chemistry alchemy. They never, by any chance, themselves reasoned rightly. How then can they teach us to do so, or how can their theory of reasoning be of the slightest value?

This traditional logic did not, however, originate in the middle ages, but with Aristotle at a comparatively early period in the history of the Greek mind, when it was fully developed upon the artistic side, but before it had done much in the way of discovering truth. Aristotle's logic was a mere first essay; it did not fairly represent Aristotle's own methods of thought, —far less those of the early Greek mathematicians. It did not enjoy, in antiquity, the immense renown which has since attached to it; although it did faithfully mirror the distrust which the ancients generally entertained of Observation as a foundation for scientific truth. Those who did believe in observation, the Epicureans, — like Roger Bacon in the 13th Century, —were hostile to the Aristotelean logic.

In modern times, logic has very naturally had a bad reputation. First, there was in the 16th Century a pretended reform of the science in the interest mainly of literary elegance, by Peter Ramus. He imported the Dilemma from rhetoric into logic. Early in the 17th Century came two important books, the *Discours de la méthode* of Descartes, and the *Novum organum* of Francis Bacon. The former represents, in a very vague sketch, but with great perspicuity, the methods of thinking of modern metaphysicians, of those who draw their convictions from within. The latter is an eloquent and majestic assertion of the dignity of nature and the littleness of man, who can only attain knowledge by observation. All the greatest steps in the progress of modern science have involved improvements in the art of reasoning. Harvey's discovery of the circulation [of] the blood, Kepler's researches on the orbits of the planets, Galileo's development of the principle of inertia, and many other such works contained lessons in logic, —which were perhaps even more valuable to the world and contributed even more to the progress of civilization than their more special teachings. This has been particularly the case with discoveries in mathematics. Pure mathematics, indeed, is nothing but an art of drawing conclusions of a particular description. But the great discoveries in mathematics have carried with them very important improvements in the methods of thinking concerning nearly every subject. In ancient times, the discovery by Euclid that the whole of Geometry can be deduced from a few elementary principles suggested the idea of metaphysics and profoundly modified all subsequent science and philosophy. In modern times, several mathematical discoveries have added even more momentous consequences. The Coördinates of Descartes are now constantly applied to every subject of reasoning with great advantage; while the ideas of the infinitesimal calculus have penetrated everywhere, forming, for example, an important not to say the principal factor in Ricardo's political economy.

The doctrine of chances is a direct contribution made by modern mathematicians to the general principles of logic. It may be called the modern logic for it decidedly outweighs all that was known of reasoning before this invention.

The present century has produced important treatises upon the theory of logic, which have all come either from Germany or from Great Britain. Two different schools have prevailed in the two countries, which have not in the least understood one another. The German school, of which Hegel is the type, have approached the subject from the side of theology and metaphysics; the English school, represented, for example, by Mill, have viewed the matter from the side of modern science. But there is good reason to trust that the breach between those two schools, which is but the continuation of a dispute existing almost since the birth of philosophy, is now in the process of being closed and healed.

CHAPTER III
THE MODUS PONENS

When the validity of the simple consequence, "*P*, therefore, *C*," is challenged, we are led to search for the principle upon which the inference proceeds; and perhaps we may come to recognize this in the judgment that if *P* is true, *C* is true. It would be a mistake to suppose that is always even tacitly assumed in the simple consequence; for it may be the manner in which the premise, *P*, presents itself to my notice, and not its bare truth alone, which causes my judgment that *C* is true. [In] any case, in drawing the simple consequence, "*P*, therefore *C*," I do not consciously judge that "If *P*, then *C*." But in the after criticism of the inference, I shall be apt to recognize this proposition as its principle; and from that moment my reasoning ceases to be a simple consequence, because the proposition "If *P*, then *C*," now becomes a second premise. In short, I now reason in the form

> If *P* is true, *C* is true;
> But *P* is true;
> Hence, *C* is true.

This form of inference is called the *modus ponens* (or positing mode, to distinguish it from the *modus tollens*, or removing mode, which will be noticed hereafter). *P* is called the *antecedent*, *C* the *consequent*. The hypothetical proposition is called the *consequence*.

For example, a little girl might reason,

> If I am good, my mamma will love me;
> Now, I will be good;
> And so my mamma will love me.

We have seen that even in drawing a simple consequence, I have a vague consciousness of a body of possible inferences to which the inference actually drawn belongs. In the *modus ponens* the conception of the Possible emerges more clearly. We have here, as one of the premises, a judgment, "If *P*, then *C*," which is not *categorical*, that is, does not relate to the real world alone, but is *hypothetical*, that is, is meant to apply to a whole universe of possibilities. To say that a fact is *possible* means primarily that it is not known not to be true, that our knowledge leaves room for it. But there are, besides, kinds of possibility, which are determined not by the knowledge which we happen just now to have, but by every conceivable state of information. Thus, we say, Napoleon did not win the battle of Waterloo, but he might possibly have done so. We use the past tense, *might*, because the fact supposed is now no longer consistent with what we have learned; but we mean that a knowledge of all the previous conditions, capable of being known beforehand, would leave room for such an event. We say that it is physically possible that a needle should be so balanced as to rest upright upon its point. We mean that if we knew only the laws of physics and not various familiar facts, we should not know that this did not happen. We say that it is impossible for anything in nature to happen otherwise than it does happen. We mean that if we knew all past facts and all the laws of nature and all that could be deduced therefrom we should know everything that was about to happen. There is therefore a sense in which only the actual is possible. At the furthest extreme from that is the logically possible. A proposition is logically possible when if we knew no facts at all, but only the meanings of words, we could not reject the proposition. It is true that such a state of knowledge is itself in a certain sense impossible, like a geometrical line or surface; but it is none the less a very useful conception.

A supposed fact which would be true in *some or all* the states of things for which an assumed condition of knowledge leaves room is said to be *possible*; one which is true in none of these states is said to be *impossible*;

one which is true in all is said to be *necessary*; one which is true in a part or none is said to be *contingent*.

To say that "If *P* is true, *C* is true," means that in an assumed condition of information, every possible state of things in which *P* would be true is a state of things in which *C* would be true.

Every addition or improvement to our knowledge, of whatsoever kind, comes from an exercise of our powers of perception. In necessary inference my observation is directed to a creation of my own imagination, a sort of diagram or image in which are portrayed the facts given in the premises; and the observation consists in recognizing relations between the parts of this diagram which were not noticed in constructing it. Different persons no doubt construct their logical diagram in different ways; many probably very oddly; but every person must construct some kind of a diagram or its equivalent, or he could not perform necessary reasoning, at all. It is a part of the business of logic to teach useful ways of constructing such diagrams. The whole range of logical possibility may be represented by an imaginary sheet of paper occupying the whole field of vision. Every point on this sheet is to represent some conceivable state of things, of which the real state of things is one, undistinguished from the rest. Everything learned cuts off and removes from this sheet some part and leaves a range of possibility less than the whole range of logical possibility, but still containing the unknown point which represents the actual state of things. To form a diagram of the truth of the hypothetical proposition "If *P* is true, *C* is true," we must suppose that all the points which represent states of things in which *P* is true are gathered together in one area, and that a line is drawn around this area; and that the same is done with the points representing states of things in which *C* is true. If then the boundary of the area of the truth of *P* is imagined to lie wholly within the area of the truth of *C*, the hypothetical proposition is represented as true. If this happens before the original sheet has been cut down at all, the proposition is represented as true by logical necessity; but if the sheet has been cut down it is only represented to be true for some degree of knowledge not defined.

Let us now suppose that, in addition to the hypothetical proposition, we are certain that the antecedent, *P*, is true. Then, in our state of knowledge it is not possible for *P* to be false; and to represent our knowledge, we must cut down the sheet of possibility so as to leave only a part lying wholly within the area of *P*. We perceive that this will also lie within the area of *C*, and therefore we shall be certain that the consequent is true.

CHAPTER VI
THE LOGICAL ALGEBRA OF BOOLE

The actual state of things is represented by a single point in the field of possibility; and the reasonings above treated have been concerned with this single point. Categorical propositions, it is true, in one sense refer to whole classes of individuals, but they only take these classes *distributively*, that is to say, they only consider characters as belonging to any one or some one individual of such a class, not as relations between several individuals.

Every qualitative reasoning about single individuals may be analyzed into syllogisms and dilemmas; but it is beyond the power of ordinary language to state complicated arguments of this kind with perspicuity or precision. In the case of such complicated reasoning,—and indeed wherever the logic of our inferences requires to be analyzed,—there is a great saving of trouble and gain of accuracy in employing a logical "algebra," or perfectly systematic written language or body of symbols in which the premises being expressed, the conclusion may be obtained by transformations of the expressions according to formal rules.

I speak of a body of symbols, but in point of fact, syllogism and dilemma, every qualitative reasoning about individuals may be expressed by use of one single symbol besides the expressions for the facts whose relations are examined. This symbol must signify the relation of antecedent to consequent. In the form I would propose for it, it takes the shape of a cross placed between antecedent and consequent with a sort of streamer extending over the former. Thus, "If a, then b," would be written

$$\overline{a} + b.$$

"From a, it follows that if b then c," would be written

$$\overline{a} + \overline{b} + c.$$

"From 'if a, then b,' follows c," would be written

$$\overline{\overline{a} + b} + c.$$

To say that a is false, is the same as to say that from a as an antecedent follows any consequent we like. This is naturally shown by leaving a blank space for the consequent, which may be filled at pleasure. That is, we may write "a is false"

$$\overline{a} + ,$$

implying that from a every consequence may be drawn without passing from a true antecedent to a false consequent, since *a* is not true. Instead of the blank space we might use a special symbol, say a circle, and write

$$\overline{a} \mathbin{\overline{+}} \circ.$$

By means of this sign of illation, or *copula*, we may express the most complicated relations without the slightest ambiguity. Thus, we may write

$$\overline{\overline{\overline{b \mathbin{\overline{+}} c \mathbin{\overline{+}} m \mathbin{\overline{+}} n \mathbin{\overline{+}} \overline{a \mathbin{\overline{+}} b \mathbin{\overline{+}} l} \mathbin{\overline{+}} m \mathbin{\overline{+}} \overline{a \mathbin{\overline{+}} c \mathbin{\overline{+}} l}}} \mathbin{\overline{+}} n}$$

Let, for instance, *a* mean that any object is either Enoch or Elijah, *b* that that object is a man, *c* that that object is mortal, *l* that Enoch and Elijah were not snatched up to heaven, *m* that the bible errs, *n* falsity. Then what is written signifies that if it be false that all men's being mortal would imply that the Bible errs, and if the Bible would err were the fact of Enoch and Elijah being men to imply that they were not snatched up to heaven, then it is false that the fact of Enoch and Elijah being mortal implies that they were not snatched up to heaven. Thus, by some strain we can express such a proposition in ordinary language. But to express the *general* proposition about *a*, *b*, *c*, *l*, *m*, *n*, we should have to say

> If {If [If (If *a*, then *c*), then *m*], then *n*},
> then (if {if [if (if *a*, then *b*) then *l*], then *m*},
> then if [if (if *a*, then *c*) then *l*], then *n*.).

But when we write the algebraic formula and desire to pronounce it, we may do so in this way,—the rule being to pronounce "if" at the beginning of each streamer and "then" at each cross. The braces, brackets, and parentheses, are quite unnecessary; and the *ifs* and *thens* may be pronounced rapidly and without pause, because the symbol exhibits the meaning.

Here then we have a written language for relations of dependence. We have only to bear in mind the meaning of the symbol $\overline{+}$ (not by translating it into *if* and *then*, but by associating it directly with the conception of the relation it signifies), in order to reason as well in this language as in the vernacular,—and, indeed, much better.

So far, we have a *language* but still no *algebra*. For an algebra is a language with a code of formal rules for the transformation of expressions, by which we are enabled to draw conclusions without the trouble of attending to the meaning of the language we use.

Our algebraic rules must enable us to prove the two propositions,

> If $\overline{a} \mathbin{\overline{+}} b$, then if *a* then *b*; and
> If from *a* follows *b*, then $\overline{a} \mathbin{\overline{+}} b$.

Any rules which will prove these propositions will evidently enable us [to] prove every conclusion. But that is not enough; for we require that the rules should enable us to dispense with all reasoning in our proofs except the mere substitution of particular expressions in general formulae. We do not as yet demand rules which shall enable us to dispense with difficult reasoning in discovering the truth and in inventing modes of proof,—that would be demanding more than an algebra, namely a *calculus*,—but we do require that in the proofs themselves nothing but simple substitutions shall be called for. The first of the above propositions, however, namely, that if a and $\overline{a} + b$ then b, being nothing but the *modus ponens* , might stand as a fundamental rule. The second proposition may be divided into two, each of which follows from it while it follows from them dilemmatically. They are

> If not a, then $\overline{a} + b$; and
> If b, then $\overline{a} + b$.

In making our code of laws, we shall first want a rule to show what substitutions can be made; next, we must give the properties of denial; and finally we need a general test of necessary truth.

I. The general rule of substitution is that if $\overline{m} + n$, then n may be substituted for m under an even number of streamers (or under none), while under an odd number m may be substituted for n.

This may be proved true by the peculiar kind of reasoning invented by Fermat. Any expression in which m is under an even number of streamers may be written in one of the forms of the two infinite series here exhibited

$$
\begin{array}{ll}
m & \overline{s + m} \\
\overline{m + A + a} & \overline{s + m + A + a} \\
\overline{m + A + a + B + b} & \overline{s + m + A + a + B + b} \\
\text{etc.} & \text{etc.}
\end{array}
$$

The rule holds for the first form of the first series by the *modus ponens*, and for the first form of the second series by syllogism. If in either series the rule holds from the first member up to any form for which we may write M, then let this form after the substitution of n for m become N. Then according to the rule, $\overline{M} + N$. The next form of the series after M will be say $\overline{M + I} + i$. Now by dilemmatic reasoning, we have

$$
\begin{array}{l}
\overline{M} + N \\
\overline{\overline{M + I} + i}
\end{array}
$$

Hence, $\overline{\overline{N + I} + i}$.

But this conclusion is the result of substituting n for m in $\overline{M \barwedge I} \barwedge i$. Thus we see that the rule is true of the first form of each series, and if true of any form true of the next in the same series; consequently, it is true of all forms which can be built up in this way.

In precisely the same way, any proposition in which n is under an odd number of streamers, may be written in one of the forms

$$\overline{n \barwedge p}$$
$$\overline{\overline{n \barwedge p} \barwedge A} \barwedge a$$
$$\overline{\overline{\overline{n \barwedge p} \barwedge A} \barwedge a \barwedge B} \barwedge b$$
etc.

$$\overline{s \barwedge n \barwedge p}$$
$$\overline{\overline{s \barwedge n \barwedge p} \barwedge A} \barwedge a$$
$$\overline{\overline{\overline{s \barwedge n \barwedge p} \barwedge A} \barwedge a \barwedge B} \barwedge b$$
etc.

The rule holds of the first form of the first series by syllogism. It holds of the first form of the second series, because

$$\overline{s \barwedge m \barwedge s} \barwedge m$$

is true identically; and taking this as a premise of a syllogism of which the other is $\overline{m} \barwedge n$, we have the conclusion

$$\overline{s \barwedge m \barwedge s} \barwedge n,$$

and this taken with $\overline{\overline{s} \barwedge n} \barwedge p$ gives by another syllogism

$$\overline{s \barwedge m} \barwedge p.$$

If the rule holds for any form N, let this after the substitution be M, and we have by the rule $\overline{N} \barwedge M$. The next form of the same series is $\overline{N \barwedge I} \barwedge i$ and by the same dilemma as before we prove that $\overline{M \barwedge I} \barwedge i$, and consequently that the rule is correct.

II. For the circle which is the symbol of falsity (or that all propositions are true), we have the general formula

$$\overline{\bigcirc} \barwedge a$$

whatever a may be. The falsity of a, usually written $\overline{a} \barwedge$ is really equivalent to $\overline{a} \barwedge \bigcirc$.

The formula of identity $\overline{a} \barwedge a$, for which we may write ∞ for short, has properties conjugate to those of the circle. Namely, we have the general formula

$$\overline{b \barwedge \overline{a} \barwedge a} \barwedge \quad \text{or} \quad \overline{b} \barwedge \infty,$$

for this is a case under the general principle that if B, then $\overline{A} \barwedge B$, since $\overline{a} \barwedge a$ is necessarily true. The truth of b may be written

$$\overline{\infty} \barwedge b;$$

for if b is true, so is this by the general principle just cited, and if this is true, so is b by the *modus ponens*.

III. If a proposition is true when the circle is substituted for one of its letters or terms, and also when ∞ is substituted for the same term, then the proposition is true in its original form; for the term replaced must be either true or false.

Besides these fundamental rules, there are a number of others which may be deduced from them. Among these are the following:

IV. The circle may be substituted for any term under an odd number of streamers, and ∞ for any term under an even number or none.

V. If a proposition has its antecedent false or its consequent true, it is true; and conversely, if its antecedent is true while its consequent is false, the proposition is false. That is to say,

> If b, then $\overline{a} + b$
> If $\overline{a} +$, then $\overline{a} + b$
> If $\overline{a} + b$, then either $\overline{a} +$ or b.*

(*It must be carefully borne in mind that all our discourse in this part of the book is about individuals. Give $\overline{a} + b$ a categorical form, and the proposition stated seems false. Namely, it does not follow that because all men are mortal, therefore either every object is a non-man or else that every object is mortal; but it does follow that each single individual is either not a man or is a mortal.)

The first proposition follows by the rule of substitution from $\overline{a} + \infty$ and $\overline{\infty} + b$, which are true by Rule II. The second proposition follows from $\overline{a} +$ o and $\overline{o} + b$. The third may be proved by Rule III, as follows. First, put $a = \infty$, $b = \infty$, and the proposition becomes

> If $\overline{\infty} + \infty$, then either $\overline{\infty} +$ or ∞, i.e. $\overline{a} + a$

Now $\overline{a} + a$ is proved by Rule III, for both

> $\overline{\infty} + \infty$ and $\overline{o} +$ o

are true by Rule II. Second, put $a =$ o, $b =$ o, and the proposition becomes

> If $\overline{o} +$ o, then either $\overline{o} +$ (that is, $\overline{o} +$ o) or o,

and the first alternative is true, by Rule II. Third, put $a =$ o, $b = \infty$, and the proposition becomes

> If $\overline{o} + \infty$, then either $\overline{o} +$ o or $\overline{a} + a$.

Fourth, put $a = \infty$, $b =$ o. In this case, the proposition becomes

> If $\overline{\infty} +$ o, then either $\overline{\infty} +$ o or o.

Now if $\overline{\infty} + \mathrm{o}$ then $\overline{\infty} + \mathrm{o}$ is a case under $\overline{a} + a$.

VI. We have necessarily

$$\overline{a + \overline{b} + c} + \overline{b} + \overline{a} + c,$$

so that, by the *modus ponens*,

$$\overline{a + \overline{b} + c} = \overline{b} + \overline{a} + c$$

and antecedents can be transposed. This [is] proved by Rule III, for

$$\overline{\mathrm{o} + \overline{\mathrm{o}} + c} + \overline{\mathrm{o}} + \overline{\mathrm{o}} + c \text{ is true, since } \overline{\mathrm{o}} + c \text{ is true;}$$

$$\overline{\mathrm{o} + \overline{\infty} + c} + \overline{\infty} + \overline{\mathrm{o}} + c \text{ is true for the same reason;}$$

$$\overline{\infty + \overline{\mathrm{o}} + c} + \overline{\mathrm{o}} + \overline{\infty} + c \text{ is true, since } \overline{\mathrm{o}} + \overline{\infty} + c \text{ is true;}$$

$$\overline{\infty + \overline{\infty} + c} + \overline{\infty} + \overline{\infty} + c \text{ is true by identity.}$$

VII. We have necessarily

$$\overline{\overline{a + \overline{b} + c} + d} + \overline{a} + \overline{b} + c + d,$$

so that by the *modus ponens*,

$$\text{If } \overline{a + \overline{b} + c} + d, \text{ then } \overline{a} + \overline{b} + c + d,$$

and the ends of two streamers terminating together can be cut off to any point together. This is conveniently proved by a *reductio ad absurdum* by means of Rule V. The proposition can only be false if

(1) $\overline{a + \overline{b} + c} + d$ is true, while

(2) $\overline{a} + \overline{b} + c + d$ is false.

The second can be false only if

(3) a is true, while

(4) $\overline{b} + c + d$ is false.

The fourth can be false only if

(5) $\overline{b} + c$ is true, while
(6) d is false.

But if (1) is true while d is false,

(7) $\overline{a + \overline{b} + c}$ is false, and therefore
(8) $\overline{a} + b$ is true, while
(9) c is false.

But if c is false, while (5) is true,

(10) b is false.

And if b is false, while (8) is true, a is false, which contradicts (3).

VIII. Suppose two propositions such that when any term or terms are replaced by the circle, one proposition becomes false, while when the same terms are replaced by identity, $\overline{a} + a$, the other proposition becomes true, then the latter follows from the former, both in their original forms. For let A be the proposition which becomes $\bigcirc$, and let A' be what it becomes when identity is substituted. Let B be the proposition which becomes true, and let B' be what it becomes when the circle is substituted. Then the proposition $\overline{A} + B$, becomes in the two cases $\overline{\bigcirc} + B'$ and $\overline{A'} + \infty$, both of which are true; so that $\overline{A} + B$ is true by Rule III.

IX. We have necessarily

$$\overline{A + B + C} + \overline{\overline{C} + L} + \overline{A} + \overline{B} + L,$$

so that from

$$\overline{A} + \overline{B} + C \text{ follows } \overline{\overline{C} + L} + \overline{A} + \overline{B} + L.$$

For if either A or B is replaced by the circle the proposition is true. If both are replaced by ∞ and c by the circle, $\overline{A} + \overline{B} + C$ is false, and again the proposition is true. If C is ∞ and L is $\bigcirc$, $\overline{C} + L$ is false and the proposition is true, while if L is true the proposition is true.

X. Any two premises a and b may at once be united in the form $\overline{\overline{a} + \overline{b} +}$. For if the two premises are true, this form becomes $\overline{\infty + \infty + \bigcirc} + \bigcirc$ which is true because $\overline{\infty} + \overline{\infty} + \bigcirc$ is false.

Rule I

If a and b, then $\overline{\overline{a} + \overline{b} +}$.

This is the *rule of combination*. It is plain that we cannot reason, unless we can combine different premises; and $\overline{\overline{a} + \overline{b} +}$ expresses no more than that that a and b are both true at once. To show this, we use a diagram invented by Mr. Venn [Fig. 1].

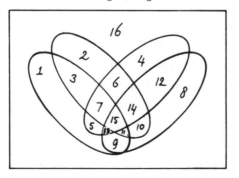

Fig. 1

The rectangle represents the whole field of possibility. The oval containing the odd numbers represents the area of possibility of a. The oval containing the even numbers not divisible by 4 and the odd numbers one less than [a multiple of] 4 (2, 6, 10, 14, and 3, 7, 11, 15) represents the area of possibility of b. The oval containing the numbers 4 to 7 and 12 to 15 inclusive represents the area of possibility of another fact c ; and finally the oval containing the numbers from 8 to 15 inclusive represents the area of possibility of a fourth fact d. The proposition $\overline{b} + c$ is true wherever b is not true or c is true, that is everywhere except in the compartments 2, 3, 10, 11. The proposition $\overline{a} + \overline{b} + c$ is true wherever $\overline{b} + c$ is true or a is not true, that is, everywhere except in compartments 3 and 11.

Finally, the proposition $\overline{a + \overline{b} + c} + d$ is true wherever $\overline{a} + \overline{b} + c$ is not true or $[d]$ is true, that is, in compartment 3 and all those numbered from 8 to 15 inclusive. Now the last expression $\overline{a} + \overline{b} + c + d$ becomes $\overline{a} + b +$ on substituting o for c and d. That is, we must erase the ovals c, and d. Then there remain only the compartments 1, 2, 3, 16, of which 3 is that where a and b are both true, and also that where $\overline{\overline{a} + b} +$ is true.

Rule II

$\overline{a} + \overline{b} + b$ is necessarily true.

This is the *rule of identity*. To prove it, we note as above that in the diagram the proposition $\overline{a} + \overline{b} + b$ is true everywhere except in the compartments 3 and 11. Now make $c = b$, that is, erase all compartments where either extend beyond the other, that is, compartments 2 to 5 and 10 to 13. Then we have erased 3 and 11 and consequently all where $\overline{a} + \overline{b} + c$ is not true. So that $\overline{a} + \overline{b} + b$ is true everywhere.

Rule III

$\overline{a} + b + c = \overline{b} + a + c$.

This is the *rule of commutation*. We have seen that $\overline{a} + \overline{b} + c$ is true everywhere except in compartments 3 and 11, which are symmetrically placed with reference to the ovals of a and b. Consequently, a and b can be transposed in the proposition without change of meaning.

Rule IV

If $\overline{a} + \overline{a} + c +$, then c.

This is the rule of the *modus ponens*. The proof of the rule of combination, by putting $\overline{a} + c$ in place of b, shows that $\overline{a} + \overline{\overline{a} + b} +$ expresses the conjoint truth of a and $\overline{a} + c$. Consequently, by the *modus ponens*, c follows from it.

Rule V

If $\overline{\overline{a + b} + \overline{b + c} + }$, then $\overline{a} + c$.

This is the *rule of syllogism*. It is proved simarly to Rule IV.

Rule VI

If $\overline{b} + $, then $\overline{b} + c$.

This is the *rule of contradiction*. It serves to state that negation is equivalent to saying that every consequent follows from the fact denied. It is proved by the diagram which shows that $\overline{b} + c$ is true over the whole field of possibility except a part of the area of b. If therefore the actual state of things is a point outside of b, it is within the area of the truth of $\overline{b} + c$.

Rule VII

If $\overline{\overline{a + b} + c + \overline{a + c} + }$, then c.

This is the *rule of dilemma*. It is proved like Rules V and VI.

We now have a complete algebra for qualitative reasoning concerning individuals. But it is not yet a very commodious calculus. To render it so, we introduce certain abbreviations which make it identical with the logical algebra of Boole as modified by Jevons and Mitchell.* (*Other modifications by Mr. Mitchell relate to the logic of relatives.) Namely, we first separate the streamer of the sign of illation from the cross, and in place of

$$\overline{a + } b \quad\quad \text{write} \quad\quad \overline{a} + b.$$

Second, wherever the sign of illation is followed by a blank we omit the cross, and thus

in place of $\overline{a} + $, write $\overline{a}$.

Third, as the sign of the simultaneous truth of a and b, instead of writing $\overline{\overline{a} + \overline{b} + }$, or $\overline{\overline{a} + \overline{b}}$, we write simply ab. This last is a superfluous sign, adopted for the sake of abbreviation. Our Seven Rules now take the following form.

    I. *Rule of combination.* If a and b, then $\overline{\overline{a} + \overline{b}} = ab$.

    II. *Rule of identity.* $\overline{b} + \overline{a} + a$.

III. *Rule of commutation.* $\bar{a} + \bar{b} + c = \bar{b} + \bar{a} + c$, or $\bar{a} + \bar{b} = \bar{b} + \bar{a}$.

IV. *Rule of the modus ponens.* If $a(\bar{a} + c)$, then c.

V. *Rule of Syllogism.* If $(\bar{a} + b)(\bar{b} + c)$, then $\bar{a} + c$.

VI. *Rule of contradiction.* If $\bar{b}$, then $\bar{b} + c$.

VII. *Rule of dilemma.* If $(a\bar{b} + c)(\bar{a} + c)$, then c.

MEANING (PRAGMATISM) (622)

I shall not attempt now any sketch of the awakening of the modern mind to the need of logical investigations, such as I shall endeavor to prepare for my large work upon logic. Suffice it to say that already before Mill some works had appeared that would have been of great importance for the education of the modern mind in the domain of reasoning, had they attracted the attention they deserved; and that shortly after his work there appeared a number of investigations that must be taken into account in attempting to describe the state of modern thought about reasoning. In 1837 Bernard Bolzano, a Catholic priest of Buda-Pest, produced a treatise on logic in four volumes. I deeply regret that I have never had an opportunity to read this work; for it is quite certain that its author did confer a signal and very singular benefit upon humanity of such a nature that it seems more likely than not that his logic contains matter of high value. Namely, that which he did was to propound a very simple and obviously correct definition of the relation of a greater collection to a lesser one. It was applied by mathematico-logicians, especially by Dr. Georg Cantor, and the result has been an opening up to clear thought of the nature of infinite quantity with an amazing system of infinite quantities, of which as yet little or no use, it is true, has been made, but which is, nevertheless, a great enlargement of our conceptions.

In 1847, an eminent mathematician of great strength and originality, George Boole, published a book of 82 pages which treated syllogistic reasoning from a mathematical point of view. I can state its two fundamental ideas in a few words, though I shall not present them exactly as Boole did, since first attempts to set forth fertile conceptions can always be improved upon, and I think that the way I shall state these two conceptions does them more justice than did Boole's first wrestling with them in the dark.

A proposition may have more *value* or less. A true assertion has more value than a false one. Let v (the initial of *verum*) stand for the value of a true proposition, or assertion, and let f stand for the value, or want of value of a false (*falsum*) one. Let us treat these as numbers, as they may very well be, since we can mark two numbers on things to signify anything we choose; and Leibniz had the habit of doing so with remarkable effects. Of course, we have at once the inequality

$$v \neq f$$

that is, v is not equal to f. Consequently, if x, the unknown value (as being true or as being false) of a proposition of whose truth or falsity we know nothing, be equal to v it cannot be equal to f; or

If $x = v$, then $x \neq f$.

The principle may be expressed in countless mathematical forms, but always as an *inequality*, never by an equation. Logicians call this the "*principle of contradiction.*"

Moreover, no matter what the proposition may be whose value is x, it is either true or false. In order algebraically to express that principle called by logicians *The Principle of Excluded Middle* (*Principium exclusi medii*, or *tertii*, "medium" and "tertium" having nearly the same meaning in Latin), he availed himself of the fact that if a product be zero, some one or more of its factors are zero, and wrote substantially

$$(v - x)(x - f) = 0.$$

This, or its analogue, holds for every proposition whatsoever, so that he might have written

$$\sum_i (v - x_i)(x_i - f) = 0,$$

making use of the principle that if a sum of quantities, none of which can be negative be zero, every one of them must be *zero*. Now if we assume $v > f$, none of the terms of such a sum can be negative.

I have now stated the main feature of his algebra, and I will show how he made use of it to draw necessary conclusions. Let x and y be the logical "values" of two different propositions, then

(1) $(x - f)(v - y) = 0$

implies that either $x - f = 0$, or $x = f$, or $x = v$, and x is false, or $v - y = 0$, $y = v$, and y is true; or both (but I shall invariably use "or" as admitting the possibility of *both* alternatives). Thus the equation is equivalent to the assertion that if x is true, so is y. In like manner, the equation

(2) $(y - f)(v - z) = 0$

will express that if y is true so is z. Now multiply the equation numbered (1) into $(v - z)$. This, since zero multiplied into any quantity yields zero as the product gives

(3) $(x - f)(v - y)(v - z) = 0$

Then, multiplying equation (2) by $(x - f)$, we get

(4) $(x - f)(y - f)(v - z) = 0.$

Adding (3) and (4), we get as their sum

(5) $(x - f)(v - f)(v - z) = 0.$

But, *by the principle of contradiction*, $(v - f)$ is not zero. We can therefore divide by this difference, without getting an indeterminate result; namely,

(6) $(x - f)(v - f)(v - z) = 0$

which equation means and says, "If x is true, so is z," which is the conclusion we have deduced from our two premises. In the same manner, if we write m_i to mean "i (whatever that may denote) is a man," d_i to mean "i is mortal," and S_i to mean "i is Socrates," and write Σ_i before any algebraical expression involving i in any way, with the understanding that the whole shall be equivalent to the sum of all the expressions that could result from substituting throughout any one of them a sign that should be restricted to denoting any one individual person or thing in place of i; then

$$\Sigma_i(m_i - f)(d_i - v) = 0$$

will assert that if any individual thing in the universe be a man, then that thing will be mortal; and in like manner,

$$\Sigma_i(S_i - f)(m_i - v) = 0$$

will assert that if anything be Socrates, then that thing will be a man; and eliminating m_i between the two just as we did above, we get

$$\Sigma_i(S_i - f)(d_i - v) = 0$$

which means that if anything in the Universe be Socrates, that thing will be mortal. We, thus, work out a stock syllogism.

I have now set forth the more fundamental of the two ideas which governed Boole in his logical system, and I will go on to set forth the other, which is Boole's virtual answer to this question, Is there any reason for preferring particular numbers to the algebraic indeterminate v and f? I say the virtual answer because Boole never separated the questions, and never discussed them. Two particular numbers occurred to him as enabling him to work out algebraically problems in necessary reasoning, and though he was struck with the advantages of those two numbers he never thought of any others as possible. I introduce these questions as throwing great light upon the nature of Boole's method. This question of the particular numbers is a question of convenience, and depends upon the forms in which the values of different assertions are prevailingly combined. Now by far the most important forms of combination of assertion are the two which are known in logic as *composition, copulation*, or *conjunction*, and *disjunction*, or *aggregation*. Copulation is combination by "and" or some equivalent. Aggregation is combination by "or," taken in the sense in which it does not deny copulation, "or or more." The words refer to the meaning, and not to the use of the particular grammatical conjunctions, which I have used to show what I mean. Quantitatively considered, any assertion formed by combining other assertions is a "function" of those assertions, as its "variables." I need not here give an absolutely precise definition of the correlative mathematical terms "function" and "variable." Suffice it to say that when the value of a quantity, F, depends, in a fixed way, upon the values of one or more quantities x, y, z, etc. whose values are considered as being less fixed, F is said to be a *function* of x, y, z, which are termed its *variables*; while any other quantity upon which the value of the function depends, but which is considered as more fixed in value than the variables is sometimes called a *"parameter"* (a word with other meanings). For example $\sqrt{h^2 - a^2}$, where h is the hypotenuse and a a leg of a right triangle, is a function of h and a as variables; but if we regard it as $(h^2 - a^2)^p$, where p has the particular value $\frac{1}{2}$, we may call p its "parameter." The two chief functions of assertions, the copulate and the aggregate, belong to the particular class of functions that mathematicians term "means." By a *mean* mathematicians understand a function of an indefinite number of variables (that is to say, the number of variables may be increased without the mathematician regarding the function as thereby necessarily changing its character at all), which function has two characters common to all "means." Namely, in the first place, every "mean" is a "symmetric" function, that is to say, if two of the variables interchange values, that interchange can never affect the value of the function; thus $x^n + y^n + z^n$ is a symmetric function, if x, y, z, are regarded as its only variables, and $xy + yz + zx$ is another. In the second place, whenever all the variables of a "mean" take the same value, no matter what possible value this may be, the "mean" itself takes likewise that value. The four kinds of means most used by mathematicians are the "arithmetic mean," $(1/n)(x_1 + x_2 + \ldots + x_n)$; the "geometric mean," $\sqrt[n]{x_1 x_2 \ldots x_n}$; the "harmonic mean," $n/(x_1^{-1} + x_2^{-1} + \ldots x_n^{-1})$; and the "quadratic mean,"

$$\sqrt{\frac{x_1^2 + x_2^2 + \ldots + x_n^2}{n}}.$$ A sum or a continued product are not means; for though both are symmetric functions, if $x_1 = x_2 = $ etc. $= x_n$, we have, in such case, $x_1 + x_2 + \ldots + x_n = nx_1$ and not x_1, and $x_1 x_2 \ldots x_n = x_1^n$ and

not x_1. But the copulate and the aggregate of assertions are means. For if A_1, A_2, etc. are assertions, the copulate assertion, "A_1 and A_2 and A_3 and etc.", if A_1, A_2, A_3, etc. are all false, is false, and if they are all true, is true; and the same is true of the aggregate, or disjunct, assertion, A_1 or A_2 or A_3, etc.

As being means, the copulate and the aggregate are of a radically different nature from sums or products. Nevertheless, it would be permissible to regard a product as a mean in an algebraic system in which the only values were zero, unity and a positive infinity and positive infinitesimal both of the zero order; since $0 \cdot 0 = 0$, $1 \cdot 1 = 1$, $+ \infty^0 \cdot (+ \infty^0) = + \infty^0$; and it would be permissible to regard a sum as a mean in a system whose only values were zero, perhaps infinitesimals, and infinites of zero order, or, perhaps, of any even order, since $0 + 0 = 0$, and $+ \infty^0 + \infty^0 = + \infty^0$, and I do not know but a similar equation would hold for any infinite of even order. I am pretty positive that $\infty^1 + \infty^1$ is indeterminate, For $\infty^1 = \frac{1}{0}$ and therefore is equal to $- \infty^1 = \frac{1}{-0}$; and, it is evident that an endless line with an initial point may be cut off from another similar line [and] can leave a line of any length. As for objections to what I have been saying founded on Cantor's doctrine of ordinals (which doctrine I approve so far as it is pertinent to the points here considered), I will not detain the reader with discussions of them; but I am confident that, properly understood, they in no way controvert the old doctrine of orders of infinity. Since then $0 + \infty^0 = + \infty^0$ while $0 \times \infty^0 = 0$ (as appears from $\frac{1}{\infty^1} = 0$), I think that there would be no serious objection to making $v = + \infty^0$ and $f = 0$. But I doubt whether there is any other logical way of representing copulation by multiplication and aggregation by addition, as was independently proposed by Jevons, perhaps by myself, by M'Cosh, and by Schröder. If we write C for copulate and $\overset{n}{\underset{1}{C}}_i x_i$ for the copulate of x_1, x_2, etc. up to x_n inclusive, and write A for aggregate and $\overset{n}{\underset{1}{A}}_i x_i$ in a similar way, we shall have the simplest way of expressing these algebraically as means,

$$\overset{n}{\underset{1}{C}}_i x_i = \frac{(x_1 - f)(x_2 - f) \ldots (x_n - f)}{(v - f)^{n-1}} + f,$$

$$\text{and } \overset{n}{\underset{1}{A}}_i x_i = v - \frac{(v - x_1)(v - x_2)(v - x_3) \ldots (v - x_n)}{(v - f)^{n-1}}.$$

The copulate is false if any of its variables is false, while the aggregate

MEANING (PRAGMATISM) (622)

is true if any of the variables is true. If we adopt these expressions for the copulate and aggregate, no assertion need enter into any expression except in one of the forms $\dfrac{x-f}{v-f}$ and $\dfrac{v-x}{v-f}$; and this will be true of copulates and aggregates themselves; and writing $C(x_1, x_2 \ldots x_n)$ for the copulate and $A(x_1, x_2 \ldots x_n)$ for the aggregate we shall have in place of the above formulae,

$$\frac{\overset{n}{\underset{1}{C}}_i x_i - f}{v - f} = \overset{n}{\underset{1}{\Pi}}_i \frac{x_i - f}{v - f} \text{ and } \frac{v - \overset{n}{\underset{1}{A}}_i x_i}{v - f} = \overset{n}{\underset{1}{\Pi}}_i \frac{v - x_i}{v - f},$$

where $\overset{n}{\underset{1}{\Pi}}_i$ means the continued product of all the factors each of the form that follows, where in place of i are to be substituted successively all the whole numbers from 1 to n inclusive. Thus, if $n = 2$ the equation involving C is

$$\frac{C(x_1, x_2) - f}{v - f} = \frac{x_1 - f}{v - f} \cdot \frac{x_2 - f}{v - f}$$

But the same equation means that if $n = 3$,

$$\frac{C(x_1, x_2, x_3) - f}{v - f} = \frac{x_1 - f}{v - f} \cdot \frac{x_2 - f}{v - f} \cdot \frac{x_3 - f}{v - f}.$$

But if $n = 4$,

$$\frac{C(x_1, x_2, x_3, x_4) - f}{v - f} = \frac{x_1 - f}{v - f} \cdot \frac{x_2 - f}{v - f} \cdot \frac{x_3 - f}{v - f} \cdot \frac{x_4 - f}{v - f}$$

The formula relative to A is to be interpreted in a similar way. But the reader will do well to fix his attention, at first, upon the structure of the equation involving C. The truth of the copulate consists of the truth of all its members, just as $\dfrac{C-f}{v-f}$ consists in the product of all factors of this form $\dfrac{x-f}{v-f}$ where the x is any one of the members of the copulation. The suggestion of the form is too urgent to be resisted that $\dfrac{C-f}{v-f}$ represents the truth of the copulate, while $\dfrac{x_i-f}{v-f}$ represents the truth of one of its members. But is not the degree of truth of a proposition precisely what we have been meaning by its logical value? If so, and if the product means here the simultaneous truth of all its factors, we must admit that

$$x = \frac{x - f}{v - f}$$

Namely, its two possible values being v and f,

$$v = \frac{v - f}{v - f} = 1 \text{ (since } v \neq f,)$$

$$\text{and } f = \frac{f - f}{v - f} = 0.$$

This gives us precisely the values of v and f which Boole adopted for reasons that he was unable to put into distinct and appraisable form.

Let us turn to the consideration of the Disjunctive Combination, which asserts that 'Either x_1 or x_2 or x_3 etc. up to x_n is true' and goes by the rather unfortunate name of the Aggregate. The equation

$$\frac{v - \overset{n}{\underset{1}{A}}_i x_i}{v - f} = \overset{n}{\underset{1}{\prod}}_i \frac{v - x_i}{v - f} = \frac{v - x_1}{v - f} \cdot \frac{v - x_2}{v - f} \cdot \frac{v - x_3}{v - f} \cdots \frac{v - x_n}{v - f}$$

asserts that the falsity of the aggregate can only consist in the falsity of all its members. Accordingly we see that $\dfrac{v - x}{v - f}$, whatever assertion x may be, is the falsity of x. But putting $v = 1$ and $f = 0$, as we have just found them to be, the expression for the falsity of x reduces to $1 - x$. Accordingly, the aggregate itself, that is to say, its truth, is

$$\overset{n}{\underset{1}{A}}_i x_i = 1 - (1 - x_1) \cdot (1 - x_2) \cdot (1 - x_3) \ldots (1 - x_n) =$$

$$1 - \overset{n}{\underset{1}{\prod}}_i (1 - x_i).$$

I have now given a slight sketch of the theoretical foundation of Boole's system, though not at all of its beautiful developments, wherein its principal utility for ordinary minds chiefly lies. Not that I would recommend the use of this or of any other system of formal logic in ordinary reasonings. Its utility consists in its giving one a clear conception of logical relations, both of simple and of intricate kinds. My account of its theoretical foundation is not Boole's; for Boole did not distinctly apprehend the nature of his own system. Mathematics had to make the general advance that it has since made before it was likely that anybody would comprehend the matter. Its publication made considerable sensation in the intellectual world, and somewhat in proportion to the intellectuality of those who looked into the new discovery. The argument in favor of the special values of v and f that Boole adopted, $v = 1$ and $f = 0$, is a strong one although it is not conclusive. It would be an excellent exercise in reasoning for I care not whom, to formulate this precisely, to show exactly what its premises are and just what it does prove. I have not mentioned, however, the strongest reason for the choice of numerical values of v and f. This reason lies in the connection of the system with the calculus of probabilities, in which affirmative certainty is represented by 1, and negational certainty by 0, and in which the probability of an aggregate and that of a negation are just what they are in Boole's algebra of logic, the former becoming, as with him, the sum of the probabilities of the members when these are mutually *exclusive*, while the probability of a copulate, or compound event, is the product of the probabilities of the members when these are mutually independent. It would not be a bad exercise in logical analysis, by the way, accurately to make out what is the precise relation between *mutual exclusion* and *mutual independence*. But it must be noted that it is quite obviously no more necessary to make affirmative certainty 1, and negational certainty 0, in the doctrine of chances, than it is in the algebra of logic. Yet in the one field, as in the other, the advantages of doing so are decided, and in most problems very nearly conclusive. It was the application by Boole of his logical algebra to the doctrine of probabilities that proved that it was not a mere brilliant idea but was of real and great importance for logic. His treatment of probability is given in the important work in which he gave the full development of his algebra of logic. The title-page of that treatise is worthy of attention. It reads: "An Investigation of the Laws of Thought on which are founded the Mathematical Theories of Logic and Probabilities. London: Walton and Maberly. Cambridge: Macmillan & Co. 1854." It is a volume in 8vo of 424 pages. Please to note the date, to begin with. It was before the epoch of Weierstrass; that is to say, before the present styles of mathematical reasoning had come in. Riemann had put forth nothing but his then too recent Inaugural Dissertation, von Staudt had not appeared, none of Cayley's really great memoirs had been written; and in short, the mathematics of the present day was, from a logical point of view, not yet fully out of the shell. It is not surprising, therefore, that Boole's volume is largely filled up with vague metaphysicoidal matter, and that he would have been utterly unable so much as to give a correct definition of Probability. It is, on the contrary, a proof of the power of his idea that it enabled him to treat the doctrines of Probabilities in such a manner as completely to refute all those calculations of Laplace that amounted to assigning a definite probability to an inductive conclusion. If his own treatment of the problem was dense fog, nevertheless it contained the fertile egg of a comprehension of the true relation of Probability to Induction.

At the same time as Boole, another mathematician, Augustus DeMorgan, began an important series of works on logic. His first memoir was presented to the Cambridge Philosophical Society on Nov. 9, 1846. It was followed by an octavo treatise entitled "Formal Logic" in 1847, by a "Syllabus of Logic" in 1860, and by four other extensive quarto memoirs, running from 1850 to 1863. As a mathematician DeMorgan lacked the genius of Boole, but his ideas were far more lucid. He opened up the logic of relations, which is of high importance; and he gave a complete formulation of an important kind of necessary reasoning which had been entirely overlooked. Those two make up the sum of the world's obligation to him: it is not slight.

We now come to living men. Decidedly, the most profound among those whom I am to mention is, incomparably, Alfred Bray Kempe. His Memoir on Mathematical Forms, published in the Philosophical Transactions for 1886, must, when sufficient time has elapsed to prepare the mathematically thinking world for the study of it, prove of much avail. I have only, I much regret to say, by accident come across some of his other contributions to logic. But one of them treating of the relation of betweenness attacks a very fundamental question in an ingenious manner. Our American Josiah Royce has an extensive memoir on this subject. The Method of Least Squares, which, whatever its relation to the Calculus of Probabilities may be, is usually treated as an application of that method, having been considerably illuminated by the works of the late General Annibale Ferrero, of Crofton, and others and later by those [of] Dr. Karl Pearson, has been applied to questions of heredity by the last, by Francis Galton and his disciples in ways which, if they have added nothing new to the theory of logic, have much improved the practice of reasoning about such subjects. Mr. John Venn, also, wrote a book called the "Logic of Chance" and another on "Empirical Logic" which have been of use in their ways.

Another set of men who have done still more for logic by much profounder thoughts are those who have studied infinite numbers, both ordinal and those that express multitude. Dr. Georg Cantor has been the chief of them. . . .

ABSTRACTS OF 8 LECTURES
[TOPOLOGICAL BASIS OF PHILOSOPHY OF CONTINUITY]
(942)

We thus see that the bare Nothing of Possibility logically leads to continuity.

For the first step a unidimensional continuum is formed.

Logically, this step is of the nature of induction. Now induction arranges possible experience after the type of logical law. But the logical law *par excellence* is that of logical sequence. Hence, the first dimension of the continuum of quality is a sequence. A sequence is a unidimensional form in which there is a difference between the relation of A to B and of B to A. Mathematically considered, in one dimension it is a progress from a point A to a point B, where A and B are different [as in Fig. 1] or A and B may coincide [as in Fig. 2], or they may both vanish [as in Fig. 3]. Of these three forms of sequence, the first is

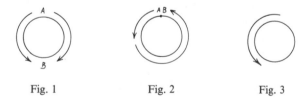

Fig. 1   Fig. 2   Fig. 3

distinctly that of logic since the ultimate antecedent and the ultimate consequent are different in logic. You cannot proceed from antecedent to consequent till you reach again your original antecedent (as in the 3rd kind of sequence, the elliptical), nor do you *tend* to such a return (as in the second, or parabolic sequence), but the two are distinct.

It follows that the first dimension of the continuum of possible quantity had to be of the nature of a hyperbolic sequence. That is to say, there is one general mode of relation, which we may name *coming after*, defined by these conditions:

1st, of any two qualities which are not entirely alike in their relations of *coming after*, one *comes after* the other;

2nd, whatever *comes after* another comes after whatever that other comes after; or otherwise stated if N comes after M then whatever, say P, comes after N also comes after M;

3rd, nothing comes after itself.

This general relation among qualities is that of *greater* intensity. This seems at first a somewhat arbitrary identification. But careful consideration will show that it is not so arbitrary after all.

But before showing this, we must ask what form the logical law takes as applied to two or more dimensions. Now the logical sequence itself is essentially unidimensional, because it is a purely internal law, and unity and interiority are inseparable. The problem of applying this law to a multidimensional continuum is therefore the mathematical problem of the *Potential*. For a *potential* is a unidimensional continuum, — or, in the usual mathematical parlance, a continuous system of real scalars, — distributed through a space or pluridimensional continuum.

There is a book about this general problem, by Neumann. There is much about it in Maxwell's *Electricity and Magnetism*. For two dimensions it is considered in the Theory of Functions. It is closely allied to the Theory of Distance in the Non Euclidean Geometry. There are different solutions. Thus, for two dimensions Maxwell suggests two general forms of solution quite independent of any metrical or even projective i.e. optical considerations. Namely, in the first place the absolute maxima and minima potential may be at points, say the maximum at the earth's North Pole, the minimum at the earth's South Pole and the parallels of latitude for equipotential lines. In the second place, the absolute maxima and minima potential may be at *lines*; say the maximum at one meridian, the minimum at another and the other equipotential lines meridians likewise.

But the law of logic calls for a different solution. The Logical sequence has from early times been considered as a *tree*. The subjective logic which I have sketched in the former lectures of this course proves that it is so. The ultimate antecedent is a *zero* without extension; the ultimate consequent is a vast manifold. Hence the continuum of possible quality in N dimensions must be in a sequence starting from a point and expanding to a final limit of $N - 1$ dimensions. Logic radiates like light.

At one end of the sequence then all the qualities come together in a *zero*. But they are separate from one another as they separate from zero. This you perceive describes the relation of varying intensity. A green with zero luminosity is the same as a red of zero luminosity, or a sound of zero loudness. It is all one zero.

Such a form of relation is equally possible in an artiad and a perissid continuum. This is a distinction which I must explain. The boundary between two bodies, say that which is common to the space a mass of ice

in the midst of water occupies and the water space about it, is a *surface*.
If the ice is only partially submerged, there is an ice-water surface, an
ice-air surface, and a water-air surface, and these three have in common
an ice-water-air line. But suppose the piece of partially submerged ice is
attached to the side of the wooden bucket. Then there are besides the
ice-water-air line, an ice-water-wood line, an ice-air-wood line, and a
water-air-wood line, and these four lines have in common,—not a
point, but *two* points, where ice, water, air, and wood all come together.
Now space is *artiad* if four solids always abut upon one another in an
even number of points, as 0, 2, 4, etc. But space is *perissid* if four solids
can abut upon one another in an *odd* number of points. In what is called
Projective Geometry–and ordinary Analytical Geometry,—space is
represented as *perissid*; while quaternions make it *artiad*. Thus according
to projective geometry, if against an endless ray or straight line you were
to place one edge of a quadrant shaped endless copper bar whose cross
section nowhere becomes a point and beside that an endless Iron bar like
it and beside that an endless Tin bar and beside that an endless Lead
bar–you would have to give the whole a half-twist to make them fit
together [Fig. 4]. For if instead of the twist, you enabled them to lie

Fig. 4

together by sawing the whole through at one cross section then at that
section Copper would abut upon Tin and Iron upon Lead, and the centre
of that cross section would be the one sole point where each abuts upon
all the rest,—thus proving the space of that theory to be *perissid*. Of
course the combined bar would not be a cylinder. Because a cylinder is a
cone with the vertex at infinity, according to projective geometry. The
bar might have an outer surface like a *phimus* or one-sheeted hyper-
boloid [Fig. 5]. Now, it would seem, at first, that the law of logic would

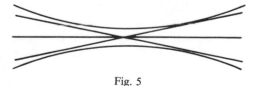

Fig. 5

not determine whether the continuum of possible quality were artiad or perissid. It would then be the nature of quality within itself which would determine that. But quality has in itself unity—from without endless manifoldness. The idea of duality is foreign to it. But an artiad continuum embodies duality; since it has *even* numbers appearing in all sorts of ways in its properties. It is stamped with duality in ever-repeating pattern like a pestering wall paper. So that the continuum of possible quality might well be assumed to be *perissid*.

Here, however, we abut upon a necessary consequence which [is], at once, apparently in direct conflict with experience and at the same time directly contrary to what has just been remarked of the non-duality of quality in itself.

To show this, it will be convenient to explain a geometrical term and conception of my own invention. You know that a *point* is an indivisible place. A *particle* is a thing which at any one instant occupies a point. A *line* is a place which a particle can occupy in the course of time. In no time, even infinite, can a particle occupy all the points of a surface. Let us use the word *filament* to mean a thing that at a single instant can occupy a line. Then a surface is a place that a filament can occupy in the course of time. Using the word *film* for a thing that in one instant can occupy a surface, a *space* is a place that a film can occupy in the course of time. Now a particle can start to move two ways along a line, forward or backward. A filament can start to move two ways on a surface so as instantly to leave the whole of its line. A film can start to move two ways in a space, and so on for higher dimensions. But a line-figure or figure composed of lines and points alone may have *singular* points from which a particle can move in less or more than two ways, namely, *isolated points*, from which the particle can move *zero* ways, *extremities*, from which it can move *one* way, and nodes of furcations from which it can move in 3 or more ways. So a surface may have singular *lines* from which a filament can move zero ways or one way (i.e. boundaries) or several ways i.e. lines of splitting. I call such places *Topically singular* places, or *singularities*. Now I say unless we suppose the continuum of possible quality to have topical singularities,—which is quite inadmissible at the present stage of development, though they may be evolved later,—then we must admit that when a quality diminishes in intensity to *zero* it can then continue the same line of change and increase in intensity in a definite quality, a *contrary quality*. And moreover owing to the perissid character of the continuum, the absolute maximum of any quality is identical with the absolute maximum of its contrary.

This, I say, appears to conflict with experience. But that need not trouble us because we know that sensation as a limen[1] which is a point of discontinuity, a boundary where the mathematical law breaks. This singularity cannot exist in the possible quality itself, but must be established at a later stage of the development.

But what does seem puzzling is that we find a duality to be intrinsic to variations of intensity. A little reflection will show this arises from the duality in the logical law itself. Logic takes its origin in duality. If there is no difference between yes and no, truth and falsehood, logic has no *raison d'être* at all. No doubt, logic encourages the idea of contrary qualities, for the antecedent is so far as it goes the negation of a similar consequent. But what logic absolutely insists upon is the absolute and ultimate difference between the consequence of contraries. That is where everybody sees at once that Hegel is wrong. It is only a long drawn out series of logical blunders upon which that peculiarity of his system is built. Hence, it is the inherent dualism of the logical law which compels us to abandon the idea of the continuum of possible quality being perissid, or if it was at first perissid must have insisted upon supplementing it with another half so as to make it artiad. This is a mathematically possible transmogrification. For instance, the gnomonic projection of the sphere, which represents every great circle by a straight line is perissid; for the parts of the plane at infinity represent a great circle (since it is an optical projection from the centre) and thus there is a straight line at infinity, which is the characteristic of a perissid plane. But this projection only shows half the sphere. To represent the whole two sheets are necessary, and these two sheets together make an artiad surface, namely the degenerate quadric surface formed by two planes cutting one another

---

[1] Among fragments the following is found: "To say that every quality has a contrary quality, that hobby horse of Bain, seems to be utterly refuted by psychology. That, however, need not trouble us much; for the absence of the contrary quality may be due to some discontinuity connected with the human mind and introduced by its peculiar constitution. And we find that in point of fact this suggestion is remarkably confirmed. For precisely at the zero of sensation we have the phenomenon of the *limen*, on one side of which the mathematical law holds good yet has fact corresponding to it on the other. Thus there certainly is a psychological discontinuity, at the very point where it would have the effect of destroying the contrary quality. And this seems to be peculiar to peripheral sensations, for internal feelings can mostly appear to have contraries. At any rate, the contrary quality is an unavoidable consequence of the non-singular continuity of quality; and that it is quite impossible to doubt. That difficulty, therefore, is not serious."

Peirce probably borrowed the concept of the *limen* from Fechner. Herbart had earlier made it familiar having had it from Leibniz. In MS. 1221 Peirce discusses the psychological principles that underlie punctuation and says, "One of these, perhaps the chief of them, is, that excitations which perhaps do not reach the *limen* of sensibility, or, if they do, produce, at any rate, severally, no sensations which any effort of attention can assure us of, may, nevertheless, in the aggregate of very numerous occurrences, produce very serious effects.

in a ray,—in this case the ray at infinity.

Thus, there is an incipient attempt of the continuum of possible quality to be perissid; but the logical law at once doubles it and makes an artiad surface.

Moreover, after having added the second sheet, it separates the two at their junction so that their section changes [as in Fig. 6].

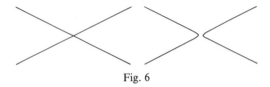

Fig. 6

And yet a relic of the original junction must remain. That I must concede to a Hegelian view. And since the sequence is hyperbolic, a singularity namely a boundary is established at the absolute maximum of quality, and thus the continuum of possible quality only occupies half of a larger continuum, of which the other half can only be a duplicate or rather a mirrored reflection, or as the mathematicians say a *perversion*, that is a looking glass reversion. It would be quite in the Hegelian spirit to admit that the infinite degree of any quality is identical with the infinite degree of its contrary.

The objection to it, however, springs from the logical law which intensity typifies. Logic takes its origin from duality. If there is no difference between *yes* and *no*, between truth and falsehood, logic has no *raison d'être* at all. The logical law supports the theory of contrary qualities, since an antecedent is as far as it goes the denial of a similar consequent. The logic of the copula makes that clear. But logic uncompromisingly insists upon the ultimate and absolute difference between the consequences of contraries. Everybody who is not dazzled by Hegel sees that it is only his personal fondness for parody and weakness of logical accuracy which has made possible that long drawn series of logical sophisms upon which his system upon a superficial view as well as in his own eyes seems to be built. But the inherent dualism of the logical law make the identity of contraries even in their infinite maxima quite inadmissible. This objection, therefore, is decisive.

It follows then that the logical law requires the maximum of quality to form an artiad continuum which is a limit separating the whole continuum of possible quality into two regions.

Now in a world of pure possibility arbitrary complications are as inadmissible as are arbitrary limitations. But if that boundary is artiad, it is simplest to suppose the whole continuum artiad, and therefore we are forced to hold it to be such.

The line of variation of a quality which leads up to its absolute maximum, must however, since it is continuous and without arbitrary limitations, extend beyond that maximum and since it does not then enter a degree of the contrary quality and cannot enter any other quality it must in the second region reënter that very same quality. It now passes away from the maximum and of course diminishes. And this diminution will continue till a zero is reached.

We are thus forced by this kind of exact reasoning about the subject to conclude that the continuum of possible quality consists of two regions which are just alike (non-metrically, of course, for measurement is as yet undeveloped); are just alike, I say, except that each one is like the other reversed or, as the mathematicians say, perverted like a looking-glass image. But these two regions thus mirroring one another are in their infinite maxima identical. Now what is the significance of this?

Every quality in itself is absolutely severed from every other. It has no relations, no parts, no degrees of intensity. It is nothing but what it is for itself; and it cannot be represented or expressed in anything else as it is for itself.

But quality generalized, as it is in the continuum of quality, is essentially *represented*. Without being represented in something else, it cannot be what it is. There is that essential feature of duality in it. The quality, or tinge of consciousness, which *seems*, and the quale-consciousness, which *feels* that quality, are now two, because the quality, being generalized, and continuity we remember is generality, is capable of entering different consciousnesses. Indeed, though it is distinguishable from consciousness by this very plurality, yet it cannot *be* in its generalized state without the possibility of being felt.

The intensity of the quality is one thing, the intensity of the feeling that represents it is another. I dimly remember when I was a boy going aboard a naval vessel with the gentleman who then and there took command of it, and a salute was fired with the brass Dahlgren guns then used in the navy. They were horribly noisy and I still have a *very faint* consciousness of a sound which I call to mind as being of *most dreadful intensity*. I do not suppose two different intensities are present to me, but it is like a painter's scale of values, where the brightest color his pigments afford represents the brightest color in nature, and the rest are scaled down correspondingly.

Quality, however, in its infinite degree is at a point of discontinuity. Every infinity is recognized by mathematicians as a discontinuity, and is truly such according to the strictest logical analysis. The infinite degree of a quality is thus broken off from the continuum. It is in no possible comparison with other degrees or other qualities. It is just that quality in and for itself which we have seen is not represented in any consciousness other than its own. That explains why the two regions of the continuum are at one in their maxima.

Let me say, by the way, that there is in the logical law this difference between the absolutely first antecedent and the absolutely last consequent, both of which are unattainable limits. The last consequent is the very reality itself. That is our very conception of reality, the essence of the word, namely, what we should believe if investigation was carried to its furthest limit where no change of belief further was possible. That is of the nature of an infinite, a true singularity of the logical continuum differing *toto caelo* from every intermediate step however near to it. I mean that it thus differs, not merely in its logical relations as leading to no consequent other than itself, but also and more particularly, as being a radically different kind of consciousness, a consciousness which is the very reality itself and no mere image seen *per speculum in aenigmate.* But the absolutely first antecedent is simply the blank ignorance, the *zero* of knowledge, although in its logical relations it is singular in leading to nothing, as a needle precisely balanced on its point will never fall, yet as a state of mind it differs indefinitely little from other states near it. Hence, though a limit as to the advance of logical development, it is not so as a mode of consciousness.

I have not yet done with the continuum of possible quality. I shall return to it in a few moments. But it will be convenient to consider at once the next step of the evolution.

Since the world of quality can be represented in a consciousness there can be a multitude of such representations, upon the same principle upon which the unity of immediate consciousness viewed from without is seen to be limitlessly manifold. There will be, indeed, upon that same principle a continuum of such representations. This continuum will lead from one feeling of the quality through others up to the very quality itself. Now will this, as final consequent and the very truth be an absolute limit? Not at all, for the reason that quality in intermediate degrees of intensity is itself of the nature of consciousness. As one of the feelings copies another and is thus a mediate copy of the quality, so the quality in intermediate degrees is only a feeling which may be regarded as copying other feelings, and there will be no limit. Every feeling of the series can be regarded as a mediate copy of itself. The continuum returns into itself in this dimension. I shall term the continuum of possible quality so enlarged by this new dimension the continuum of possible feeling.

I need hardly say that in this lecture I am merely sketching a series of theorems which are capable of a much more strict demonstration. But I do not say that there are no lacunae, no saltus, still remaining in the demonstrations which I have as yet succeeded in obtaining. I only say they are far more nearly perfect [than] I can [make] them in one lecture.

But now to return to the continuum of possible quality. Every complexus of qualities is a quality, and as such, considered by itself, is all that is in and for itself. Not only every complex of qualities but every generalization of such complexes is a possible quality. Every such quality makes a dimension of the continuum of quality. But in this way, the dimensions of the continuum of quality ought to exceed every discrete multitude. In short, they should form a continuum. But there is no such thing as a continuum of dimensions. It is impossible. Hence, these dimensions of complex quality are only abstractly possible. They cannot have simultaneous being in the world of potentialities.

What then does the logic of events require? What is required, as an objectively hypothetic result, is that an arbitrary selection of them should crowd out the others. This is *existence*, the arbitrary, blind, reaction against all others of accidental combinations of qualities. Here we have very nearly Hegel's definition of existence as the immediate unity of reflection into self and reflection into another.

All objective hypothetic results involve an arbitrary element, and thus offend against current notions of logic. But it is I believe most true and a proposition which will force itself upon anyone who considers the subject long enough that individual existence depends upon the circumstance that not all that is possible is possible in conjunction.

But existence is continuous as far as the nature of the case admits. At every point of it, it reunites all qualities each in some degree.

The thisness of it consists in its reacting upon the consciousness and crowding out other possibilities from so reacting.

It has often been said that the difference between the real world and a dream is that the real world coheres and is consistent. Undoubtedly this is the principal characteristic. The real events conspire as it were against the unreal ones, because there is not room for all. But observe that the logical analysis of this statement is that existence has its root in *pairing*. As soon as duality appears, as in contrast between the quality and the feeling of the quality, there is already an adumbration or prophetic type of real. But it is the composite of pairing in exclusion of another pair with the banding or pairing together of such exclusive pairs which produces *thisness*.

The word *individuality* applied to *thisness* involves a one-sided conception of the matter, as if unity and segregation were its characteristic. But this is not so. Segregate unity belongs to immediate consciousness, to quality; and wherever it appears that is its real origin. The true characteristic of *thisness* is duality; and it is only when one member of the pair is considered exclusively that it appears as *individuality*.

But what is commonly in our minds when we speak of individuality is a positive repugnance to generality. Our thoughts are so impregnated with generality, that we look at everything from its standpoint. Instead of thinking of *thisness* as it is in itself and for itself, we think of it in its relation to generality. But then we so exaggerate the importance of *feeling* or immediate consciousness, that we are accustomed to think of generality as characterized by *unity*, instead of by mediation, and its positive contrary, thisness, we think of as also characterized by unity,— which is logically absurd. Positive anti-generality is not unity, but duality,—the setting of objects over against one another with a great gulf between, instead of conceiving them as cases joined by a continuous medium or perpetual thirdness.

The quality, by and for itself, is not general. But it is not positively repugnant to generalization. Its unity is not lost by the generalization. The generalized quality is still *one*. But when duality is subjected to generalization, the twoness disappears from the continuum. Instead of having either *A* or *B*, we have a merging and blending of the two into a pervasive mediation. Thus, morality divides actions by a sharp line into *honest* and *dishonest*; but a generalizing view, whatever the advantages and disadvantages may be, conceives of all degrees of openness and of concealment as forming such a continuum that no sharp line can be drawn.

The *thisness* of the accident of the world of existence is positively repugnant to generality. It is so because of its intrinsic duality; and if you call it *individual* you are forgetting one term of the pair.

For example, a *this* is an object, but it only is so, by virtue of being in reaction with a subject. A *this* is accidental; but it only is so in comparison with the continuum of possibility from which it is arbitrarily selected. A *this* is something positive and insistent, but it only is so by pushing other things aside and so making a place for itself in the universe.

Thisness, in short, is *reaction*. Whatever reacts against something else is a *this*; and every this so reacts. Reaction is duality. All duality is like reaction in the world in which the duality subsists. If the duality is merely mental,—my own arbitrary pairing of things,—then the one idea reacts upon the other in my arbitrary thought. But if the reaction belongs to an arbitrary system of pairs, which insist upon being paired,—and which system sets itself off an emphatic second to the world of possibility, then we say it is real reaction.

Every reaction is antigeneral. It is *this* act. It is act not power. Second-ness not firstness.

Although reaction is antigeneral, blind, and brutal, in itself, it is nevertheless more or less won over into a general system.

Each accident of reaction is in itself, in its own blindness and brutality, absolutely unrelated to any other. But just as the qualities, which as they are for themselves, are equally unrelated to one other, each being mere nothing for any other, yet form a continuum in which and because of their situation in which they acquire more or less resemblance and contrast with one and other; and then this continuum is amplified in the continuum of possible feelings of quality, so the accidents of reaction, which are waking consiousnesses of pairs of qualities, may be expected to join themselves into a continuum.

Let me remind you that three things do not necessarily make a genuine triad or have what I call *threeness*, except in a modified and reduced sense. By a genuine triad I mean three things in a genuine triple relation which is not a mere mixture of dual relations. Hence when I say that accidents of reaction are pairs of qualities, I do not mean to deny that they may be compounds of more than two qualities, since in general every quality is concerned, but merely that the mode of combination is strictly com*bin*ation, such a relation as two things can have. So there may be a large family of brothers and sisters, but then relationship is nothing but the mixture of the relations of the different pairs.

Since, then an accidental reaction is a combination or bringing into special connection of two qualities, and since further it is accidental and antigeneral or discontinuous, such an accidental reaction ought to be regarded as an adventitious singularity of the continuum of possible quality, just as two points of a sheet of paper might come into contact. Topically the mere bending of the sheet is no change at all. But the instant that bending brings two points together, there is a discontinuous singularity.

But although singularities are discontinuities, they may be continuous to a certain extent. Thus the sheet instead of touching itself in the union of two points may cut itself all along a line. Here there is a continuous line of singularity.

In like manner, accidental reactions though they are breaches of generality may come to be generalized to a certain extent.

They will all extend into that dimension of successive copies which distinguishes the continuum of possible feelings from the continuum of possible quality. They necessarily must do so, since in the order of logical development that dimension results subsequently to the ac-cidental reactions themselves and they will be copies along with the body of the continuum of possible quality. But though in that dimension the continuum of possible quality comes round into itself with no absolute limit, there is no reason why these singularities should do so. Accidental as they are, you would not expect them to have that perfect smoothness and featurelessness of monotonous regularity. Indeed, it is quite impossible that they should so come round. For an accidental reaction, to use the metaphor of time again, is not anything eternal like a possibility, but something that happens. There are, therefore, reactions which it cannot determine but which only determine it. Observe by the way that the determination I speak of is not any exercise of power or intelligence on the part of the reactions in themselves, which are quite blind and brutal, but are conferred upon them by the continuity, in the same manner in which two qualities are not in themselves like or unlike but only have likeness and unlikeness imputed to them from their situation in the continuum of possible quality. But that remark is merely parenthetic. The law of logic strictly requires of the extensions of the different accidental reactions into the dimension of successive copies that they should have one limiting absolute first antecedent and one limiting last consequent.

The dimension of successive copies of feeling, so far as it applies to accidental reactions, I identify with Time.

With the emergence of Time, the first book, as it might be called, of an Historical *Encyclopädie*, what Hegel terms "Logic" is brought to a close and we pass to what he calls "Naturphilosophie."

But shall I not rather say, the first day of Creation is done? It would not be fair, nor paying our just dues, not to remind ourselves of that Babylonian philosopher from whom it would seem the first chapter of Genesis was cribbed. It is remarkable that though subconsciously yet he has perceived the need of every element which was needed for the first day. His *tohu wabohu, terra inanis et vacua* is the indeterminate germinal Nothing. His *Spiritus Dei ferebatur super aquas* is consciousness. His *Lux* is the world of quality. His *fiat lux* is an arbitrary reaction. His *divisit lucem a tenebris* is the recognition of the necessary duality. His *vidit Deus lucem quod esset bona* is the waking consciousness. Finally, his *factumque est vespere et mane, dies unus* is the emergence of Time. The Chaldeans had a native turn for scientific investigation decidedly surpassing that of the Greeks. But it took them over 15 centuries, or say 50 generations, to work out that account of the first day. Let a great mind devote himself intently and long to a problem, and it is wonderful how he will bring forth expressions of thought of which any distinct apprehension is beyond his reach. I have often remarked the phenomenon in reading Kant. With more truth than falsity this phenomenon might be called Inspiration.

This first day of creation was all in the very first moment of time.

But suppose that I were to undertake to tell you the History of Man from the beginning till now. I might say Man had a hand. He made himself tools. Thinking of his tools so developed his brain that he became able to invent language and then writing. He found out a number of laws of nature and some things about the constitution of the mind, and made some progress toward regulating his conduct in a purposeful way and toward developing conscious and rational purposes. To say that I should thus be narrating the history of man in *infinitely* more detail than I have narrated the history of the first moment of time would not be hyperbole at all; but it would be quite unspeakably understating the case. Were I to develop the history of that first moment in a book as large as the Chinese Encyclopedia the same thing would be *not one bit* less true.

That first moment of time was of course infinitely long ago. But more than that, although it was but one moment, it was infinitely longer than any number of ages. It contained as great a multitude of ages as there are points upon a continuous line. In one sense this continuum was not time, it is true, because it all occupied but a moment of time. But it was not only strictly analogous to time, but it gradually and continuously developed into time; so that it was of one continuous nature with time. All that follows from the principles of continuity.

Well begun is half done; and it may truly be said that when that first moment was over the main work of creation was already accomplished.

We place ourselves, then, at the beginning of time. Qualities are already possible. Actual existence has begun. Accidental reactions are taking place. Several continua are established. A tendency toward generalization is operative.

But as yet no *thing* can be said to exist; much less any personal consciousness. The accidental reactions are purely accidental, unregulated in any degree by law, the work of blind and brutal chance.

But now the tendency toward generalization which is already operative, and which indeed is more ancient than actual existence itself, begins to group the accidental reactions into fragmentary continua. Into continua, because such is the logical nature of generalization. Into fragmentary continua because the tendency to generalization has to fight the lawless brutality of chance with its youthful freakishness, and ebullient vivacity. At first, during the early eternities, those generalizations and continua are smashed as soon as they appear. For they are themselves of a haphazard kind with no vitality. But in this endless haphazard shindy between generalization and chance this generalization happens to come about, namely a limited but still a general tendency toward the formation of habits, toward repeating reactions that had already taken place under like circumstances. Now the difference between this generalization, this tendency toward law, and the rest was that this was one which by its own law was always tending to grow stronger. Therefore, although this was doubtless smashed like the others billions upon billions of times, to use a hyperbole of stating matters infinitely weaker than I really mean, yet still, it was often springing from its ashes, and on the whole was tiring out the lawlessness, until at length,—of course after an infinite lapse of time subsequent to the first moment, although infinitely long ago, there came to be a decided and so to say a sensible degree of tendency in nature to take habits.

This was the earliest of the laws of nature and was and still is continually strengthening itself. A habit of acquiring habits began to be established, and a habit of strengthening the habit of acquiring habits, and a habit of strengthening that habit, and so on *ad infinitum.*

The acquiring [of] a habit is nothing but an objective generalization taking place in time. It is the fundamental logical law in course of realization. When I call it objective, I do not mean to say that there really is any difference between the objective and the subjective, except that the subjective is less developed and as yet less generalized. It is only a false word which I insert because after all we cannot make ourselves understood if we merely say what we mean.

Now the question is, given that tendency to acquire habits, how would the world physical and psychical necessarily be developed under that tendency.

Everybody today is evolutionist. This is said to be the day of evolutionism. But in truth every important philosopher from Pherecydes down has always been evolutionist. Above all, everybody who talks of the plan of creation is evolutionist. That side of evolution, the *plan*, has I am very confident, much more real importance than the last generation has seen in it, and much closer an affinity for evolutionism. It is doubtful whether any consistent philosophical position other than an evolutionist position is possible. If you adopt an atheistic, Epicurean hypothesis, that we well know leads to evolution. If you adopt a theistic hypothesis you must either say that God does not think, and therefore, does not *plan*, or else his thought will result in evolution just as surely as that of the author of a book gives that book a gradual development. If you say that God does not think you seem to be led to a kind of agnosticism which gives up the whole task of philosophy and will not allow even science much longer to survive.

The main difference between, say Herbert Spencer's theory and mine is, that he restricts evolution to certain elements of the universe, and gives it a merely secondary position as a corollary of physical law of the "persistence of force" thus making that law something absolutely inexplicable and inscrutable, and barricading the roadway of inquiry by pronouncing at the outset that nothing can be found out about that, notwithstanding that that law presents intellectual relations which stimulate scientific curiosity in the highest degree; while I on the other hand say let us try, let us inquire, and see whether or no we cannot find out how that law came about with its curious intellectual and mathematical character, and how all the laws of the universe as well as all the facts of universe came to be as they are. Logic calls upon us to do this. We must either throw away scientific methods or we must apply them where they have not been applied before, for that is precisely what the scientific method prescribes for us and what the scientific spirit implores us to do.

The first thing which our inquiry demands is that we shall bring to it a clear and distinct idea of what habit and the acquiring of habit are. This we can only learn by studying these things where we see them in formation in the human mind. In doing this I am not much afraid of specializing too much and of assuming that the universe has characters which belong only to nervous protoplasm in a complicated organism. For we must remember that the organism has not made the mind, but is only adapted to it. It has become adapted to it by an evolutionary process so that it is not far from correct to say that it is the mind that has made the organism.

The distinction between the inner and the outer worlds antedates Time. I do not mean by the inner world that human consciousness which Baldwin and Royce have lately so forcibly reminded us is a social development and therefore very recent, only now in fact in process of taking a shape which has not yet been attained. The inner world that I mean is something very primitive. The original quality in itself with its immediate unity belonged to that inner world, a world of possibilities, Plato's world. The accidental reaction awoke it into a consciousness of duality, of struggle and therefore of antagonism between an inner and an outer. Thus, the inner world was first, and its unity comes from that firstness. The outer world was second. The social world was logically developed out of those two and the physiological structure of man was brought to forms adapted to that development.

That is why I make bold to go to the human mind to learn the nature of a great cosmical element. At any rate, although here and there in physics we may pick up a useful fact or two about habit, we really are obliged to go to the mind for the bulk of our information about it.

But even from the human mind we only collect external information about habit. Our knowledge of its inner nature must come to us from logic. For habit is generalization.

Habits appear to be formed in the human organism as a part of the process of nutrition. We are continually being "born again" by nutrition, and in being "born again" we are born into a second nature. Nutrition itself probably takes place only when waste takes place in the course of exercise; and that is why it is that the second nature which we acquire in nutrition is a natural tendency to act as we have acted before, as we were acting in the exercise which made the waste that nutrition repairs.

In this way habits are formed by exercise.

According to the Lamarckian theory, it is exercise and the consequent growth which by imperceptible steps has transformed the Moner into Man. But according to Darwin's hypothesis exercise has nothing to do with it. The whole gulf has been bridged by imperceptible variations at birth. But if we can trust to the lessons of the history of the human mind, of the history of habits of life, development does not take place chiefly by imperceptible changes but by revolutions. For some cause or other trade which had been taking one route suddenly begins to take another. In consequence merchants bring new goods; and new goods make new habits. Or some invention like that of writing, or printing, or gunpowder, or the mariner's compass or the steam engine, in a comparatively short time changes men very profoundly. It seems strange that we who have seen such tremendous revolutions in all the habits of men during this century should put our faith in the influence of imperceptible variations to an extent that no other age ever did. Is it because we have so little of Asiatic immovability before our eyes that we do not realize now what the conservatism of old habit really is?

That habit alone can produce development I do not believe. It is catastrophe, accident, reaction which brings habit into an active condition and creates a habit of changing habits.

To learn is to acquire a habit. What makes men learn? Not merely the sight of what they are accustomed to, but perpetual new experiences which throws them into a habit of tossing aside old ideas and forming new ones.

The most striking difference between the state of society today and that of Dr. Johnson's and Horace Walpole's time is the readiness of people nowadays to adopt new ways of life and new ways of thinking. And think what it was before the invention of printing! This plastic habit which we have acquired has not been brought about by imperceptible degrees but by a succession of revolutions, printing, the Reformation, Newton, the French Revolution, steam, and so on.

Consider the changes which take place in man's inmost habits, in their character and purposes. Are they brought about by imperceptible changes? Only in the way of decay. Conversions and reformations, — which after all are not unknown phenomena, — are always consequent upon impressive experience.

Habits are not for the most part formed by the mere slothful repetition of what has been done, but by the logical development of the potential germinal nature of the man, generally by an effort, the accident of having done this or that merely having an adjuvant effect.

Take the history of physical science, — say the history of mechanics, which has been so well narrated by Ernst Mach, — it would certainly be very false to say that it was a mere evolution of the Ichheit. Yet even that would be less false than the prevalent idea that it was a mere compend of facts. It is formed by the interaction of the two elements, a $[\,\ldots\,]$ mind of common origin with the universe, and facts which are selected by that mind as its suitable pabulum.

A habit then is not so much a mere tendency to repeat any action you happened to perform, though there is a certain tendency of this kind as it is the adoption of something which you happened to do because it happened to afford an outlet for the logical development of your germinal nature.

Therefore in considering how the universe would develop under the influence of a tendency to take habits, we must not content ourselves with merely calculating by the doctrine of chances what sort of accidents would happen the oftenest but while not neglecting this factor we must also take into account the logical development of tendencies already in germ, which will make one appropriate accident outweigh millions of ordinary ones.

The tendency to take habits having gained some strength, three processes of habit taking would be certain to commence and would go on about simultaneously, though one would be in advance of the others. I propose, however, to speak of only two of these just now.

As soon as casual reactions took place the objective logic of Hypothetic inference would create for each of them two subjects. Time and habit taking having established their reigns those subjects would take permanence and become *substances*. Of course, I use the word substance in the old Aristotelian sense, not in the chemical sense which has driven the other out of common parlance. We have seen that the essence of actual existence is reaction, and its *thisness* comes from the fact that a reaction is anti-general. This reaction which confers actual existence upon the substances must for that purpose be kept up through time. Since reaction is antigeneral, and rebels against law, no doubt that in the infancy of time, before habit had consolidated its throne, a good many of the casual reactions did not do this, and the substances were annihilated; and in like manner, others sprang into being. In fact, it must be that the same thing happens now. For we have seen that induction can do no more than establish a ratio of frequency, and therefore it can never forbid abnormal occurrences but can only insist that their ratio of frequency shall be zero. But you will ask what phenomena exhibit those everlasting reactions which I maintain confer existence upon the substances of the universe. I admit that this question is a pertinent one; for it is true that nothing can have actual existence without manifesting itself in phenomena. Indeed, Idealism, in the sense in which Objective Logic, as I understand it, is Idealism, may be defined as the doctrine that nothing exists but phenomena and what phenomena bring along with them and force upon us, that is Experience, including the reactions that experience feels and all that logically follows from experience by Deduction, Induction, and Hypothesis. And by *us*, we mean our neighbors, all that are embraced in the community, or society, very indefinite to our apprehension of which you and I are, as it were, histological cells. Here I may just remark that an objection to tychism, or the doctrine that absolute chance plays a part in the universe, might be based on this definition. For it may be said that an event cannot be *seen* by immediate perception to be casual, nor can any logical hypothesis infer that an event is casual, since it can only infer *explanations*, that is a fact from which what is observed would happen according to a law, while the chance event would not happen according to law. My reply to that objection has two parts, which must be united in the mind in order that the force of the argument may be understood. On the one hand, I say that although the fact that a given event is causal cannot be inferred, yet the operation of chance itself is a part of the regularity of nature and can be inferred. The variety and multiplicity of the universe proves it. On the other hand, I say that the single sporadic event in point of fact does not exist; because regularity is an essential element of existence. But it *begins* to exist; and after it has begun to exist it may continue to exist by becoming somewhat regular. Ever so little regularity is sufficient, and then not being thoroughly regular it may become extinct. The points of discontinuity at its origin and extinction are not full existence; and therefore are according to my form of Idealism not obliged to be fully phenomenal. For I hold that potentialities have a being, though they are not actual; so that in that respect, as I understand it, I am more of an evolutionist than Hegel who, I believe, does not admit this germinal being. I stand in the Aristotelian ranks here. I maintain that we directly contemplate this ideal world and when we open our eyes we perceive in the world about us that which corresponds to the freedom of the ideal world. It is true that reflection is required to enable us to recognize it. But that reflection *recognizes* it, and assures us that we *saw* it from the very first impression of sense.

But this is a digression. I go back to your question, what phenomena exhibit those everlasting reactions which I maintain confer existence upon the substances. I reply that in the first place there are some special phenomena which I shall speak of later. But, in the second place, even if there were none, those substances are manifested by the clustering about them of other habits of reaction, and not only of *habits* of reaction, for that would not be enough to preserve their identity, if that identity were preserved in no other way, but actual reactions continuing through time. For notwithstanding the *thisness* of reaction, it can continue in time, by virtue of the continuity which time confers. Casual reaction does not stop with those pretemporal reactions which bring existence to the substances. These we may call *substantial reactions.* But after the substances are formed there are additional casual reactions submitting themselves at length to habit; and these we may term *adventitious reactions.* These adventitious reactions do not create subjects because they occur between pairs of substances already existing. That is, though the substances are themselves of the nature of reaction, as their *thisness* testifies, yet that does not prevent their entering into additional reactions against one another. And this element once introduced spreads until every substance reacts against every other in manifold ways, and throughout time. All these reactions are phenomenal. That habit of reaction, together with that continuity of reaction, would suffice to maintain the identity and thus the existence of each substance were there no special phenomena effecting the same results.

So much for the first of the processes which commences as soon as habit has gained some strength. The second process concerns the substantial reactions. In these reactions, every substance reacts against every other, and reacts differently against all the others. For originally, in their antigenerality there is no likeness at all between the reactions, except so far as you may call the reaction of A against B and the reaction of A against C alike in that A is concerned in both. But in fact A simply reacts against the universe and it is only a logical distinction, —though not the less real for that, —that analyzes this into A against B, A against C, etc. all of which reactions have originally no further likeness than that they concern A. But the generalization of habit-forming brings all these into a continuity, thereby conferring upon them resemblances and differences; and this continuity, like every law, transcends all that exists, which is only casual and non-general, and it may therefore be called the *continuity of modes of substantial reaction*. The mode of reaction of any pair of substances must vary incessantly throughout time. If they did not do so, their reaction would cease. Indeed, it is a familiar proposition of mathematics that two variables cannot be equal throughout a portion of a continuum and not be equal throughout the whole of the same dimensions unless there is some breach of a continuous law in their variation. Now were the reaction between a pair of substances constant throughout time it would cease to have the *thisness* of an event; and however the power of habit may become predominant it can never from its very nature absolutely nullify the *thisness* of duality that belongs to reaction. For this last reaction, too, however much alike the modes of reaction between *A* and *B* and *A* and *C* may become, they never can be *exactly* alike, not only not through any time, but *not at any single instant*. The antigenerality of reaction may and will yield to habit in a measure, but rebels against a complete surrender. This is so because no induction can be absolute. It is in every respect approximate only.

The law of logic causes a sequence in this continuum as in others. But since this is a continuum of pairs, the sequence is not absolute but is only the distinction of *more or less* in reactions absolutely alike. It is true that no reactions will be absolutely alike, for a single instant even abstracting from the distinction of more or less; but the continuum must necessarily allow them to be so, because it places no limit upon the degree of their likeness. Hence, by the mathematical principle of the limit, as far as the continuum is concerned, two reactions might be exactly alike, at any rate except as to more or less.

I need hardly say that this continuum of the possible modes of substantial reaction is Space. The bodies must be Boscovichian particles.

How many dimensions would space have according to this theory? It may perhaps at first have had a great multitude. But a continuum of dimensions is not possible. There was therefore no reason at the outset why there should be one multitude rather than another. It must have been fixed by accident, that is, by casual reactions. But habit which is the agency which formed the continuum is essentially assimilative and would have reduced the dimensions as much as possible. The *thisness* of reaction would not, however, permit the dimensions to be as few as two, since in that case two bodies *B* and *C* moving about *A* would have their *radii vectores* brought into one line, that is, the reactions of *A* against *B* and *A* against *C* would differ only as more or less, contrary to what we have seen would be the effect of *thisness*. Accordingly there must have been at least three dimensions and exactly three unless there was some cause to make the number greater.

But it seems to me that such a cause did exist and that the number of dimensions must originally have been four and that it is probably four still for atoms in a chemical molecule. I mean that there are four dimensions, but space is shrunk in one dimension in the tendency to reduce the number, the room in that direction is too small to allow molecules to pass one another, although very likely atoms can still do so. There are certain facts of chemistry which seem to me to point to that hypothesis, although owing to the infant state of chemical theory, it is little more than a surmise.

The cause which it seems to me must have made the number of dimensions four is not the *thisness* of reaction, but was, on the contrary, habit in its intellectual character. Generalization as it seems to me, must have acted to make the modes of motion from any one position the same as from any other, just as undoubtedly it did act to make all the different parts and directions of space alike in other respects. I do not mean that generalization would act merely to make the modes of motion from any one position equal in variety to those from any other, but that it would act to make every mode of motion from one position definitely identifiable with some mode of motion from each other position. Now that is not the case in three dimensions. Take for instance a pencil standing vertical. It can fall toward the north, or toward the east, or in intermediate ways. But suppose it lies horizontal north and south. It can turn up or turn toward the east, or turn in intermediate ways. Its turning up is a mere continuation of the falling of the vertical pencil toward the north, and must therefore be identified with that mode of motion. But we cannot identify its turning toward the east with the falling of the vertical pencil toward the east. For if we did, we should be obliged to say that a pencil lying horizontal east and west had two identical modes of turning, which is absurd. For one of its modes of turning is a continuation of the turning of the north and south pencil toward the east.

But in four dimensions the modes of turning are perfectly identifiable. . . .

LECTURES ON PRAGMATISM
LECTURE II (302)

If I were asked to give a young gentleman a liberal education in 100 lessons, I should devote 50 lessons to teaching some small branch no matter what thoroughly,—say perhaps to boiling an egg,—or at any rate so nearly thoroughly that the young man should begin to know what thoroughness really means,—and should never thereafter be guilty of the ridiculous conceit of fancying that he knew English, for example. The other fifty I would distribute as follows: three lessons should teach the science of mathematics, one esthetics, two ethics, one metaphysics, one psychology, one the living and dead languages, one history, geography, and statistics ancient and modern, one dynamics and physics, one chemistry and biology, one astronomy, geology, and physical geography, one law, divinity, medicine and the other applied sciences, and the remaining 36 should be devoted to logic. Thereupon I would give him a certificate to the effect that he was a more truly educated man than two thirds of the doctors of philosophy the world over; and this certificate would have the singularity of being strictly true.

For thirty-six lessions into which the teacher should throw his whole soul would suffice really to teach the leading principles of logic, and logic, the ability to think well, constitutes about three eighths of a truly liberal education. But it is a mighty important three-eighths. A liberal education ought to be a living organism and logic may truly be said to be the heart of it. But I do not say that six lectures on a fragment of logic will have the same proportionate value. It will be like cutting out five sixths of a man's heart and leaving him the remainder. Or rather it would be that if lectures were *lessons*, which they certainly are far from being.

Now, gentlemen, every minute counts. There remain about fifty in which to present to you glimpses of those conceptions of mathematics which have any relation to our problem.

Pure Mathematics is the study of pure hypothesis regardless of any analogies they may have in our universe. The simplest possible hypothesis would be that there is a single element

A

and nothing more. The mathematics of this consists in a single proposition, as follows:

There is nothing that can be said of *A*.

The next branch of mathematics we come to supposes two elements, or, as analogy suggests that we should call them, two *values*. We might denote them by

B and *M*,

the initials of *bonum* and *malum*. They are different. In regard to anything, *x*, we may inquire whether under any assumption it is *B* or *M*. We know that it cannot be both. We also know that it is one or the other; for that is our hypothesis.

The resulting mathematics if developed by means of arrays of letters with conventional signs to signify relations between them will constitute the Boolian algebra of logic. It may equally be developed by means of diagrams composed of lines and dots, and this in various ways of which the *Eulerian diagrams* form one example while my *Existential Graphs* and *Entitative Graphs* are others.

Of this mathematics under its original limitations confining its applicability [to] non-relative logic no masterly presentation has ever been given. The nearest approach to such a thing in print is contained in the first two chapters of my paper in Vol III of the *Am. Jour. Math.*; and I may mention that Schröder's criticism of my definitions of aggregation and composition there given, although at first I assented to it, is all wrong and that the demonstration which Schröder professes to demonstrate cannot exist does exist and is perfect.* (*I value Schröder's work highly and he was a highly sympathetic man whom it was impossible to know and not to like even more than the great merit of his work justified. But he was too mathematical, not enough of the logician in him. The most striking thing in his first volume is a fallacy. His mode of presentation rests on a mistake and his second volume which defends it is largely retracted in his third [and] is one big blunder. There are some very fine things in his third volume and his posthumous volume I hope will contain still better. He was a growing man.) But the whole thing is bad; first, because it does not treat the subject from the point of view of pure mathematics, as it should have done; and second because the fundamental propositions are not made out. I follow too much in the footsteps of ordinary numerical algebra. And the sketch of the algebra of the copula is very insufficient.

I devoted some months last year to attempting a strict presentation and found it an extremely difficult job. I have the thing all typewritten but I am far from being satisfied with it. I shall try again one of these days, if I can find where with all to keep body and soul together while I am doing the work.

This kind of mathematics is rather poverty-stricken as to valuable ideas as might naturally be expected. Nevertheless, there is one. The relation expressed by the copula of inclusion the fundamental importance of which I was the first to discover and to demonstrate in 1870 is a matter on which I regret not having time to discourse at large. It is the relation which B has to itself which M has to itself and which M has to B but which B has not to M.

This is the first germination in mathematics of that wonderful conception of greater and less and of all systems however complicated having dimensions, each in itself linear.

I call this kind of mathematics which rests on the hypotheses of two objects, elements, or values, *Dichotomic Mathematics*.[1]

In 1870 I made a contribution to this subject which nobody who masters the subject can deny was the most important excepting Boole's original work that ever has been made.

I think it was in 1883 that I printed at my own expense a brochure presenting the pure mathematical aspect of this, —not by any means as well as I could now do but still tolerably. When it was done and I was correcting the last proof, it suddenly occurred to me that it was after all nothing but Cayley's theory of matrices, which appeared when I was a little boy. However, I took a copy of it to the great algebraist Sylvester. He read it and said very disdainfully —Why it is nothing but my *umbral notation*. I felt squelched and never sent out the copies. But I was a little comforted later by finding that what Sylvester called "*my* umbral notation" had first been published in 1693 by another man of some talent, named Godfry William Leibniz. He himself speaks of it as "une overture assez extraordinaire." You will see it in its original French in Muir's admirable *Theory of Determinants in the Historical Order of its Development*. Sylvester's name *umbra* which is the only distinctive name the thing has ever received, must, I fear, be retained, although *ion* or *radicle* would be far better. For who ever heard of two shadows combining together to form a substance!

They are things that do not exist. That is to say they do not belong to the universe of the fundamental hypothesis, being neither B nor M, in the dichotomic mathematics. In other mathematics, they have no existence in the universe of quantity. But joined together in sets, they do. They are

---

[1] See Section 16, a (p. 285 of this volume).

just like chemical radicals, each having a certain number of unsatisfied wants. When each of these is satisfied by union with another, the completely saturated whole has an existence in the universe of quantity. Surely the word *umbra* utterly fails to suggest all that; while the word *radicle* gives the idea exactly.

The mathematics which results from following out this idea of Leibniz which I rediscovered for myself and applied to dichotomic mathematics is, in mathematics taken generally, now most usually called the theory of matrices.

But I do not think that the icon of a matrix exhibits the idea quite so well as the idea of a chemical radicle does.

The application of this idea to logic gives the *exact logic of relatives*. DeMorgan had before me developed to some extent the logic of relatives. Schröder thought I greatly exaggerated the importance of DeMorgan's work. But Schröder greatly exaggerated the merits of that particular algebra of mine to the study of which his third volume is mainly devoted.

This exact logic of relatives and even in some degree DeMorgan's development, simply dynamites all our traditional notions of logic and with them Kant's Critic of the Pure Reason which was founded upon them.

But I must tell you that all that you can find in print of my work on logic are simply scattered outcroppings here and there of a rich vein which remains unpublished. Most of it I suppose has been written down; but no human being could ever put together the fragments. I could not myself do so. All I could do would be to make an entirely new presentation and this I could only do in five or six years of hard work devoted to that alone. Since I am now 63 years old and since all this is matter calculated to make a difference in man's future intellectual development, I can only say that if the *genus homo* is so foolish as not to set me at the task, I shall lean back in my chair and take my ease. I have done a great work wholly without any kind of aid, and now I am willing to undergo the last great effort which must finish me up in order to give men the benefit of what I have done. But if I am not in a situation to do so but have to earn my living, why that will be infinitely the more comfortable way of completing the number of my days; and if anybody supposes that I shall regret missing the fame that might attach to the name of C. S. Peirce,—a name that won't be mine much longer,—I shall only say that he can indulge that fancy without my taking the trouble to contradict him. I have reached the age when I think of my home as being on the other side rather than on this uninteresting planet.

Taking leave of dichotomic mathematics, I may mention that trichotomic mathematics which starts from the fundamental hypothesis of three elements [Fig. 1] has never received any development at all, to

Fig. 1

speak of, although it would certainly be extremely interesting and a field in which there would be soil for the growth of great and wonderful works of genius.

Let me call your attention to the circumstance that there is only one way in which 3 things can be arranged. *ABC* and *CBA* are different provided you recognize the difference of shape of *B* and *C* toward their right and their left sides. But if you do that, you are dealing with more than 3 objects. You are dealing with *A*, *B* viewed from the left, *B* viewed from the right, *C* viewed from the left, *C* viewed from the right.

If *ABC* are mere designations of dots on a line [Fig. 2], the arrangement *CBA* is merely the arrangement *ABC* viewed from the other side. That is you introduce a fourth and fresh object which is the pair of objects implied in the idea of passing through one way or the other. We may represent this idea of passing through the series one way or the other by two additional dots, which we may call *I* and *J*. Then indeed [Fig. 3] and [Fig. 4] are different.

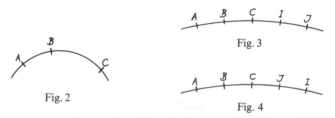

Fig. 2

Fig. 3

Fig. 4

But it may be asked whether *ABC* and *BCA* (that is, *ACB*) are not different arrangements. I reply No. They are so when you conceive that in *ABC* from *A* you can pass to *B*, and from *B* to *C* but cannot in the same way pass from *C* to *A*. But this is substantially to suppose a fourth object that puts a stop to the passage. For if I have *ABC* as dots on a line [Fig. 5] nothing prevents my completing the oval [as in Fig. 6] and then

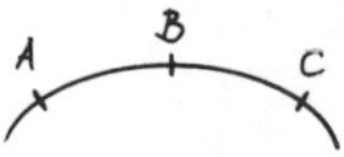

Fig. 5

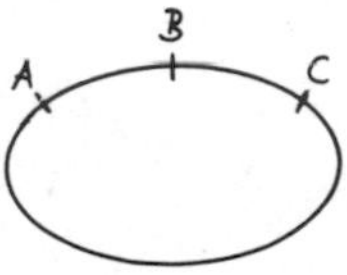

Fig. 6

I can as well begin at *B* as at *A* and describe the arrangement as one in which if I pass from *B* to *C* I may go on beyond *C* to *A*. This I can do unless a stop say *Ω* is inserted [as in Fig. 7 and Fig. 8]. Then of course [these two figures] are different.

Fig. 7   Fig. 8

But it may be asked whether three objects *ABC* may not be conceived to be so related that *AB* have a particular connection with one another while *C* stands apart from them. To that I reply that that particular mode of connection is either described and made definite or it is not. In the latter case, if it be merely that there is *some* respect in which *AB* are connected together while *C* stands unconnected, then this not merely *may be* the case, but it *must be* the case. Namely *A* and *B* are any way specially connected in that they are members of the special pair *AB* while *C* stands apart in that respect. So this will not be the ground of any distinction between one arrangement of three things and another.

But if you describe that respect in which *A* and *B* agree, then that constitutes a fourth object which you introduce, a sort of hyphen between *A* and *B* [Fig. 9]. So the theorem stands that there is but one possible arrangement of 3 objects.

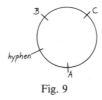

Fig. 9

How about 4 objects. Could they be arranged in more than one way. You will tell me perhaps that I have just shown that there are three arrangements [Fig. 10]. But if anybody [thought] there were three arrangements of 4 objects, it must have been you. I certainly did not.

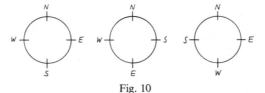

Fig. 10

Four objects *on one line* are *in reference to that line* in one or other of three orders. But any four objects can have three different lines drawn through them so as [to] be at once in all three arrangements [Figs. 11 and 12]. This shows that unless the particular mode of drawing the line is specified, in which case there are *at least* five objects; four objects can only be arranged in one way.

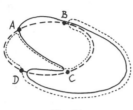

Fig. 11

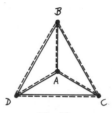

Fig. 12

Following out this idea, we soon see that no number of objects can be *in themselves* arranged at all. Indeed, when this has once been proved, it seems to be self-evident. But here is another self-evident truth. Let those who are taking notes, note this; because axioms are a highly distinctive class of truths, as everbody will admit. I proceed then to enunciate this:

It is self-evident that every truth of pure mathematics is self-evident if you regard it from a suitable point of view.

And conversely, nothing that has not been expressly enunciated in your original hypothesis can be self-evident unless you regard it from a suitable point of view.

That will be worth thinking over when you get home because it is an *aperçu* into the nature of pure mathematics.

As to arrangements, we see that an arrangement is neither an ordinary object nor is it equivalent to any finite number of ordinary objects.

An arrangement is a relation.

A relation is an object; but it is not an ordinary object.

Let us return to the chemical idea I call an object with one unsatisfied valency a *monad*, with two a *dyad*, with three a *triad*, etc. with more than two a *polyad*. With none at all a *medad*.

In our very simplest hypothesis of but one object, that one is plainly a *medad*.

As there is nothing whatever that can be said of it, it is just the same as blank nothingness. So much truth there is in the proposition that *pure being* is *blank nothingness*.

But in dichotomic mathematics we distinguish between the two objects, *M* and *B*. Then we *ipso facto* suppose not merely those two objects but also sixteen different relations between them. Or if we exclude the idea of a thing being related to itself there are 4. These relations are *dyads*. And the two objects since they can enter into these relations are *monads*. If they had been medads they would be perfectly indistinguishable and would have been merely one.

The different relations of the two to one another have themselves a system of relations between them. How many? To write down the number of them would require 78 figures.

These again have their relations and I could not tell you even how many figures would be required to write the number. The number of figures required merely to write down the number of figures would be too great.

The infinite series now begins to get its second wind.

These however are merely the dyadic relations which we have been considering. There are also triadic, tetradic, etc. relations which are vastly more numerous.

LECTURE II (303)[2]

I feel that I must not waste time; and yet an investigation, in order to be really solid, must not confine itself too closely to a question set beforehand. Let us not set our thoughts on pragmatism but survey the whole ground and let the evidences for or against pragmatism or in favor of a modification of it come when they will without being teased.

Pure mathematics is the study of pure hypotheses regardless of any analogies that they may present to the state of our own universe. It would be wild to deny that there is such a science, as actively flourishing and progressive a science as any in the whole circle, if sciences can properly be said to form a circle. It certainly never would do to embrace pragmatism in any sense in which it should conflict with this great fact.

Mathematics, as everybody knows, is the most ancient of the sciences; that is, it was the first to attain a scientific condition. It shows today many traces of its ancient lineage; some of which are excellences while others are unfortunate inheritances. The Greeks were very fine reasoners. Throughout the XVIIIth century, the opinion prevailed among mathematicians that the strictness of Greek reasoning was unnecessary and stood in the way of advances in mathematics. But about the middle of the XIXth century it was found that in important respects the Greek understanding of geometry had been truer than that of the moderns. Gradually, beginning we may say with Cauchy early in the XIXth century a vast reform has been effected in the logic of mathematics which even yet is not completed. This has quite revolutionized our conception of what mathematics is, and of many of the objects with which it deals, as well as of the logical relations between the different branches and the logical procedure. We are now far above the Greeks; but pure mathematics as it exists today is a decidedly youthful science in such an immature state that any student of logical power may very likely be in possession of important *aperçus* that have not yet become common property.

---

[2] Peirce writes: "Rejected. No time for this. And it would need two if not three lectures."

That mathematical reasoning is by no means confined to quantity is now generally perceived, so that it now becomes an extremely interesting logical question why quantity should play so great a part in it. It is recognized that the main business, if not the only business, of mathematics is the study of pure hypotheses and their consequences, or, as some say, the study of the consequences of pure hypotheses only. The Greeks approached this conception without attaining to it.

In the procedure of *all* mathematics whatsoever, the observation of diagrams plays a great part. That this is true in geometry was shown, though rather vaguely, by Stuart Mill in his logic. It is not so obvious that algebra makes use of the observation of visual images; but I do not think there ought to be any doubt of it. Arrays of letters are observed, although these are mixed with conventional signs with which we associate certain so-called "rules," which are really permissions to make certain transformations. There is still some question how far the observation of imaginary, or artificial constructions, with experimentation upon them is logically essential to the procedure of mathematics, as to some extent it certainly is, even in the strictest Weierstrassian method, and how far it is merely a psychological convenience. I have sometimes been tempted to think that mathematics differed from an ordinary inductive science hardly at all except for the circumstance that experimentation which in the positive sciences is so costly in money, time, and energy, is in mathematics performed with such facility that the highest inductive certainty is attained almost in the twinkling of an eye. But it is rash to go so far as this. The mathematician, unless he greatly deludes himself, the possibility of which must be considered, reaches conclusions which are at once enormously and very definitely general and yet, but for the possibility of mere blunders, are absolutely infallible. Anybody who fancies that inductive reasoning can achieve anything like this has not made a sufficient study of inductive reasoning.

Induction is no doubt generalization and mathematicians,—especially mathematicians of power,—are so vastly superior to all other men in their power of generalization, that this may be taken as their distinctive characteristic. When we are dealing with the real world cold water gets dashed upon any generalizing passion that is not well held in check. There are very few rules in natural science, if there are any at all, that will bear being extended to the most *extreme cases*. Even that invaluable rule that the sum of the angles of a plane triangle is equal to two right angles shows signs of breaking down when by the aid of photometric considerations and that of the numbers of stars of different brightness, we compare statistically that component of stellar proper motions that is due wholly to the real motions of the stars with that component that is partly due to the motion of the solar system.[3] But when we come to pure mathematics we not only do not avoid the extension of principles to extreme cases, but on the contrary that is one of the most valuable of mathematical methods. In regard to any ordinary function for example, if we only know for what values of the variable it becomes zero and for what values it becomes infinite, we only need to know a single finite value to know all there is to be known about it.

No minute analysis of any piece of characteristically mathematical reasoning has ever appeared in print. There are numbers of attempts which profess to be successful. There have even been professed representations of the reasoning of whole books of Euclid in the forms of traditional syllogistic. But when you come to examine them, you find that the whole gist of the reasoning, every step in the progress of the thought which amounts to anything, instead of being analyzed logically is simply stated in the form of a premiss. The only attempts that are in any important degree exceptions to this are Mill's analysis of the *pons asinorum* which has its value, but which relates to too slight a bit of reasoning to teach much and Dedekind's little book on the foundation of arithmetic with Schröder's restatement of it. This is certainly very instructive work. Yet it is open to both the same criticisms. In the first place, the mathematics illustrated is, most of it, of too low an order to bring out in strong colors the real peculiarities of mathematical thought; and in addition to that, the real mathematical thinking is, after all, only stated in pretty much the old fashion of all mathematical writers, that of abridged hints. It is not really analyzed into its logical steps. Every now and then the intelligent reader will say, "I wonder how he got the idea of proceeding so and so." But it is just at these points that the fine mathematical thinking comes in. It is left undissected.

When I first got the general algebra of logic into smooth running order, by a method that has lain nearly twenty years in manuscript and which I have lately concluded that it is so impossible to get it printed that it had better be burned,—when I first found myself in possession of this machinery I promised myself that I should see the whole working of the mathematical reason unveiled directly. But when I came to try it, I found it was the same old story, except that more steps were now analyzed. About the same amount as in Dedekind's and Schröder's subsequent attempts. Between these steps there were unanalyzed parts which appeared more clearly in my representation than in theirs for the

[3]   On 21 Dec. 1891, Peirce was writing to Simon Newcomb of his desire to investigate in these terms the possibility of the negative curvature of space. See "The Charles S. Peirce-Simon Newcomb Correspondence" by the editor, in *Proceedings of the American Philosophical Society* 101:5.

reason that I attempted to analyze a higher kind of mathematics. I was thus forced to the recognition in mathematics of a frequent recurrence of a peculiar kind of logical step, which when it is once explained is so very obvious that it seems wonderful that it should have escaped recognition so long. I could not make any exact statement of it without being led into technical developments that I desire to avoid and which are besides precluded by my being obliged to compress what I have to say into six lectures.[4] But I can give you a general idea of what the step is so that you will be able by subsequently studying over any piece of mathematics to gain a tolerable notion of how it emerges in mathematics.

Let me at this point recall to your minds that the correlatives *abstract* and *concrete* are used in two very little connected senses in philosophy. In one sense, they differ little from *general* and *special* or *particular*, and are for that reason hardly indispensable terms; though that is the usual meaning in German, which is so to say pushed to an extreme in Hegel's use of the word. The other sense in which for example *hard* is concrete and *hardness* is abstract, is more usual in English than in other languages for the reason that English is more influenced by medieval terminology than other languages. This use of the words is fully as well authorized as the other if not more so; and this is the sense in which I shall exclusively employ the words. Hard is concrete, hardness abstract.

You remember the old satire which represents one of the old school of medical men, — one of that breed to whom medicine and logic seemed to be closely allied sciences, — who asked why opium puts people to sleep answers very sapiently "because it has a dormitive virtue." Instead of an explanation he simply transforms the premiss by the introduction of an *abstraction*, an abstract noun in place of a concrete predicate. It is a poignant satire, because everybody is supposed to know well enough that this transformation from a *concrete predicate* to an abstract noun in an oblique case, is a mere transformation of language that leaves the thought absolutely untouched. I knew this as well as everybody else until I had arrived at that point in my analysis of the reasoning of mathematics where I found that this despised juggle of abstraction is an essential part of almost every really helpful step in mathematics; and since then what I used to know so very clearly does not appear to be at all so. There are useful abstractions and there are comparatively idle ones; and that one about *dormitive virtue*, which was invented with a view to being as silly as one could, of course does not rank high among abstractions. Nevertheless, when one closely scrutinizes it and puts it under a magnifying glass,—one can detect something in it that is not pure nonsense. The statement that opium puts people to sleep may, I think, be understood as an induction from many cases in which we have tried the experiment of exhibiting this drug, and have found, that if the patient is not subjected to any cerebral excitement, a moderate dose is generally followed by drowsiness, and a heavy dose by a dangerous stupor. That is simply a generalization of experience and nothing more. But surely there must be some explanation of this fact. There must be something, say to fix our ideas, perhaps some relation between a part of the molecule of morphine or other constituent of opium which is so related to some part of the molecule of nerve-protoplasm as to make a compound not so subject to metabole as natural protoplasm. But then perhaps the explanation is something different from this. Something or other, however, there must be in opium some peculiarity of it which if it were understood would explain our invariably observing that the exhibition of this drug is followed by sleep. That much we may assert with confidence; and it seems to me to be precisely this which is asserted in saying that opium has a dormitive virtue which explains its putting people to sleep. It is not an explanation; but it is good sound doctrine, namely that *something* in opium must explain the facts observed.

Thus you see that even in this example which was invented with a view to showing abstraction at its very idlest, the abstraction is not after all entirely senseless. It really does represent a step in sound reasoning.

Before going on to consider mathematical abstractions, let us ask ourselves how an *abstraction*, meaning that which an abstract noun denotes, is to be defined. It would be no proper definition of it to say that it is that which an abstract noun denotes. That would not be an analysis but a device for eluding analysis, quite similar to the old leach's offering *dormitive virtue* as an explanation of opium's putting people asleep. An abstraction is something denoted by a noun substantive, something having a name; and therefore, whether it be a reality or whether it be a figment, it belongs to the category of *substance*, and is in proper philosophical terminology to be called a *substance*, or thing. Now then let us ask whether it be a real substance or a fictitious and false substance. Of course, it may chance to be false. There is no magic in the operation of abstraction which should cause it to produce only truth whether its premiss is true or not. That, then, is not in question. But the question is whether an abstraction *can* be real. For the moment, I will abstain from giving a positive answer to this question; but will content myself with pointing out that upon pragmatistic principles an abstraction may be, and normally will be, *real*. For according to the pragmatistic maxim this must depend upon whether all the practical consequences of it are true. Now the only practical consequences there are or can be are embodied in the statement that what is said about it is *true*. On pragmatistic principles *reality* can mean nothing except the *truth* of statements in which the real thing is asserted. To say that opium has a dormitive virtue means nothing and can have no practical consequences except what are involved in the statement that there is some circumstance connected with opium that explains its putting people to sleep. If there truly be such a circumstance, that is all that it can possibly mean, — according to the pragmatist maxim, — to say that opium really has a dormitive virtue. Indeed, nobody but a metaphysician would dream of denying that opium *really* has a dormitive virtue. Now it certainly cannot *really* have that which is pure figment. Without, then, coming to a positive decision as yet, since the truth of pragmatism is in question, we shall if we incline to believe there is something in pragmatism also incline to believe that an abstraction may be a real substance. At the same time nobody for many centuries — unless it was some crank, — could possibly believe that an abstraction was an ordinary primary substance. You couldn't load a pistol with dormitive virtue and shoot it into a breakfast roll. Though it is in opium, it is wholly and completely in every piece of opium in Smyrna, as well as in every piece in every joint in the Chinatown of San Francisco. It has not that kind of existence which makes things *hic et nunc*. What kind of being has it? What does its reality consist in? Why it consists in something being true of something else that has a more primary mode of substantiality. Here we have, I believe, the materials for a good definition of an abstraction.

An abstraction is a substance whose being consists in the truth of some proposition concerning a more primary substance.

By a primary substance I mean a substance whose being is independent of what may be true of anything else. Whether there is any primary substance in this sense or not we may leave the metaphysicians to wrangle about.

By a *more* primary substance I mean one whose being does not depend upon all that the being of the less primary substance [does] but only upon a part thereof.

Now, then, armed with this definition I will take a shot at the abstractions of geometry [and] endeavor to bring down one or two of them.

We may define or describe a point as a place that has no parts. It is a familiar conception to mathematicians that space may be regarded as consisting of points. We shall find that it is not true; but it will do for a rough statement.

We may define a *particle* as a portion of matter which can be, and at every instance of time is, situated in a point.

According to a very familiar conception of matter,—whether it be true or not does not concern us,—every particle is supposed to exist in such a sense that all other matter might be annihilated without this particle ceasing to exist. Supposing that to be the case the particle is, so far at least as matter is concerned, a primary substance.

Now let us imagine that a particle moves. That is, at one instant it is in one point at another in another. That may pass as a concrete description of what happens but it is very inadequte. For according to this, the particle might be now here now there without continuity of motion.

But the geometer says: The place which a moving particle occupies on the whole in the course of time is a *line*. That may be taken as the definition of a line. And a portion of matter which at any one instant is situated in a line may be called a *filament*.

But somebody here objects. Hold, he says. This will not do. It was agreed that all matter is particles. What then is this filament? Suppose the objector is told that the filament is composed of particles. But the objector is not satisfied. What do you mean by *being composed*? Is the filament a particle? No. Well then it is not matter, for matter is particles. But my dear sir the filament is particles. Then it is not one but many and this single filament you speak of is a fiction. All there is is particles. But my dear sir, do you not understand that although all there is is particles yet there really is a filament because to say that the filament exists is simply to say that particles exist. Its mode of being is such that it consists in there being a particle in every point that a moving particle might occupy.

Thus you see that if the particles be conceived as primary substances the filaments are abstractions, that is, they are substances the being of any one of which consists in something being true of some more primary substance or substances none of them identical with this filament.

A *film* or that portion of matter that in any one instant occupies a surface will be still more abstract. For a film will be related to a filament just as a filament is related to a particle.

And a solid body will be a still more extreme case of abstraction.

Atoms are supposed to have existences independent of one another. But in that case according to our definition of an *abstraction*, a collection of atoms, such as are all the things we see and handle are *abstractions*.

They are just as much abstractions as that celebrated jack-knife that got a new blade and then a new handle and was finally confronted with a resurrected incarnation of its former self.

There is no denying this I believe, and therefore I do not think that we need have any further scruple in admitting that abstractions may be real,—indeed, a good deal less open to suspicion of fiction than are the primary substances. So the pragmatistic decision turned out correct in this instance, though it seemed a little risky at first.

That a *collection* is a species of abstraction becomes evident as soon as one defines the term *collection*. A *collection* is a substance whose existence consists in the existence of certain other things called its *members*.

An abstraction being a substance whose existence consists in something being true of something else, when this truth is a mere truth of existence the abstraction becomes a collection.

When we reflect upon the enormous rôle enacted in mathematics by the conception of *collection* in all its varieties, we can guess that were there no other kinds of abstractions in the science (instead of the hosts of them that there are), still the logical operation of abstraction would be a matter of prime importance in the analysis of the logic of mathematics.

I have so much more to say than I have of time to say it in that all my statements have to be left in the rough and I know I must produce an impression of vagueness and haziness of thought that would disappear upon close examination. I shall be obliged to presume that after leaving the lecture room you will do some close thinking on your own accounts.

I have no time to speak further of the interesting and important subject of the reasoning of mathematics. Nor can I discuss Dedekind's suggestion that pure mathematics is a branch of logic. It would I think be nearer the truth (although not strictly true) to say that necessary reasoning is not one of the topics of logical discussion. I am satisfied that all necessary reasoning is of the nature of mathematical reasoning. It is always diagrammatic in a broad sense although the wordy and loose deductions of the philosophers may make use rather of auditory diagrams, if I may be allowed the expression, than [of] visual ones. All necessary reasoning is reasoning from pure hypothesis, in this sense, that if the premiss has any truth for the real world that is an accident totally irrelevant to the relation of the conclusion to the premiss; while in the kinds of reasoning that are more peculiarly topics of logical discussion it has all the relevancy in the world.

But I must hurry on to the consideration of the different kinds of mathematics, a subject of which the slightest sketch keeping close to what is wanted for the study of pragmatism ought in itself to occupy three good lectures at least, and would be much more interesting so.

The different branches of mathematics are distinguished by the different kinds of fundamental hypotheses of which they are the developments.

The simplest conceivable hypothesis is that of a universe in which there is but one thing, say A, and nothing else whatever of any kind. The corresponding mathematics consists of a single self-evident proposition (that is it becomes evident by logical analysis simply) as follows:

Nothing whatever can be predicated of A and it is absolutely indistinguishable from *blank nothingness*.

For if anything were true of A, A would have some character or quality which character or quality would be something in the universe over and above A.

The next simplest mathematics seems to be that which I entitle *dichotomic mathematics*. The hypothesis is that there are two things distinguished from one another. We might call them B and M these being the initials of *bonum* and *malum*. Then the problem of this mathematics will be to determine in regard to anything unrecognized, say x, whether it is identical with B or identical with M. It would be a mere difference of phraseology to say that there are countless things in the universe, x, y, z, etc. each of which has one or other of the two values B and M. The first form of statement is preferable for reasons I cannot stop to explain.

The Boolian algebra of logic is a mere application of this kind of pure mathematics. It is a form of mathematics rather poverty-stricken as to ideas. Nevertheless, it has some features which we shall find have a certain bearing upon the foundations of pragmatism.

In the first place, although the universe consists of only two primary substances, yet there will *ipso facto* be quite a wealth of abstractions. For in the first place there will be the universe of which M and B are the two parts. Then there will be three prominent *relations*. Namely 1st, the relation that M has to B and that M has to nothing else, and that nothing but M has to anything. 2nd, there will be the converse relation that B has to M and to nothing else and that nothing but B has to anything, and 3rd, the relation that B has to M and to nothing else and [that] M has to B and to nothing else, without counting the absurd relation that nothing has to anything. That third relation is the self-converse relation of *otherness*.

Those four relations are *dyadic relations*. That is, considered as abstractions their existence consists in something being true of two primary substances. Thus to say that M is in the relation of otherness to B is to say the M is other than B which is a fact about the two primary substances M and B.

But there are also *triadic relations*. It is true that owing to there being but two primary substances, there is no triadic relation between three different primary substances. But there can be a relation between three different dyadic relations.

There are dyadic relations between dyadic relations. Thus the relation of M to B is the *converse* of the relation of B to M and this relation of *converseness* is a dyadic relation between relations.

As an example of a triadic relation between relations take the relation between the 1st, 2nd, and 3rd relations between M and B. That is the relation between first the relation of M only to B only, of B only to M only, and of otherness of M to B and of B to M. This triadic relation is a case of the general triadic relation of *aggregation*. To say that Z stands in the relation of aggregation to X and Y is to say that Z is true wherever X is true and wherever Y is true and that either X or Y is true wherever Z is true.

Another important kind of triadic relation between dyadic relations is where R is in the relation of relative product of P into Q where P, Q, R are dyadic relations. This means that if anything A is in the relation R to anything C there is something B such that A is in the relation P to B while B is in the relation Q to C and conversely if A is not in the relation R to C then taking anything whatever β either A is not P to β or else β is not Q to C.

Applying this idea of a relative product we get the conception of *identity* or the relation which M has to M and to nothing else and that B has to B and to nothing else.

I have only noticed a few of the most interesting of these abstractions. But I have not mentioned the most interesting of all the dyadic relations, that of *inclusion*, the great importance of which, now generally recognized, was first pointed out and demonstrated by me in 1870.

It is the relation that M has to M and to B and that B has to B but that B does not have to M.

It is the connecting link between the general ideas of logical dependence and the idea of the sequence of quantity.

All these ideas may be said to have virtually existed in the form of the Boolian algebra originally given by Boole. But in 1870 I greatly enlarged and I may say revolutionized the subject by the virtual introduction of an entire new kind of *abstractions*.

9

TYPES OF REASONING (441)

I fear you got little good from the conversation at the end of the last lecture; for I am long dishabituated to talking philosophy. But upon my side I found it very delightful to be assured of so intelligent and critical an auditory.

I am truly vexed to find myself obliged this evening to plunge into the dry and dreary subject of formal logic. But you know as well as I do that there is a part of the philosophical world whose unlucky convictions force them to base metaphysics upon formal logic. To that unfortunate party I appertain. We have at our head three men of might, Aristotle, Duns Scotus, and Kant. Among our most redoutable antagonists there were in the ancient world Pythagoras and Epicurus, and in the modern world Descartes, Locke, and I must add Hegel. The general question of which party is right I shall not argue. The scheme of classification of the sciences which I laid before you last time serves to define my position that metaphysics must draw its principles from logic and that logic must draw its principles neither from any theory of cognition nor from any other philosophical position but from mathematics. The defence of this view must rest upon the fruits that it can exhibit.

In order to enliven the deathly dullness of Formal Logic, I will throw what I have to say in the form of a narration of my own mental history, so that you may at least see how a man might keep up a lively interest in this subject during many years. Besides, it will perhaps not be altogether unprofitable for young thinkers to contemplate the picture of an unusually sustained train of thought along a narrow pathway, and to note the advantages and disadvantages of such systematized thinking. For that it has both advantages and disadvantages is I think not to be denied. At the same time I shall take care not to weary you by over-accuracy of autobiographical detail, the main thing being to sketch the formal logic, using the narrative form merely for embellishment.

Having been bred in a highly scientific circle, I entered upon the study of philosophy, not at all for the sake of its teachings about God, Freedom, and Immortality, concerning the practical value of which I was very dubious from the outset, but moved rather by curiosity in regard to Cosmology and Psychology. In the early sixties I was a passionate devotee of Kant, at least of that part of his philosphy which appears in the Transcendental Analytic of the Critic of the Pure Reason. I believed more implicitly in the two tables of the Functions of Judgment and the Categories than if they had been brought down from Sinai. But even then it seemed to me that formal logic ought not to be made to rest upon psychology. For the question whether a given conclusion follows from given premises is not a question of what we are able to think, but of whether such a state of things as that set forth in the premises was able to be true without the state of things related to it as that set forth in the conclusion is to that set forth in the premises being true likewise.

But Kant, as you may remember, calls attention to sundry relations between one category and another. I detected some additional relations between the categories, *all but* forming a regular system, yet not quite so. Those relations seemed to point to some larger list of conceptions in which they might form a regular system of relationship. After puzzling over these matters very diligently for about two years, I rose at length from the problem certain that there was something wrong with Kant's formal logic.

Thereupon I fell to reading every book on logic that I could procure. At first, I had a senseless prepossession in favor of German books; but I soon found myself forced away from it. The German logics, for example, since Kant, are almost unanimous in dividing propositions into the categorical, the conditional (which they wrongly call the hypothetical) and the disjunctive. This was a division introduced by Kant because he rightly or wrongly imagined that the traditional one would spoil his table of categories. The traditional definition of hypothetical, that is, of compound, propositions was into the conditional, the disjunctive, and the copulative. As an example of a copulative proposition, take this: It lightens and it thunders. But in truth, conditional propositions merely form a special variety of disjunctive propositions. Thus, "you will either take care where you tread or you will wet your feet" is disjunctive. "If you march equably, you will wet your feet" is a conditional proposition which comes to the same effect. But you will find it impossible to express the same idea in a copulative form. Thus, hypotheticals are in truth either disjunctive or copulative. The same thing is equally true of categoricals. The universal propositions are disjunctive, the particular propositions are copulative. Thus, to say that every man who is without sin may fire a stone is as much as to say of any man you may select that he is either sinful or is at liberty to stone the person. On the other hand to say that some swan is black, is the same as to say that an object can be found of which it is true that it is a swan *and* that it is black.

Cicero informs us that in his time there was a famous controversy between two logicians, Philo and Diodorus, as to the signification of conditional propositions. Philo held that the proposition "if it is lightening it will thunder" was true if it is not lightening or if it will thunder and was only false if it is lightening but will not thunder. Diodorus objected to this. Either the ancient reporters or he himself failed to make out precisely what was in his mind, and though there have been many virtual Diodorans since, none of them have been able to state their position clearly without making it too foolish. Most of the strong logicians have been Philonians, and most of the weak ones have been Diodorans. For my part, I am a Philonian; but I do not think that justice has ever been done to the Diodoran side of the question. The Diodoran vaguely feels that there is something wrong about the statement that the proposition "If it is lightening it will thunder," can be made true merely by its not lightening.

Duns Scotus, who was a Philonian, as a matter of course, threw considerable light upon the matter by distinguishing between an ordinary *consequentia*, or conditional proposition, and a *consequentia simplex de inesse*. A *consequentia simplex de inesse* relates to no range of possibilities at all, but merely to what happens, or is true, *hic et nunc*. But the ordinary conditional proposition asserts not merely that here and now either the antecedent is false or the consequent is true, but that in each possible state of things throughout a certain well-understood range of possibility either the antecedent is false or the consequent true. So understood the proposition "If it lightens it will thunder" means that on each occasion which could arise consistently with the regular course of nature, either it would not lighten or thunder would shortly follow.

Now this much may be conceded to the Diodoran, in order that we may fit him out with a better defence than he has ever been able to construct for himself, namely, that in our ordinary use of language we always understand the range of possibility in such a sense that in some possible case the antecedent shall be true. Consider, for example, the following conditional proposition: If I were to take up that lampstand by its shaft and go brandishing the lamp about in the faces of my auditors it would not occasion the slightest surprise to anybody. Everybody will say that is false; and were I to reply that it was true because under no possible circumstances should I behave in that outrageous manner, you would feel that I was violating the usages of speech.

I would respectfully and kindly suggest to the Diodoran that this way of defending his position is better than his ordinary stammerings.

Still, should he accept my suggestion I shall with pain be obliged to add that the argument is the merest *ignoratio elenchi* which ought not to deceive a tyro in logic. For it is quite beside the question what ordinary language means. The very idea of formal logic is, that certain *canonical forms* of expression shall be provided, the meanings of which forms are governed by inflexible rules; and if the forms of speech are borrowed to be used as *canonical forms of logic* it is merely for the mnemonic aid they afford, and they are always to be understood in logic in strict technical senses. These forms of expression are to be defined, just as zoölogists and botanists define the terms which they invent, that is to say, without the slightest regard for usage but so as to correspond to natural classifications. That is why I entitled one of the first papers I published, "On the Natural Classification of Arguments." And by a *natural* classification, we mean the most *pregnant* classification, pregnant that is to say with implications concerning what is important from a strictly logical point of view.

Now I have worked out in MS. the whole of syllogistic in a perfectly thoroughgoing manner both from the Philonian and from the Diodoran point of view. But although my exposition is far more favorable to the Diodoran system even than that of DeMorgan in his *Syllabus of Logic*, which is much the best presentation of the Diodoran case ever made by an adherent of it, yet I find that the Philonian system is far the simpler,—almost incomparably so. You would not wish me to take you through all those details. This general statement is all that is appropriate for this brief course of lectures.

Be it understood, then, that in logic we are to understand the form "If *A*, then *B*" to mean "Either *A* is impossible or in every possible case in which it is true, *B* is true likewise," or in other words it means "In each possible case, either *A* is false or *B* is true."

That being granted, I say that there is no logical difference between a categorical and a hypothetical proposition. A universal categorical is a disjunctive hypothetical and a particular categorical is a copulative hypothetical. This is my position today just as it was in 1867. If any distinction could be admitted, it would be necessary to declare that categorical propositions are of more complicated structure than hypotheticals. Indeed, the most convenient order of exposition will be to begin by asserting provisionally that this is the case, making afterward the slight correction that is necessary to establishing the perfect identity between the categoricals and hypotheticals. I say, then, that but for a consideration of secondary importance, a categorical proposition is of an essentially more complex structure than a hypothetical proposition.

This sounds like a paradox. A hypothetical proposition is by definition a compound proposition. Of what, then, is it compounded? To that question I reply that it may be compounded of simple propositions that for the present may be regarded as non-categorical. But no matter of what it be compounded. My point is that categorical propositions are equally compound; only they are not *merely* compound, but are affected by a peculiar complication from which hypotheticals may for the present be regarded as free.

The truth is that the appearance of paradox is due merely to the preconceptions which we imbibe from the European languages. These languages do not represent the nature of thought in general, nor even that of human thought. They form one small class among half a dozen large classes, and in many respects are quite exceptional. In other respects, the Indo-European and Shemitic languages taken together are exceptional. The idea that a categorical proposition such as "All men are rational animals," is so simple is an idea suggested by a language in which the *common noun* is a distinct part of speech. Now *Proper names* must, of course, exist in all languages. But the common noun seems to be nowhere else developed as it is in the Indo-European languages. In the Shemitic languages common nouns are met with; but by far the greater part are mere verbal inflections. They are not felt as independent words, nor for the most part entered in the vernacular dictionaries. The other languages of mankind, that is to say the great bulk of human speech, are almost devoid of the part of speech. Take, for example, a language with which in these days of tourists a good many people acquire some smattering, I mean the Ancient Egyptian. I believe there is no such thing in that language as a word which can only be used as a common noun and cannot be used as a verb. A language in which the noun is as fully developed as in our own certainly cannot express the idea, "Every man is a rational animal" without the verb *is*. But even in a language so closely allied to ours as Greek this verb may be omitted,—a fact which shows that even so near home as that, the common nouns still retain some verbal life. Outside of the Indo European branch, a language which requires an *is* in such a sentence is a distinct rarity. The Old Egyptian often uses a copula, it is true; but what is the nature of that copula? According to LePage Renouf and I think Brugsch it is not a verb at all; it is a *pronoun* whose function is to show that the man and the rational animal refer to the same object.[1]

---

[1] Sir Peter LePage Renouf (1822–1897), the British Egyptologist whose Hibbert Lectures for 1879 were published by Scribner's Sons (1880) under the title *The Origin and Growth of Religion as illustrated by the Religion of Ancient Egypt*. Heinrich Kare Brugsch (1827–1894) was an eminent German Egyptologist.

We thus see that the truth of the matter is this. There are in the first place certain propositions, such as *fulgurat* and *pluviet*, which can be expressed by a single word. Those propositions are certainly not distinctively categorical. They express no inherence, no character of any object. They do not convey any definite idea of a *thing*. In the second place, these propositions or any propositions may be *denied*. In the third place, any propositions may be logically compounded either by way of disjunction or by way of conjunction. Thus, we get *hypotheticals*. In the fourth place, to these last there may be attached as in Egyptian relative pronouns; or other linguistic devices may be employed to signify that the objects to which the different clauses refer are identical. It is this last complication which makes a proposition distinctly *categorical*.

Thus far, the categorical proposition appears as something more complex than the hypothetical. Thus, Every man is a rational animal means, under a Philonian interpretation, If there be a man, then *he* feels and thinks. It is the identifying pronoun "he" that distinguishes this preposition as categorical. If you object that the categorical proposition asserts the existence of men, I reply that this is a Diodoran interpretation. But now it is time to *correct* the view that categoricals are more complex than hypotheticals by showing that hypotheticals equally involve something equivalent to the pronoun. In my paper of 1867 on the Natural Classification of Arguments, I had already reached the Knowledge of the perfect identity of Categoricals and Hypotheticals; but I could not then have given the account of the matter which I am now about to borrow from my long subsequent studies of the logic of relatives. We must in the first place consider such propositions as these: "This is beautiful, — I thirst, — The sky is blue." In "This is beautiful," we have the pronoun, but it does not identify two common nouns; it only identifies beautiful with the object which the pronoun *This* directly points out. The demonstrative and personal pronouns, *this, that, I, you, we*, etc. have very peculiar powers. They enable us to convey meanings which words alone are quite incompetent to express; and this they do by stimulating the hearer to look about him. They refer to an experience which is or may be common to speaker and hearer, to deliverer and interpreter. They are thus quite *anti-general*, referring to a *hic et nunc*. The proper name "the sky" is of similar effect. It is that which if you look up, you will see. Any proper name is of the same nature. "The Bible" could not mean to a Japanese what it means to us. For in order to discuss with an Afghan the character of George Washington, you must begin by making him partake in some of the experience you have yourself. It is not the same thing when you talk to the man blind from birth about colors. It is no *definite experience* in that case that is lacking, but only a power of imagining. In the case of George Washington, the experience is not *quite* definite in its outlines; still it is approximately so.

Allow me just here to make a parenthetical observation. It is far more true to say that such a name as George Washington is a feeble substitute for a *this* or *that* which should spread the very experience referred to before the interpreter's eyes, than it is to say that "*this*" and "*that*" are *pronouns*, or *substitutes for nouns*. They are not substitutes for nouns at all; they are *stimulants to looking*, like the bicyclist's bell, or the driver's "hi! there," or "mind your eyes." Duns Scotus clearly saw this and composed a new definition of a pronoun which was accepted by all grammarians down to the time of the reformation of the universities in the 16th century. Then, the hatred of the reformers for the Dunces was such that they reverted to the absurd old definition; and this was continued down to our times. But I see that of late years the new edition of Allen and Greenough and some other grammars have substantially returned once more to the Scotistic definition.

Of a somewhat different nature, yet involving the same element of hæcceity, are the selective pronouns, *some, any, every, whoever*, etc. Boole imagined that the proposition "Some swan is black" could be represented by writing, "An undefined kind of swan is black." The very first paper on logic I ever published corrected this error. It is *in any case* true that "Some kind of a swan is black," namely, "every black swan is black"; but that is not what the proposition "Some swan is black" asserts. It declares that here in this actual world *exists* a black swan. There is a reference to a *here and a this*; only the interpreter is not told precisely where among *this here* vast collection of swans the one referred to is to be found. "*Every*" or "*any*" is of a similar nature. When I say "every man dies," I say you may pick out your man for yourself and provided he belongs to *this here* world you will find he will die. The "some" supposes a selection from *this here* world to be made by the *deliverer* of the proposition, or made in his interest. The "every" *transfers* the function of selection to the *interpreter* of the proposition, or to anybody acting in his interest.

Now, then, in the sense in which the words "some" and "every" are demonstrative, that is refer to an examination of an experience common to the deliverer and interpreter, in that sense the same element is to be found even in hypotheticals. "If it lightens, it will thunder," means select from *this here* range of possibility any case *you* like, and *either* it will not be a case of present lightening or else it *will* be a case of thunder shortly following. So "It may lighten without thundering," means: *I* could find you in *this here* range of possibilities, a case in which there *should be* lightening *without* any thunder.

Having thus satisfied ourselves of the perfect logical equivalence of categorical and hypothetical propositions, we may for the sake of convenience speak of them as expressed in categorical form. The copula, understood as it must be according to this analysis, is called the *copula of inclusion*. It is opposed to the *copula of identity* which is employed by those who advocate the quantification of the predicate, if any such logicians still linger on the scene after their doctrine has been thoroughly exploded.

There are various other modes of logically analyzing propositions, which are equally exact. Two of them are exemplified in the papers by Mrs. Franklin and Prof. Mitchell in the Johns Hopkins Studies in Logic. So there are innumerable different systems of coördinates in geometry all equally correct. And there are different modes of space-measurement, the elliptic, the parabolic, and the hyperbolic all equally correct. But of all the different kinds of geometrical coördinates, *rectangular point-coördinates alone* correspond to the *principles of mechanics*; and consequently this kind of coordinates must be taken as the *basic coordinates* in the study of mechanics. So it is with the different kinds of space-measurement. The *parabolic alone* corresponds to the principles of *Euclidean geometry*; and nobody would dream of using any other, unless he intended to suppose Euclidean geometry to be false. In complete analogy with these illustrations, of all the methods in which propositions *may be* analyzed and analyzed correctly, *that one* which uses the copula of inclusion *alone* corresponds to the *theory of inference*. And *this* it does, inasmuch as it makes the relation of subject to predicate of a categorical proposition *precisely the same* as the relation of antecedent to consequent of a conditional proposition; for this latter is manifestly no other relation than that of a premise to its conclusion. The greatest scholastic doctors, who were unquestionably the most exact reasoners there have ever been *except* the mathematicians, always called the minor premise the antecedent and the conclusion the consequent.

In my paper on the Classification of Arguments, I made use of this scholastic doctrine of the *Consequentia* in order to get a ποῦ στῶ from which to start a doctrine of inference. I reasoned in this way. Suppose we draw a conclusion. Whether it be necessary or probable I do not care. Let *S* is *P* represent this conclusion. Now we certainly never can be warranted in drawing any conclusion about *S* from a premise, or set of premises, which does not relate in any way to *S*. If the inference is drawn from more than one premise, let all the premises be colligated into one copulative proposition. Then this single premise must relate to *S*; and in that sense, it may be represented thus: *S* is *M*. I do not, of course, mean that S need appear formally in this premise as a subject, far less as the sole subject. I only mean that "S is M" may in a general sense stand for any proposition which virtually relates to S. The inference, then, appears in this form

> *Premise* S is M
> *Conclusion* S is P.

But, *whenever we draw a conclusion*, we have an idea, more or less definite, that the inference we are drawing is only an example of a whole class of possible inferences, in each of which from a premise more or less similar to the actual premise there would be a sound inference of a conclusion analogous to the actual conclusion. And not *only* is this idea present to our consciousness,—as is shown by our thinking that the premise *leads to* the conclusion,—but, what is still more important, there is a principle *actually operative* in the depths of our minds,—a *habit*, natural or acquired, by virtue of which we really *should* draw that analogous conclusion in each of those possible cases. This operative principle I call, after the logician Fries, the *leading principle* of the inference. But now *logic* supposes that reasonings are *criticised*; and as soon as the reasoner asks himself what *warrant* he has for concluding from S is M that S is P, he is driven to *formulate* his leading principle. Now in a very general sense we may write as representing that formulation, M is P. I write M is P instead of P is M because the inference takes place from M to P, that is M is antecedent while P is consequent. So that the reasoner in consequence of his selfcriticism reforms his argument and substitutes in place of his original inference, this *complete* argument:

> Premises $\{$ M is P
> S is M
> Conclusion S is P.

I do not mean that the formulation of the leading principle necessarily takes the form M *is* P is any *narrow sense*. I only mean that it must express some general relation between M is P, which not merely in reference to the special subject, S, but in all analogous cases will warrant the passage from a premise similar to S is M to a conclusion analogous to S is P.

This second argument has certainly itself a leading principle, although it is a far more abstract one than the leading principle of the original argument. But you might ask, why not express this *new* leading principle as a premise, and so obtain a *third argument* having a leading principle *still more* abstract? If, however, you try the experiment, you will find that the third argument so obtained has no more abstract a leading principle than the second argument has. Its leading principle is indeed precisely *the same* as that of the second argument. This leading principle has therefore attained a *maximum degree* of abstractness; and a leading principle of maximum abstractness may be termed a *logical principle*.

It is thus proved that *in an excessively general sense* every complete argument, i.e. every argument having a leading principle of maximum abstractness, is an argument in the form of *Barbara*.

The purpose of this whole investigation was to ascertain what the principal types of reasoning were; and my plan of proceeding was this. Since all reasoning is in an excessively general sense of the form of *Barbara*, if I could only find a kind of reasoning which should be of the same form of *Barbara* in a more special sense, and yet not in so special a sense as to prevent my discerning clearly its different species, then I might try whether reasoning in general being compared to that kind of reasoning would not show itself to be divided into *analogous classes*.

The first figure of syllogism is of the form of *Barbara* in a special sense; but the difficulty was that its four moods appear to be mere accidental variations of essentially the same kind of inference. But demonstrative syllogism *in general* may be considered as belonging to the form of *Barbara*. For every mood of such syllogism can be reduced to the first figure by the aid of immediate inferences which seem to be inferences in form only—mere transformations of statements. Thereupon, the question arose whether any differences in the mode of reasoning in demonstrative syllogism could be detected with certainty. Kant in his brochure *Über die falsche Spitzfindigkeit der vier syllogistischen Figuren* had maintained that there were no such differences. But then Kant had had in mind substantial differences, and was not on the lookout for mere *forms* of difference, which might nevertheless be quite unmistakable. Moreover, Kant had seriously tripped in his reasoning in that little book in *two particulars, not to mention* less important errors. In the first place, he had argued that because the second, third, and fourth figures *might be* resolved into reasonings in the first figure together with immediate inferences, as indeed all logicians since Aristotle had *fully explained*, therefore those figures *must* be regarded as so compounded, which was very much as if he had argued that because a force could be resolved into two components therefore it could not be regarded as really composed in any other way. In the second place, Kant had never thought of inquiring what the *nature* of those *immediate inferences* might be; but had hidden the *question* from himself by a verbal device consisting in calling them *Folgerungen* and denying to them the title of *Sch[l]üsse*. The immediate inferences in question are in principle

two, one being the inference

> No M is P
> $\therefore$ No P is M,

while the other is the inference

> Some S is M
> $\therefore$ Some M is S.

Now it is quite true that No M is P and No P is M are nothing more than different expressions for the same fact; so that *in substance* there is no inference at all; and the same thing is true of Some S is M and Some M is S. But *then*, I bethought me that it frequently happens *in geometry* that we have to distinguish between two forms which in *themselves considered* are *just alike*, but which differ *toto caelo* when we consider each as a limiting case of a continuous series of other forms. For example, we say that two coincident straight lines cross one another at two points, which may be any two points on their length. What is the sense in saying that, or how can two coincident lines differ from a single line? The answer is that if we draw different conics having their vertices at the same two points on a straight line, the nearer those conics come to the line, the nearer will the tangents to the conics from any given point in the plane come to passing through those two fixed points. [See Fig. 1.] Is

Fig. 1

there, then, I asked myself, any distinction at all analogous to this between No M is P and No P is M. Suppose that instead of saying absolutely *No M is P* and absolutely no P is M, we talk of excessively small proportions. It is one thing to say that of men who start writing poetry an excessively small proportion will turn out to have powers like those of Dante; but it is quite another thing to say that of men with powers like those of Dante an excessively minute proportion will ever start to write poetry.

To cut a long matter short, I found that each of those immediate inferences could be thrown into syllogistic form. Namely the mood *Cesare* of the second figure runs thus:

> No M is P
> Any S is P
> $\therefore$ No S is M

Now change the S to P, and this becomes

> No M is P
> Any P is P
> $\therefore$ No P is M

And here is the inference from No M is P to No P is M.
So likewise the mood *Datisi* of the third figure runs thus:

> Any S is P
> Some S is M
> $\therefore$ Some M is P

Now change the P to S and this becomes

> Any S is S
> Some S is M
> $\therefore$ Some M is S.

Here is the immediate inference from Some S is M to Some M is S.

Now I *found*, that the reduction of the second figure *always requires that immediate inference* which thus shows itself to be of the form of the second figure, and requires no other; while the reduction of the third figure always requires that immediate inference which shows itself to be of the form of the third figure and requires no other. As to the fourth figure, that turned out to be of a mixed character.

Confining ourselves then to the first three figures, their forms are as follows:

1st Figure

Any $M \therefore \begin{smallmatrix} \text{is} \\ \text{is not} \end{smallmatrix} P$

$\begin{smallmatrix} \text{Any} \\ \text{Some} \end{smallmatrix} S$ is M

$\therefore \begin{smallmatrix} \text{Any} \\ \text{Some} \end{smallmatrix} S \begin{smallmatrix} \text{is} \\ \text{is not} \end{smallmatrix} P$

2nd Figure

Any $M \begin{smallmatrix} \text{is} \\ \text{is not} \end{smallmatrix} P$

$\begin{smallmatrix} \text{Any} \\ \text{Some} \end{smallmatrix} S \begin{smallmatrix} \text{is not} \\ \text{is} \end{smallmatrix} P$

$\therefore \begin{smallmatrix} \text{Any} \\ \text{Some} \end{smallmatrix} S$ is not M

3rd Figure

$\begin{smallmatrix} \text{Some} \\ \text{Any} \end{smallmatrix} S \begin{smallmatrix} \text{is} \\ \text{is not} \end{smallmatrix} P$

$\begin{smallmatrix} \text{Any} \\ \text{Some} \end{smallmatrix} S$ is M

$\therefore$ Some $M \begin{smallmatrix} \text{is} \\ \text{is not} \end{smallmatrix} P$

It will be seen that the second figure is derived from the first by interchanging the minor premise and conclusion while changing them both from affirmative to negative or *vice versa*; while the third figure is derived from the first by interchanging the major premise and conclusion while changing both from universal to particular or *vice versa*. This does not include the two moods *Darapti* and *Felapton* because these are not valid in the Philonian system.

Having thus established this real, although merely formal, distinction between the three figures of syllogism, the *next* question was, whether I might not find that this corresponded to a *more substantial* distinction between three types of inference in general, just as we *usually find* in geometry that merely *nominal* differences in *degenerate* forms correspond to highly *important* distinctions between the genuine forms of which those degenerate forms are the *limiting cases*. Demonstrative inference is the *limiting case* of probable inference. Certainty *pro* is probability 1. Certainty *con* is probability 0. Let us start, then, with probable syllogism in the first figure. This runs as follows:

> The proportion *r* of the *M*s possess *Π* as a *haphazard* character;
> These *S*s are *drawn at random* from the *M*s;
> ∴*Probably and approximately*, the proportion *r* of the *S*s possess *Π*.

Four explanatory remarks are here called for:

1st, the phrase "probably and approximately" in the conclusion, means, loosely speaking, that the probability of the proportion approximating to the value *r* is *greater* the *wider* the limits of approximation are chosen. Were the conditions of the premises *exactly fulfilled*, the phrase "probably and approximately" would imply agreement with the law of the *probability curve*. Those conditions need not, however, be *exactly fulfilled* and that law has to be modified accordingly. I might occupy a whole lecture on this point.

2nd, The proportion, *r*, has not necessarily any *precise* numerical value. It may stand for "more than half," or for "nearly all or none" or for "near any simple ratio," and so forth. In short, we are simply to conceive that all possible values from 0 to 1 are distributed in any way whatever into two parcels, and that the statement is that the value of the ratio is contained in a specified parcel.

3rd, the condition that the *S*s shall be "taken at random" means *nearly enough* at random; and the more nearly the closer will the approximation be. But the nearer the proportion *r* approaches *all* or *none* the less important is the randomness of the selection, until when *r* is *quite* all or none, the randomness becomes *altogether indifferent*. It remains to define perfect randomness, which is very important. The *S*s are drawn *quite at random* provided they are drawn according to a method such that if the drawing were continued indefinitely no *M* would escape getting drawn.

But this cannot be, if the multitude of Ms exceeds that of the whole numbers,—as does for example, the whole collection of real irrational numbers. In regard to such a collection it is impossible directly to reason in this way. I complete the definition of this condition by noting that it is only requisite that the Ss should be random with respect to Π, that is, that the Ss that are Π and the Ss that are not Π should be drawn with the same relative frequencies.

4th, the condition that Π must be taken *haphazard* is analogous to the condition that the Ss must be taken *at random*. Qualities, however, are *innumerable*, and consequently cannot be taken strictly at random. Yet an object may be said to be taken *haphazard* from a collection, when we so far exert our wills that the object taken shall belong to the collection, while refraining from any further interference, we leave it to the *course of experience* to determine what particular object presents itself. In the case of qualities, this course of experience is not the course of outward experience but is the course of our inward thoughts. If the course of experience should not present any general character, as we hope it will, the effect will *not* be to vitiate the reasoning, but merely to prevent any conclusions from being drawn. Should the course of experience approximate toward the fulfillment of a general rule but not *precisely* fulfil any rule, the effect will be that we shall never be able to conclude any quite precise proposition. It is to be noted that every quality like every mechanical force, is resolvable into innumerable others in innumerable ways. Hence it is that I have written Π in the singular number; for in any event it will have innumerable elements. According to the present requirement, Π must be composed of elements which the course of thought, left to itself, naturally throws together. It must not be a recondite or artificially composed character. It must not be suggested by the manner in which instances are presented, nor by the characters of those instances. The safest way will be to insist that Π shall be settled upon before the Ss are examined.

This is so *important* a rule of reasoning and is so *frequently violated*, that I think I had better give illustrations of the effect of neglecting it. Here is one example:

> The immense majority of small objects which fall to the bottom of the deep sea will never be seen by mortal eye;
> Here are some specimens of Challenger Expedition dredgings drawn at random from small objects that had fallen to the bottom of the deep sea;
> ∴ Probably the immense majority of these specimens will never be seen by mortal eye.

Such reasoning is manifestly absurd. Here is another example. I wish to draw at random a few names of eminent persons. In order that the randomness of the selection may be above suspicion I have simply taken the first name on those pages of Phillip's Great Index of Biography whose numbers end in two zeros, preceded by an odd number, that is on pages 100, 300, 500, 700, 900. Here are the names:

Francis Baring	Born 1740	Died 1810 Sep 12
Vicomte de Custine	1760	1794 Jan 3
Hippostrates (Of uncertain age)		
Marquis d'O	1535	1594 Oct 24
Theocrenus	1480	1536 Oct 18.

Now suppose we reason in disregard of the present rule.

> One man in every ten is born in a year ending in a cipher, and therefore probably not more than one of these men. In fact, all but 1.
> One man in four dies in autumn. Probably therefore about one of these. In fact, all but one.
> One man in three dies on a day of the month divisible by 3. Probably therefore 1 or 2 of these. In fact everyone.
> One man in ten dies in such a year that its number doubled and increased by 1 gives a number whose last figure is the figure in the ten's place of the date itself. Probably therefore not more than one of these. In fact, every one.

I have not stated these four rules with the utmost possible accuracy but only with so much accuracy as I thought you might be able to carry away from a single lecture. The subject ought to be spread over a number of lectures. But it is necessary for the development of what I have to say later that I should sketch it out as best I may. We see here the first figure of probable reasoning which, were the rules stated accurately, would be seen to embrace all necessary reasoning as a special case under it. It is in fact Deduction.

I next proceed to form the 3rd figure of probable reasoning. This is derived from the first by interchanging the major premise and conclusion, while denying both. You will remember that r merely means that the true ratio is in one or other of two parcels. Let Z denote any ratio contained in the parcel to which r does not belong. Then the denial of the proposition

$$\text{The proportion } r \text{ of the } \left\{ \begin{matrix} Ms \\ Ss \end{matrix} \right\} \text{ have the character } \varPi$$

will be

$$\text{The proportion } \rho \text{ of the } \left\{ \begin{matrix} Ms \\ Ss \end{matrix} \right\} \text{ have the character } \varPi.$$

The *form* of the denial is the same as that of the assertion. Hence, the third figure will be

> These Ss are drawn at random from the Ms;
> Of these Ss the proportion ρ have the haphazard character Π;
> ∴Probably and approximately, the proportion ρ of the Ms have the character Π.

This is the formula of Induction. I first gave this theory in 1867. Subsequently, I remarked that it was substantially Aristotle's theory. Since that, many logicians have expounded it. The word *inductio* is Cicero's imitation of Aristotle's term ἐπαγωγή. It fails to convey the full significance of the Greek word, which implies that the examples are arrayed and brought forward in a mass. Aristotle in one place calls the reasoning ἡ ἀπὸ τῶν καθ' ἕκαστον ἐπὶ τὰ καθόλου ἔφοδος the assault upon the generals by the singulars.

Induction, then, is probable reasoning in the Third Figure. This theory is distinguished from all others by two attributes, 1st, it derives the rules of induction as necessary corollaries from the definition of Induction, and 2nd, the rules so derived are much more stringent than those of any other theory. For example, Mill and others quite overlook the necessity of the rule that the character Π must not be suggested by an examination of the sample itself.

It follows as a consequence of this rule that induction *can never make a first suggestion.* All that induction can do is *to infer the value of a ratio, and that only approximately.* If the Ms are infinite in number, the proportion of Ms that are P may be strictly as 1 to 1, and *still* there may be exceptions. But we shall see later that in fact induction *always* relates to infinite classes,—though never to such as exceed in multitude the whole numbers. Consequently, induction can never afford the *slightest reason to think that a law is without exception.* I may remark that this result is admitted to be true by Mill.

I will refrain from a minute and tedious formalistic discussion of what the Second Figure of probable reasoning will be; but will content myself with informing you that such a discussion would lead to the following result, namely that the second figure reads:

> Anything of the nature of M would have the character Π, taken haphazard;
> S has the character Π;
> ∴Provisionally, we may suppose S to be of the nature of M.

Still more convenient is the following conditional form of statement:

> If μ were true, Π, Π', Π'' would follow as miscellaneous consequences —
> But Π, Π', Π'' are in fact true;
> $\therefore$ Provisionally, we may suppose that μ is true.

This kind of reasoning is very often called *adopting a hypothesis for the sake of its explanation of known facts*. The explanation is the *modus ponens*

> If μ is true, Π, Π', Π'' are true
> μ is true
> $\therefore \Pi$, Π', Π'' are true.

This probable reasoning in the second figure is, I apprehend, what Aristotle meant by $\dot{\alpha}\pi\alpha\gamma\omega\gamma\dot{\eta}$. There are strong reasons for believing that in the chapter on the subject in the Prior Analytics, there occurred one of those many obliterations in Aristotle's MS. due to its century long exposure to damp in a cellar, which the blundering Apellicon, the first editor, filled up with the wrong word. Let me change but one word of the text, and the meaning of the whole chapter is metamorphosed in such a way that it no longer breaks the continuity of the train of Aristotle's thought, as in our present text it does but so as to bring it into parallelism with another passage, and to cause the two examples, like the generality of Aristotle's examples, to represent reasonings current at his time, instead of being, as our text makes them, the one utterly silly and other nearly as bad. Supposing this view to be correct, $\dot{\alpha}\pi\alpha\gamma\omega\gamma\dot{\eta}$ should be translated not by the word *abduction*, as the custom of the translators is, but rather by reduction or *retroduction*. In these lectures I shall generally call this type of reasoning *retroduction*.

I first gave this theory in 1867, improving it slightly in 1868. In 1878 I gave a popular account of it in which I rightly insisted upon the radical distinction between Induction and Retroduction. In 1883, I made a careful restatement with considerable improvement. But I was led away by trusting to the perfect balance of logical breadth and depth into the mistake of treating Retroduction as a kind of Induction. Nothing, I observe, is so insidious as a tendency to suppose symmetry to be exact. The first anatomist must have been surprised to find the viscera were not as symmetrical as the hands and features. In 1892 I gave a good statement of the rationale of Retroduction but still failed to perceive the radical difference between this and Induction, although earlier it had been clear enough to my mind. I do not regard the present lecture, into which I have been obliged to cram too much matter, as a formal restatement;

but it suffices to indicate that I am now ready to make such a fresh restatement which shall correct some former errors and I hope will stand as satisfactory for a good many years.

For the sake of brevity I have abstained from speaking of the argument from analogy, which Aristotle terms παράδειγμα. I need hardly say that the word *analogy* is of mathematical provenance. This argument is of a mixed character being related to the others somewhat as the fourth figure of syllogism is related to the other three.

Now let us remove the scaffolding of syllogistic forms which has served as our support in building up this theory and contemplate our erection without it. We see three types of reasoning. The first figure embraces all Deduction whether necessary or probable. By means of it we predict the special results of the general course of things, and calculate how often they will occur in the long run. A definite probability always attaches to the Deductive conclusion because the mode of inference is necessary. The third figure is Induction by means of which we ascertain how often in the ordinary course of experience one phenomenon will be accompanied by another. No definite probability attaches to the Inductive conclusion, such as belongs to the Deductive conclusion; but we can calculate how often inductions of given structure will attain a given degree of precision. The second figure of reasoning is Retroduction. Here, not only is there no definite probability to the conclusion, but no definite probability attaches even to the mode of inference. We can only say that the Economy of Research prescribes that we should at a given stage of our inquiry try a given hypothesis, and we are to hold to it provisionally as long as the facts will permit. There is no probability about it. It is a mere suggestion which we tentatively adopt. For example, in the first steps that were made toward the reading of the cuneiform inscriptions, it was necessary to take up hypotheses which nobody could have expected would turn out true,— for no hypothesis positively likely to be true could be made. But they had to be provisionally adopted,—yes, and clung to with some degree of tenacity too,—as long as the facts did not absolutely refute them. For that was the system by which in the long run such problems would quickest find their solutions.

10

REASON'S CONSCIENCE:
A PRACTICAL TREATISE ON THE THEORY OF DISCOVERY
WHEREIN LOGIC IS CONCEIVED AS SEMEIOTIC (693)[1]

CHAPTER I

Life is not nearly so long, compared with all one wishes to do with it, as it appears to the young to be; and it ought to be economized. It is idle to read a book, far more so to study it, without a somewhat definite and reasonable expectation of getting from it enough amusement, exercise of mind, useful information, or moral benefit, to recompense one for the time spent upon it. It is, therefore, fair that I should say just what I am prepared to promise that a faithful study of this book will do for the student, and what he ought to be warned that it cannot do for him.

It will fill a certain gap in his preparation for life. That he should beforehand appreciate what its importance will be, so as neither to underrate it nor overrate it, is an impossibility. But by the aid of some reflections, which will be useful in themselves, a tolerable idea of what help the book will render, may be gained. Every young person needs to learn certain things in order to excel in the special occupation of his or her life. They are of three kinds. First, he has to learn to *discriminate* certain special things. A person who intends to be a butcher, or a banker, or a doctor, must learn to distinguish one kind and quality of meat from another and know the relative value of each, or to distinguish one kind and quality of security from another, or to distinguish one kind and degree of illness from another. In the second place, he has to learn some special kind of *skill*, the one to buy meat, cut it up, to keep it untainted and to dispose of it judiciously; the second, to perform calculations very swiftly, and to take prompt advantage of every impor-

---

[1] This manuscript consists of six "volumes" as follows:

vol. 1, pp. 1–80   vol. 4, pp. 250–278
vol. 2, pp. 82–164   vol. 5, pp. 278–370
vol. 3, pp. 166–248   vol. 6, pp. 370–442
The editor is using therein only pp. 2–30; 70–140; 216–276; 278–342 — observing Peirce's instructions to reject certain parts and substitute others.

tant event; the third, to perform surgical operations, keep his patient in a healthy frame of mind, and make the family recognize the value of his services. In the third place, he has to acquire certain special *infor-mation*, quite extensive in every occupation.

If a person cares for nothing but just to live, he had better content himself with such special education as that. I knew of a brewer who sold out his business for ten million dollars; but there was a good deal of trouble in ascertaining how much it was worth, because he had been a poor boy, and was not only unable to read and write, but did not even know the ten figures of arithmetic; so that he had kept no accounts. Everybody despised him, and he was too ignorant to enjoy anything except locking up more money every day. He had been immensely suc-cessful, however. He had become rich; and that was all he had aimed at. But if one desires to enjoy the society of the most delightful people, to accomplish great and useful things, and to read in that wonderful book, the universe of matter and mind, then, even more important than the in-dispensable special training is that general education which develops all the powers of body, heart, and mind, and which ought to be more nearly alike than the special training for all superior men and women.

General education likewise involves training of three kinds some-what similar to those of special education; that is to say, 1st, in general power, *discrimination*; 2nd, in *vigor of action*; 3rd, in *finding out the truth*. There is such a thing as *discrimination in vigor*; we call it *Skill*. There is a *skill in finding out the truth*: we call it *reasoning power*.

The purpose of this book is to aid the student in improving his reasoning power. We have to consider how far it can be expected to do this.

Pedagogy, properly speaking, is a branch of rhetoric, namely, the art of teaching, and only concerns the teacher; but the word is often extend-ed to the art of learning, a branch, or fruit, of logic, which is usually and properly taught to pupils in schools. This art will form no part of the subject of this book; but it is pertinent to the question just put to remark that the power of discriminative attention, the power of well-coördi-nated exertion, and the power of readily acquiring habits, can all three be enormously strengthened by well-planned exercises steadily persisted in. Some of the points to be particularly attended to are that the exercises should more and more gradually have their difficulty enhanced, that they should be so devised as to teach only one thing, —or at any rate, very little,—at a time, and that opportunity should be afforded for that unconscious operation by which the lesson soaks into the mind while this is in a passive state. These are facts which it concerns the learner to know almost as well as the teacher knows them.

Skill in the performance of an action consisting of various parts which may on different occasions be varied,—say, for example, bridge-building, where we have our choice of an arched bridge, or a simple suspension-bridge, or a trussed-suspension-bridge, or a plain truss bridge, or a cantilever bridge, or a bridge of boats, or some other structure,—is best acquired by making a study of what is essential to the achievement of the purpose of the action, of the advantages and disadvantages of every way of carrying it out, and of the relation of these to the different general features of different circumstances under which we may need to perform the action. Such an account to the why and wherefore of an action or of an artificial thing is called the *theory* of it. If an action, although complicated, has very often to be performed, and is almost always performed in nearly the same way, it frequently happens that we have an *instinct* for performing it. The action of walking is an example; the action of throwing a stone is another. Now instinct is remarkable for its great accuracy, as well as for its adaptedness to its purpose; and it would usually be unwise in the extreme to attempt to perform such an act under the guidance of theory; for theories have to be studied very long and very deeply before they can be entirely freed from error; and even then the application of them is laborious and slow.

Now this book sets forth the theory of finding out the truth; but I shall call it a practical treatise because I aim, not only at giving the theory in the briefest adequate form, but also at explaining how the theory can be conveniently applied in practice.

Many of our reasonings, however, are performed instinctively, and it must not, for an instant, be supposed that I should recommend that such modes of action be given up in favor of theoretical procedures, except to compare theory with practice or for some other peculiar and quite theoretical purpose. Other reasonings, although not exactly instinctive, have become so habitual as to resemble instinctive actions. In many cases, the habits have come to us from tradition. Now the long course of experience in the practice of these modes of reasoning, which is precisely what has made them habitual or traditional, has involved their being put to the test very often indeed; and if these tests had not been favorable to them, on the whole, the habits or traditions would have fallen into disrepute and would have been broken up. In these cases, therefore, which embrace most of our everyday reasonings, it is usually better to adhere to the habitual or traditional ways, although there are cases that are exceptional in this particular,—cases where the theory has been worked up to the last degree of perfection, and where it can be shown without danger of error that the old ways have been wrong. In the majority of such everyday reasonings, however, it would be false to pretend that a book like this can be of any real service.

On the other hand, there are habitual and traditional ways of thinking which never have been put to the test, or not very often. In such cases, their being habitual or traditional affords no guarantee of their being sound. In cases such as these, a single careful and cautious application of the theory may save us from error on a great number of closely similar occasions.

Many cases arise where reasonings are called for of kinds that are not traditional nor habitual; and in such cases disputes frequently arise as to what is and what is not sound reasoning. The less simple and more intellectual a man's life is, the higher his aims, and the oftener his ambition or enterprise carries him into unwonted situations, the oftener will the theory of finding the truth be needed, to be applied to his practice in steering his difficult way through unfamiliar tortuosities of thought.

So, then, this book is to be a practical treatise on the theory of finding out the truth, or a practical treatise on Logic.

At the same time, I do not intend to be deaf to the hunger and thirst of the reader who adores Truth for her own sake,

> Whose profit is greater than the profit of silver,
> And the gain thereof than fine gold.
> She is more precious than pearls,
> And all the things thou canst desire are not to be compared
>   unto her.

CHAPTER II.   THE SCIENCES

Men, the principal occupation of whose lives is finding out the truth, are called *scientific men*; and their occupation is called *Science*, although this word is used in several other senses, like most common words.

Men who spend their lives in finding out similar kinds of truth about similar things understand what one another are about better than outsiders do. They are all familiar with words which others do not know the exact meaning of, they appreciate each other's difficulties and consult one another about them. They love the same sort of things. They consort together and consider one another as brethren. They are said to pursue the same *branch* of science. But the word *branch* here means sometimes a trunk branch, sometimes a branch of a branch, sometimes a bough, sometimes a mere twig. The word branch is applied to all. . . .

The ladder of the sciences, as well as I have been able to work it out, is as here exhibited:

A. SCIENCE OF DISCOVERY, embracing:

    I. MATHEMATICS, embracing:
       a. Dyadics,
       b. Arithmetic,
       c. Synectics.

    II. PHILOSOPHY, embracing:
       a. Phenomenology
       b. Normative Science; consisting of
          i.   Esthetics
          ii. Ethics
          iii. Logic, composed of
             1. Speculative Grammar,
             2. Speculative Critic,
             3. Methodeutic.
       c. Metaphysics, including
          i.   Ontology
          ii. Physical Metaphysics, including
             1. Cosmology
             2. The Doctrine of Time and Space
             3. The Doctrine of Matter
          iii. Religious Metaphysics.
             1. Metaphysical Theology
             2. The Theory of Freedom
             3. The Doctrine of another Life

   III. IDIOSCOPY.
      a. Physiognosy
         i.   General Physics, embracing
            1. Dynamics, including
              α General and Rigid Dynamics
              β Hydrodynamics
              γ Dynamics of Multitudinous Systems
            2. Physics of Special Forces
              α Molar Physics—Gravitation
              β Molecular Physics—Elaterics and Thermotics
              γ Etherial physics—Optics, Electrics.
            3. Physics of the Constitution of Matter

    ii. Classificatory Physiognosy
      1. Chemistry
        α Physical Chemistry
        β Organic Chemistry
        γ Inorganic Chemistry
      2. Crystallography
      3. Biology
        α Physiology
        β Anatomy
   iii. Descriptive Physiognosy
      1. Astronomy
      2. Geognosy
        α Geography
        β Geology

b. Psychognosy
  i. General Psychology
    1. Introspectional Psychology
    2. Experimental Psychology
    3. Phsyiological Psychology
    4. Genetic Psychology
  ii. Classificatory Psychics
    1. Special Psychology
      α Individual Psychology
      β Psychical Heredity
      γ Abnormal Psychology
      δ Mob Psychology
      ε Race Psychology
      ζ Animal Psychology
    2. Linguistics
      α Phonetics
      β Word Linguistics
      γ Grammar
      δ Forms of Composition
    3. Ethnology
      α Ethnology of Social Developments
        x Customs
        xx Laws
        xxx Religion
        xxxx Traditions and Folk Lore
      β Ethnology of Technology

iii. Descriptive Psychics
   1. History
      α Monumental
      β Ancient and Early

      γ Modern

      α′ Political History
      β′ History of Science and Art
      γ′ History of Social Developments

   2. Biography
   3. Criticism
      α Literary Criticism
      β Criticism of Art.

B. SCIENCE OF REVIEW.
C. PRACTICAL SCIENCE.

I proceed to give some explanations of this scheme. It is intended to exhibit the most important relations between the sciences considered as so many businesses of groups of men in the present state of our civilization.

Science of Discovery is that science which is pursued simply to find out the truth, regardless of what is to be done with the knowledge. *Science of Review* is that science which endeavors to form a systematized digest of the whole or some part of human knowledge, using whatever the science of discovery has brought to light and filling up its lacunae for its own purpose by investigations of its own. *Practical Science*, or the theory of the arts, is that science which is selected, arranged, and further investigated in details as a guide to the practice of an art.

Mathematics is the study of what is or is not logically possible, without undertaking to ascertain what actually exists. *Philosophy* is that science which limits itself to finding out what it can from ordinary everyday experience, without making any special observations. *Idioscopy* is that science which is occupied in making new observations and which uses these to find out what further it can by inference.

Dyadics is that branch of mathematics which finds out what impossibilities are connected from something being connected in a particular way with some objects and not with others. *Arithmetic* is that branch of mathematics which ascertains the impossibilities which result from each one of a given collection of objects being connected with some single one of an endless series of objects that are related to one another as the whole numbers are related to one another. *Synectics* is that branch of mathematics that finds out the impossibilities resulting from supposing possibilities which the whole numbers would not suffice to distinguish.

Phenomenology is that branch of philosophy which endeavors to describe in a general way the features of whatever may come before the mind in any way. *Normative science* is that science which considers any kind of excellence, and endeavors to formulate the conditions under which an object would possess that excellence, without undertaking to say whether given objects possess that excellence or not. *Metaphysics* is that branch of philosophy which inquires into what is real, that is, what has anything true of it regardless of whether anybody thinks it is true or not.

Physiognosy is that branch of idioscopy which observes by the senses and reasons upon what it observes, without the exercise of those faculties by which we recognize persons and moral influences. *Psychognosy* is that branch of idioscopy which observes the evidences of beings like ourselves and reasons upon them. There is no third branch here. But perhaps this is because nobody occupies himself with observing the workings of ideas like Truth, Humanity, etc. which, it is true, appear only in persons, just as persons appear only in the material objects which they mould, but which appear to be as distinct from the persons as the persons are from matter.

Esthetics is that normative science which studies the conditions of that kind of excellence which objects may possess in their presentation, or appearance, regardless of their relations. *Ethics* is that normative science which studies the conditions of that excellence which may or may not belong to voluntary action in its relation to its purpose. *Logic* is that branch of normative science which studies the conditions of *truth*, or that kind of excellence which may or may not belong to objects considered as representing real objects.

Ontology is that branch of metaphysics which endeavors to determine the nature and kinds of reality. *Physical Metaphysics* studies the foundations of physiognosy. *Religious Metaphysics* studies metaphysics in the interest of man's spiritual aspirations.

General physics is the science of the ubiquitous phenomena of the physical universe, first discovering them, then ascertaining their laws, and finally measuring their constants. In its development it tends toward metaphysics. *Classificatory physiognosy* describes and classifies the kinds of matter and seeks to [identify] them by the discoveries of general physics, with which in its development it finally tends to coalesce. *Descriptive physiognosy* describes individual physical objects,—the Heavens and the Earth,—and endeavors to explain their phenomena by means of the discoveries of the other two branches of physiognosy. It ultimately tends to become classificatory.

General psychology discovers the general elements of mental phenomena and ascertains their laws. *Classificatory physics* classifies manifestations of mind and endeavors to explain them by general psychology. *Descriptive psychics* describes individual works of mind and explains them by the principles of the other branches of psychognosy.

The subject of this book is to be the science of Logic; and the above table of the ladder of the sciences suggests that Logic ought to be found to be immediately dependent upon Ethics; and further that, as a normative science, it ought to be found to repose upon Phenomenology; and still further, that as a branch of philosophy it ought to be found making appeal to Mathematics. I now propose to show that in fact the problems of logic cannot be solved without taking advantage of the teachings of Mathematics, of Phenomenology, and of Ethics.

The literature of logic very frequently shows its students adopting fundamental propositions of logic upon the authority of Metaphysics, of Psychology, and of Linguistics. But the suggestions of the Table are that Metaphysics immediately depends upon Logic, so that Logic could not reciprocally depend upon Metaphysics; and that Psychology and Linguistics as branches of idioscopy, would naturally depend upon Philosophy, no branch of which could, therefore, depend upon either of them. Accordingly, after I have shown that Logic must depend upon Mathematics, Phenomenology, and Ethics, I shall proceed to show that it can find no secure foundation for any of its doctrines either in Metaphysics, in Psychology, or in Linguistics.

Mathematics used to be defined as the science of quantity; but that definition is now entirely given up by philosophical mathematicians. In fact, a mere glance at the historical origin of that definition will suffice to deprive it of all authority. Namely, at the time that definition came into vogue, in the schools of the Roman Empire, the three words *mathematics*, *science*, and *quantity*, all had entirely different meanings from any they have today. Namely, by "mathematics" was then meant, astronomy, music, geometry, and a certain wordy doctrine that went by the name of arithmetic; while computation was excluded as no part of mathematics. Hence when the men of that day called mathematics the science of quantity, they meant by "science of quantity" something that would be true of astronomy but would not be true of what we call arithmetic. But the truth is that by "quantity" they meant anything capable of being measured; so that if there had been any science of *lager beer* or *kerosene*, that would have been a branch of mathematics. There was, however, no such science in their sense of this word, since by "science" they meant the comprehension of anything from first principles,—a notion which we now recognize as futile.

If you imagine any five rays, or unlimited straight lines, to lie in one plane, no three of them meeting at the same point, and if you letter them, *a*, *b*, *c*, *d*, *e*, and employ square brackets enclosing two of these letters to denote the point of intersection of the two rays those letters denote, and if we denote the ray through [*ae*] and [*bd*] by *c'*, the ray through [*ad*] and [*bc*] by *e'*, and the ray through [*ac*] and [*be*] by *d'*, then the three points [*cc'*], [*ee'*] and [*dd'*] will always be on one ray (that is to say, a ray may be drawn through them all). This is a proposition in geometry, which is a recognized branch of mathematics; yet it involves not the slightest reference to quantity. The most interesting and important branch of geometry (except topical geometry of which what I am saying is equally true) is entirely occupied with propositions that have no reference of quantity.

Accordingly, modern mathematicians recognize as the truly essential characteristic of their science, that, as stated above, it concerns itself with pure hypotheses without caring at all whether they correspond to anything in nature or not, or at least, disregards such correspondence entirely after its hypotheses are once formed.

From this it follows that whenever a student of any science has occasion to argue, that supposing certain definite hypotheses to be true, then, no matter what else may be true or false, a certain conclusion will inevitably be found true, he is taking the place of the mathematician whose essential function it is to determine whether what that student is saying be true or not. Thus, every other science would have a mathematical part as many sciences are universally recognized as having.

Now logic, as a normative science, entirely disregards what the particular state of things may be, and undertakes [to] show what procedure must lead to the discovery of truth, whatever that truth may be. Some logicians, it is true, contend that their science has to assume that nature is uniform, and although others deny this, yet even they must concede that logic must assume *something* to be true. Still, as to what the particular state of things may be, beyond certain definite assumptions, logic must permit it to be whatever it may, or else logic would not be able to recommend any procedure as leading *invariably* to the truth, in whatever strange situation the reasoner might find himself. In order, then, to *prove* that the procedure he recommends really is of this nature, the logician has to show that as long as certain definite assumptions are supposed true, then, no matter what may be the case besides them, the line of reasoning he recommends will be the speediest road to the truth. But this is reasoning from pure hypothesis, which is the essential business of the mathematician. Thus logic must appeal to mathematics, or else, what amounts to the same thing, must invade the domain of mathematics, in order to make certain of the truth that it essentially seeks.

Descartes narrates, in the fourth part of the *Discours de Méthode* and in his *First Meditation*, that he one day "resolved" to doubt everything, and that he found there was but one thing that he could not doubt, namely the fact that he doubted. But therein he deceived himself most sadly, and committed the grand central mistake of his system. There were ever so many other things that he did not doubt. How did he know he doubted them? It seemed to him that he did; and he thought that was conclusive. There was where he was mistaken. He ought to have put himself to the test, and then he would have found that many of his doubts were purely imaginary. He did not see why he should not doubt everything except that he doubted; and he was trying to doubt all he could; and so he assumed, without any assurance of the matter at all, that he did doubt everything else, — a most stupendous piece of credulity on the part of this gentleman who fancied himself to be such an utter sceptic.

What happens to any veracious young thinker is that on the first day on which his thoughts are turned that way, he remarks that there are a multitude of things that he does not doubt; and it will not be long before he finds that what he does not doubt, he cannot doubt, until he meets with some definite reason for doubting it. Let him fancy that he doubts whether he will be alive the next day, and though this is far from being certain, yet he will soon catch himself making preparations for his next day's life without the smallest misgiving on the subject. He may fancy he doubts whether what he sees before his eyes be not a hallucination; but he will probably find himself very much indisposed to ask another person whether or not he sees it too, much as it concerns him to know whether it be a hallucination or not.

When he meets with a surprise, that is, finds that some belief of his has been false, in case he does not know where the mistake had its origin, then he will truly doubt and will doubt about more things than one. Should he really experience a hallucination, he may be led to doubt what no ordinary person doubts. But there are few minds who are able to maintain the equilibrium of doubt. What ought to cause them doubt will throw them over into the contrary belief; and if such a person experiences hallucinations he will soon be slumping about in such a slough of absurd beliefs, that he will lose all control over his reason.

But a normal person, able to take care of his affairs, will be found to have certain beliefs similar to those of all other such persons which he never can overcome, and which indeed he can only under peculiar circumstances really try to doubt. Such for instance are beliefs that he has certain descriptions of images before his mind, although his beliefs involving these descriptions have no resemblance whatever to the images themselves. Now there is no sense in finding fault with what is entirely beyond control. And there is nothing to be done with these beliefs, but to find whether they accord with those of others, and whether they can be abolished.

Now the science of Phenomenology asserts that every man who is sufficiently intelligent to testify to the matter at all will testify that whatever is at any time before his mind has certain features which it describes, and that it is not possible to think these features are not in what is before one's own mind. If anybody finds that he cannot doubt this, as far as his own mind is concerned, — as well as he can discover, — and that he does not seem to have any real doubt of the matter's being the same in other minds, he may as well admit that, for him, there is no doubt about it.

I must say, however, that all these reflections about possible doubts originate rather with the logician when it is a question of appealing to phenomenology than in any emphatic assertions of the phenomenologist himself. Phenomenology, which, according to our Table, stands just upon the borders of the purely hypothetical science of mathematics, hardly makes any explicit assertions. What phenomenology does is to distinguish certain very general elements of phenomena, render them distinct, and study their possible modes. It does not need particularly to insist upon their universality, since this is evident to everybody, who knows by his own portion of human experience something of what human experience generally is like. The work of discovery of the phenomenologist, and most difficult work it is, consists in disentangling, or drawing out, from human thought, certain threads that run through it, and in showing what marks each has that distinguishes it from every other.

The problem of the logician is to determine the general conditions of the attainment of truth. This problem as it first comes before him, may be clothed in various forms of words, all of which are used by every man every day of his life, so that he feels himself to be a perfect master of their meaning. Yet the first thing that the logician's studies disclose to him is that he does not know what it is that they mean. What is truth? An assertion is said to be an expression which is either true or false; but what is it that makes an expression either true or false? The logician at the outset is met by a multitude of such needs of definitions, and finds that his science must be all guess-work until they are answered. But when shall he be satisfied with his definitions? When shall he be sure that its terms do not themselves require to be defined? Here the results of the phenomenologist's studies are of fundamental utility to him. When, for example, he can show what goes to the constitution of truth in terms of those universal elements which the phenomenologist has shown to be primary, he has done all that in the nature of things is possible.

Normative science must not be confounded with practical science. When a general purpose is quite fixed and settled, as, for example, that dwellings must be made warm in winter, a skill is gradually developed in accomplishing that purpose. If the operations required to accomplish the purpose are complex, and if the purpose itself is complicated with secondary considerations that vary considerably on different occasions, it becomes desirable that the skill in accomplishing it, which already involves a good deal of special information, and as such is called an *art*, should be guided by a general theory of the relation of means to the purpose in question. This theory of an art is a *practical science*. There is a practical science of discovery. It is an adjunct of logic, and has sometimes been treated by itself.

Such a practical science has to be based upon an exhaustive acquaintance with all the means actually available for accomplishing the different parts of the fixed purpose, and has quite enough to do in considering all the circumstances of the existing situation without troubling itself to inquire how a corresponding purpose could be carried out on the planet Neptune or in the island of Laputa. Nothing more widely different from this can [be] imagined than a normative science, which instead of assuming the purpose to be settled, begins on the contrary by inquiring what the purpose shall be, and then out of the very considerations which have gone to determine the purpose, with whatever other considerations may be strictly needed, proceeds to evolve the general conditions that must hold good, wherever the results of phenomenology hold, for the realization of the end.

There are but three normative sciences; they are esthetics, ethics, and logic. Esthetics is the science of the general conditions of a form's being beautiful. It has to begin by finding out what this familiar but elusive idea of the beautiful really means. It has to define it, not at all with reference to its pleasing *A*, *B*, or *C*, but in terms of those universal elements of experience that have been brought to light by phenomenology. Unless this can be done, and it can be shown that there are certain conditions which would make a form beautiful in any world, whether it contained beings who would be pleased with such forms or not, there is no true normative science of esthetics.

The business of ethics is to determine the conditions of action's being well-purposed, or virtuous. It has to begin by finding out [what] the familiar but confused idea of moral goodness really consists in. Let us assume that it finds that [that] conduct is virtuous which is adapted to an end that is esthetically good. If that be so, ethics will be largely dependent upon esthetics. At any rate, it will certainly find that moral conduct is conduct which is self-controlled so as to be steadily directed toward a sort of purpose which ethics will define. The remainder of its inquiry will be directed to such conditions of morality as will hold good of every conceivable agent who should be amenable to the moral law at all. Such must be the nature of a true normative science of ethics.

Logic, developing its own purpose in a similar way, soon finds that it is essential to the action of reasoning that it should be self-controlled: for without that, all criticism of it, as good or bad, is idle. It would, therefore, be nothing but an application of ethics to a particular kind of conduct, were it not that, as a normative science, it has problems of analysis of its own, and that its end appears in the light of phenomenology to involve an element of a higher nature than the moral end. These circumstances, however, do not prevent its gaining great advantage from the affinity between truth and right.

After some hesitation I have concluded that the possible objection that perhaps it is ethics that depends upon logic is sufficiently plausible to be answered. In the first place, it is to be remembered that although necessary reasoning might be considered as, in some aspects of it, a subject of logic, yet mathematics, which must recognize it, is quite adequate both to practice it and to deal with it formally; and certainly mathematics has never required the aid of a science of logic to determine whether any reasoning that concerned it was valid or not. It is only probable reasonings which require a special science of logic; and ethics has no need of such reasonings, but has had a long and glorious history without falling into any merely logical difficulties. In fact, as long as necessary reasoning is relegated to mathematics, where it belongs, no normative science has any occasion for an appeal to logic, for the reason that such science does not deal at all with what actually is but only with what must necessarily be the case. Ethics, therefore, has no need of logic. But all probable reasoning, the only reasoning over which logic has exclusive jurisdiction,—and under the term reasoning I include acts of observation insofar as they are controlled,—depends upon a moral virtue, that of sincerity, as will be proved in this book,— and, indeed, you will find that those scientific men that are constantly using this kind of reasoning with success are also always thinking about this virtue; and as the reasoning depends upon the virtue, so must the theory of the reasoning depend upon the theory of the virtue.

I have now shown that logic must depend upon the sciences of mathematics, phenomenology, and ethics; and I shall not deny that it depends in some measure, though indirectly for the most part, upon esthetics. I now declare that it does not depend, except it be for illustrations and perhaps here and there [for] a fact that might be replaced well enough by another drawn from the stock of ordinary every-day experience, upon any other science whatsoever. I will first show that it does not depend upon three sciences to which logicians are in the habit of continually making appeal, but as I maintain with the effect only of confusing the subject of logic. I mean, as I have said above, the sciences of metaphysics, psychology, and linguistics. I will then show of what sort of use to logic is the history of science. Finally, I will deal summarily with possible attempts to make logic depend upon other sciences. . . .

CHAPTER III. MATHEMATICS

Since it has been shown that logic ought to depend largely upon the three sciences of Mathematics, Phenomenology, and Ethics, it will be best, before we actually take up the study of logic, to devote one chapter each to developing as much of those sciences as will be required in our study of logic, so that when we once do take up our proper subject, we may be better equipped for carrying on the study, and may not be forced to make confusing halts. Accordingly, the present chapter will be given to mathematics.

Mathematics is supposed to be a very difficult science; but on the contrary, it is in fact by far the easiest of all sciences. We often meet men who are supposed to be incapable of understanding mathematics. The greater part of them are not so, but have simply been ill taught and have no training in mathematical thought. In fact, that they have been ill taught is pretty certain; for though mathematics is such an easy science to study, the difficulties of teaching it are very great. A teacher of mathematics, especially in its most elementary branches (for they once passed, it does not *so much* matter), ought to have three requirements which I name in the order of their importance, though all are indispensable. He should be a strong mathematician, a subtile logician, and an accomplished psychologist. It will also be of advantage if he should be a natural teacher and an experienced teacher of mathematics. As for those persons whose difficulty with mathematics is more genuine, it will be found that they are weak in reasoning generally, and have little command over their thoughts; but this weakness being usually combined with a fortunate disposition and keen instincts (though little general power of original observation, usually), they frequently go through life with marked success.

In reading mathematics, the student should beware of falling into a passive state. He must remember that it is he himself, and nobody else, who has to perform the entire reasoning process, for which the book merely affords some hints. There are men who never dream of this. If they are asked to look at a diagram, they will say to themselves, "Yes, we will suppose that done," without the smallest consciousness of not having done all that was asked of them. If the book says, you will see that this diagram is that of a triangle, when some error of the printer has put a circle where the triangle was said to be, these readers will never remark the discrepancy. Should anybody get it into their heads that they are required to do the reasoning themselves, they will feel as if they had been stood upon the ridge pole of a cathedral and were required to walk unaided from one end of it to the other. The scare will deprive them of all their senses.

You may remark that a boy with any turn for mathematics talks of *learning* Latin but never talks of learning mathematics. He talks of *doing* mathematics. Now you, instead of standing there eternally shivering on the brink, first carefully watch you what the book says you are to do, and having thought over how you are to do it, then without further hesitation jump you in and do it; and you will soon find it is great fun. What the book invariably does is first to describe in general terms a diagram, or array of letters, or something of the sort that is to be made. Its general terms allow considerable latitude as to how this diagram (as, for the sake of brevity, we will always entitle it) is to be formed. But you must note carefully what features precisely it is required to have, and then proceed you to make such a diagram. There will probably be a figure of such a diagram in the book. Do not copy that, but make one of your own, following exactly the general description. Now the book, which understands that you have done this, invariably goes on to speak of alterations to be made in that diagram. Thereupon, what you will usually have to do is to *copy* your own diagram as exactly as you conveniently can, and perform the described alterations in the copy. Now the book proceeds to compare the original diagram with the altered diagram, and to call upon you to remark certain exact relations between them; and you have to do so. But in doing this there will be two difficulties. On the one hand, your copy of your diagram not being absolutely exact, nor your diagram itself absolutely according with the prescription, you cannot implicitly rely upon that; and upon the other hand, your diagram shows only one way out of an infinite variety of ways in which the diagram might have been constructed, so that although the relation which the book says will exist may do so in this case, yet that does not prove that it would be so in every possible case. But here the book will come to your aid and will make it quite plain and evident to you that the relation *always will* hold exactly. The book will do this by means of the fact that you have already become fully convinced without any shadow of doubt, that certain things *always will be*. If these were propositions about any outward experience,—say, about the sun's rising in the morning,—such entire freedom from all doubt could hardly be reasonable. But the propositions about which you are so sure are *not* matters of outward experience, but relate to an imaginary world,— and *what* imaginary world? Why, *any* world you please *in which certain general conditions* are fulfilled. Now the propositions that you will be so certain of are nothing but what is plainly involved in these general conditions which determined what world we are talking about. If we were to agree to talk about an imaginary world in which the moon, instead of going through its changes as ours does, should be a brilliant pie, out of which the sun every day should cut a sixty degree sector and eat it until he had consumed the whole, when taking one day of rest, he should commence on a new moon-pie. Now of such a world, we can say with absolute certainty that the moon will never last unchanged in shape for a thousand years, since if it did, it would not be the world we agreed to talk about. However, I must correct myself if I have said that we should be absolutely certain of this,—because just as in adding up a column of figures it is possible to make a mistake, so it is possible to blunder in saying what is or is not supposed in our original convention about the kind of imaginary world to which our conversation was to relate. So we cannot strictly say we are *sure* but only that if we have not gone crazy or committed some unaccountable stupidity equivalent to insanity, then we are sure.

Now my dear reader, I am going to suppose that you are one of those persons who have a difficulty about mathematics; and yet if you have no such difficulty, you will find what I am about to say neither without profit nor without interest. One of the most famous and most continually used of all books is the Στοιχεῖα or *Elements*, of Euclid, a mathematician for whom the mathematicians of the second half of the nineteenth century learned to entertain an extraordinary respect, and to confess that he understood the fundamentals of geometry much better than they had previously been doing. He was called to the university of Alexandria ninety-nine years after the death of Socrates. His work is appropriately named, not *The Elements of Geometry*, but just "Elements,"—the beginnings of all scientific knowledge. Now the fifth proposition of Euclid acquired in the University of Cambridge the appellation of the *Asses' Bridge, Pons Asinorum*. There is no sense in applying this name, as some ignorantly do, to the Pythagorean Proposition, which is the 47th of Euclid. The reason for giving the name to the 5th was perhaps because its figure somewhat resembles a logical diagram that had been so-called in the middle ages, probably because it looked a little like a truss and because only persons quite wanting in sagacity would need it. However, another reason which was early given, is a much better one for so-calling the 5th proposition, that the angles at the base of an isosceles triangle are equal. Namely, it is said that it is extremely difficult to induce donkeys to venture upon a bridge for the first time, but that as soon as they have crossed their first bridge there is never any further such difficulty. Now this proposition is the first one in Euclid which calls for a little active mathematical thought on the part of the student; and this is so contrary to the habit of mind of one whose studies have been purely receptive,—as in their Latin grammar,—that it often happens that he has a difficulty in understanding what is demanded of him, and is paralyzed at the demand when he does understand it. But as soon as this first awkwardness is conquered he goes on to far more intricate propositions without again experiencing any particular difficulty. I will, therefore, just run over that proposition in order just to point out to you what the student's part has to be in reading mathematics. We must first glance at the parts of the book that precede the fifth proposition. The work opens with twenty-three definitions, of which two or three may be noticed. The fourth reads: Εὐθεῖα γραμμή ἐστιν, ἥτις ἐξ ἴσου ἐφ᾿ ἑαυτῆς σημείοις κεῖται which is so vague that one hesitates to attempt to imitate it in English; but say, "A straight line is whatever line lies upon an equality in respect to the points upon it." This cannot, I think, mean that the line is symmetrical all round. It seems to say that a straight line is one upon which all points upon it are situated alike. But it must have been obvious to Euclid that a circle has no point upon it differently related to it from any other. A circle however does not rest by any means in the same way upon a *pair* of points nearly opposite one another upon it, that it does upon a pair nearly upon the same side of it; and if we suppose that this was Euclid's meaning, it will save his logic. But perhaps that is saved in another way. We must suppose he meant this, or else that he only meant to say vaguely that the straight line lies "flat" on its points; and in favor of this supposition is the fact that he does not usually make subtle analyses in his definitions. The eighth, for example, defines an angle, γωνία, simply by means of the word κλίσις, inclination, although this is not very accurate. It is true that the common phrase ἐξ ἴσου usually has a more definite meaning; but that is a weak argument.

The fifteenth definition is that a circle is a plane figure surrounded by one line, all the straight lines drawn to which from one point within the figure are equal to one another.

After the definitions come five "postulates," that is, assumptions which there is no attempt to prove and which are open to doubt. The word is explained by Aristotle and others. These five postulates are as follows:

I. It is postulated that from any point to any point a straight line may be drawn.

II. And that a terminated straight line (note, then, that he recognizes the possibility of an unlimited straight line) can be continuously extended in a straight line.

III. And that with any centre and radius a circle can be drawn (in any plane through that centre).

IV. And that all right angles are equal to one another.

V. And that if two straight lines (in one plane) be cut by a third straight line making the sum of the internal angles on one side less than two right angles then those two lines will meet if sufficiently produced, on that side on which the angles are less than two right angles.

Though these are the only postulates that Euclid mentions he virtually assumes several others, as I will show.

Next follow is the MSS. nine "κοιναὶ ἔννοιαι" or "matters of common knowledge," now, as then, otherwise called "axioms," and now often supposed to be self-evident. But that was not the ancient conception. Of these axioms, there is reason to think that only five were really given by Euclid, as follows:

I. Things equal to the same are equal to each other;

II. And if equals be added to equals, the sums will be equal;

III. And if equals be subtracted from equals, the remainders will be equal;

IV. And the whole is greater than the part;

V. And things which fit over each other are equal to each other.

Then follow the first three propositions, which we may skip for the present. The fourth proposition is as follows:

"If, of two triangles, one has two sides and the angle between them respectively equal to two sides of the other and to the angle between them, then the third side of the one triangle will be equal to the third side of the other, the areas of the two triangles will be equal, and the angles opposite the equal sides will be equal."

He now proceeds to draw a diagram [Fig. 1] in accordance with the condition of that general statement, as follows (I translate freely and change the lettering): "Suppose that the two triangles $A_1B_1\Gamma_1$ and

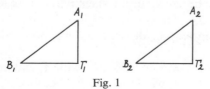

Fig. 1

$A_2B_2\Gamma_2$ be such that the side A_1B_1 of the one is equal to the side A_2B_2 of the other, and the side $A_1\Gamma_1$ of the first is equal to the side $A_2\Gamma_2$ of the second, and the angle A_1 of the first is equal to the angle A_2 of the second. Then what I am saying is that the third side $B_1\Gamma_1$ of the first of these triangles is equal to the third side $B_2\Gamma_2$ of the second of the triangles, that the area of $A_1B_1\Gamma_1$ is equal to the area of $A_2B_2\Gamma_2$ and that the angle Γ_1 of the one (opposite the side A_1B_1) is equal to the angle Γ_2 of the other (opposite the side $A_2B_2 = A_1B_1$), and the angle B_1 of the one, opposite $A_1\Gamma_1$, is equal to the angle B_2 of the other opposite $A_2\Gamma_2 = A_1\Gamma_1$."

That is what he has to prove *is* true, no matter what the triangles may be, so long as the three parts of the one are equal to the three parts of the other as is assumed; and if he proves this, he proves his general proposition which means no more than that this is true in every case.

In order to do this, he begins by remarking that if the first triangle be carried over and placed upon the second, so that the point A_1 falls upon the point A_2, the side A_1B_1 may be made exactly to cover the side A_2B_2, since it is equal to it. But this remark gives an unexpected and inadmissible meaning to what was offered as "common knowledge" that things which can coincide are equal. For it is far from being common knowledge that everything can be moved about without changing the sizes of its parts. It is true that we commonly assume that a metallic bar may be obtained such that after comparison with another in length [it] may be carried away, brought back by any other route and, observing great precaution, may be recompared with the same result. In another chapter, we shall have to consider how far such an assumption with which the results of every experiment, taken just as they come out, will conflict, is justifiable and what degree of confidence ought to be reposed in it. But whether it be true as a matter of fact or not is not the question at present. It certainly is an assumption about the nature of space which ought to be reckoned as a postulate. It ought also to be said that even if two bodies be equal in all their parts only three points of the one could with certainty be made to coincide at once with corresponding points of the other even if they could interpenetrate; because they may be rights and lefts, like our two hands. Hence, an additional assumption will be necessary; such for example as that a right and left handed body can be equal in all their parts. This assumption would patch up the difficulty. However, the imperfection of the proof is not, so far, sufficient to deprive it of all force. So let us proceed with the very free translation.

Having brought A_1 into coincidence with A_2 and made the line A_1B_1 to stretch along A_2B_2, B_1 will fall upon B_2, because A_1B_1 and A_2B_2 were equal and the magnitude is supposed to remain fixed during the motion. Now he says that if one triangle is turned over upon the other line $A_1\Gamma_1$ may be made to stretch along the line $A_2\Gamma_2$ because the angle A_1 is equal to the angle A_2 and therefore will just cover it. We see plainly that this must be so; for if A_1 being on A_2 and A_1B_1 lying along A_2B_2 if when we bring one angle, A_1, flat down upon the other, A_2, the other side of the angle A_1, that is $A_1\Gamma_1$, does not fit along the side $A_2\Gamma_2$ (it is not now a question of its length) the two angles evidently would not be equal. We assume of course as in the case of the comparison of length that there has been no change of magnitude in the motion. So, then, we have A_1 on A_2 and $A_1\Gamma_1$ stretching along $A_2\Gamma_2$; but now, under the same assumption of no change of length Γ_1 must fall on Γ_2, for if it fell short or went beyond, $A_1\Gamma_1$ would not be equal to $A_2\Gamma_2$, which is the state of things to which we are limiting all we say. So then B_1, resting all the time on B_2, and Γ_1 now falling upon Γ_2, there are two points B_1 and Γ_1 of the straight line $B_1\Gamma_1$ that are in the same places (and points are nothing but movable places) as the two points B_2 and Γ_2 of the straight line $B_2\Gamma_2$. Now, says Euclid, if the whole line $B_1\Gamma_1$ did not fit upon the line $B_2\Gamma_2$ "two straight lines would enclose a space which is impossible." He ought to have set it down as a postulate that two straight lines cannot enclose a space. Some good MSS. insert it as such, but in a second hand. Most of them make it Axiom No. 9; but the ancient doctrine given by Aristotle was that axioms being well-known, if not self-evident, it was useless to cite them in proofs; and Euclid never does so, and no such Axiom appears in the list that Martianus Capella, Boethius, and Proclus profess to copy from Euclid. So it must be set down to a negligence which Euclid would not rectify at the expense of a fresh sheet of papyrus. He goes on: Since, then, $B_1\Gamma_1$ coincides with $B_2\Gamma_2$ and just covers its whole length, these two sides, the "third sides" spoken of are equal. Furthermore, the area of the triangle $A_1B_1\Gamma_1$ fits upon the area of the triangle $A_2B_2\Gamma_2$ and must be equal to it; and the angle B_1 fits upon the angle B_2, and the angle Γ_1 fits upon the angle Γ_2. Thus, all that the general proposition said is verified, and for the reasons given would be verified in the same way in every case.

We now come to the *Pons Asinorum* which is that in a triangle two of whose sides are equal to one another, the two angles opposite those sides are also equal; and moreover, if the two equal sides are prolonged beyond the third side, the two exterior angles so formed will be equal [Fig. 2]. Take any triangle, $AB\Gamma$, that has two of its sides, AB and $A\Gamma$, equal; and let the side AB be prolonged, say to $\triangle$, and let the side $A\Gamma$ be prolonged, say to E. Then what I am saying is that the angle $A\Gamma B$ (opposite AB) will be equal to the angle $AB\Gamma$ (opposite $A\Gamma$); and moreover that the

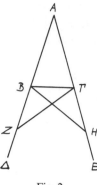

Fig. 2

external angle $B\Gamma E$ will be equal to the external angle $\Gamma B\Delta$. Euclid now takes any point, Z, on the line $A\Delta$ beyond B and marks a point H on the line AE so that AH shall be equal to AZ. Hence H will lie between Γ and E, since AH is equal to AZ which is greater than AB and hence also [than] its equal, $A\Gamma$. Proposition 3 was devoted to showing how this could be done; namely, by sticking the point of one leg of a pair of compasses into A and drawing an arc of a circle cutting $A\Delta$ and AE. It is evident to the last extreme that this will do it. He now compares the two triangles $AZ\Gamma$ and AHB by applying Proposition [4] just proved. The side AZ of the former equals the side AH of the latter and the side $A\Gamma$ of the former equals the side AB of the latter while the angle A included in both cases between the two sides is common to the two triangles. Therefore, by proposition 4, the third side of the one, that is, $Z\Gamma$, must be equal to the third side HB of the other, and the angles of the two triangles opposite the equal sides, that is to say the angle $A\Gamma Z$ opposite AZ is equal to the angle ABH opposite AH (equal to AZ); and the angle $AZ\Gamma$ opposite $A\Gamma$ is equal to AHB opposite AB (which is equal to $A\Gamma$). He now compares the two triangles $BZ\Gamma$ and ΓHB by the same principle. The angle Z of the one has just been proved to be equal to the angle H of the other, the side $Z\Gamma$ of the one has just been found equal to the side HB of the other and the other side adjacent to the angle, that is ZB is equal to $H\Gamma$ because these are the remainders after subtracting the equal lines AB and $A\Gamma$ from the equal lines AZ and AH respectively. Thus the two sides $Z\Gamma$ and ZB and the included angle $BZ\Gamma$ of the triangle $BZ\Gamma$ are respectively equal to the two sides HB and $H\Gamma$ and the included angle ΓHB of the triangle ΓHB; so that by Proposition 4, the angle $ZB\Gamma$ is equal to the angle $H\Gamma B$ and the angle $B\Gamma Z$ equals the angle ΓBH. But if we cut off these equal angles respectively from the two angles $A\Gamma Z$ and ABH which have been proved equal, you can see by inspection that

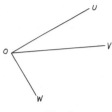

Fig. 3

the remainders are $A\Gamma B$ and $AB\Gamma$ which are therefore equal. For the angles $Z\Gamma B$ and $B\Gamma A$, *since they lie in one plane* together just cover, without overlapping the angle $Z\Gamma A$. And such will always be the case. Watch the way the letters come in the triplets [Fig. 3]. If any two angles like UOV and VOW have the same middle letter in their triplets and the first letter of one the same as the last of the other, then, *if the angles do not overlie one another, and are in one plane*, their sum is UOW with the same middle letter and the two once mentioned first and last letters. But if they lie in one plane and do overlap, like UOW and WOV, the same rule gives their difference, UOV. So, just as $Z\Gamma A$ is the sum of $Z\Gamma B$ and $B\Gamma A$ and HBA is the sum of $HB\Gamma$ and ΓBA. So since the two sums have been proved equal and since $Z\Gamma B$ and $HB\Gamma$ have just been proved equal, it must be that $B\Gamma A$ and ΓBA are equal; but these are the two angles opposite the two equal sides that I undertook to prove equal; while the external angles $ZB\Gamma$ and $H\Gamma B$ have already been proved equal and all that had to be demonstrated is demonstrated.

The *Pons Asinorum* has not the reputation of being abstruse; and no person who has any intellectual self-control can find it unsurmountable. The tradition is that this once conquered one will find nothing very redoutable in the rest of mathematics. At any rate, it is certainly true that the further one penetrates into mathematics the more intrinsically easy it becomes, as a general rule, apart from the facility that one gains by the habit of intellectual exertion. The part of mathematics which it is necessary for the student of reasoning to be thoroughly skillful in, I mean Dyadics, is far easier than the Pons Asinorum.

In addition to that branch, however, it is quite indispensable for a good reasoner to understand the doctrine of chances. This is a subject of difficulty in two ways. In the first place, the mathematics of it is a little difficult in places. But these can be smoothed down considerably. In the second place, the greatest difficulties of the theory of probabilities are not strictly speaking mathematical, or at least not what is usually recognized as mathematical, but are logical; and the discussion of these can be postponed until the reader has gained from previous parts of this book, a proper equipment for grappling with them.

Before taking up dyadics, it will be necessary, in order that the reader may have the sort of understanding of mathematics that the study of logic requires,—a somewhat different kind of understanding of it from that which he would need if he desired only to understand mathematics,—that we should make some preliminary reflections upon the nature of mathematics.

Now the feature of mathematics which separates it widely both from Philosophy and from every special science is that the mathematician never undertakes (*quâ* mathematician) to make a categorical assertion from the beginning of his scientific life to the end. He simply says what *would be* the case under hypothetical circumstances.

The higher branches of physics and of psychics,—the nomological sciences,—likewise discover and enunciate general laws which are very properly expressible in the same grammatical forms. But they only mean *at most*,—if they do generalize so far, which may be questioned,— that experience forces them to believe that such are the laws of this universe in which we find ourselves. The mathematician, however, snaps his fingers at experience and at this little universe: what he means to pronounce upon relates to any and every universe in which the antecedent of his proposition might be true. Even the normative sciences,— logic, for example,—go far beyond objective experience. The logician reasons about objects that are not even real. Still, all the normative sciences necessarily make themselves responsible for certain characters of objects of thought. The mathematician alone does not unconditionally assume or assert anything at all.

What does he rely upon, then? Ah, that is a question-begging question. He no doubt works with his imagination and his generalizing intellect; and if they were to work wrong, he would fall into error. But there are two circumstances to be taken into account, and to be taken into account, *together*. They have no relevancy to mathematics; but they are relevant to the irrelevant question that has been asked. One of these circumstances is, that in point of fact, the mathematician's imagination and intellect do *not* work wrong,—although once in a while, from inattention, he adds up a column wrongly, and says that 2 and 2 make 5. But such blunders are so promptly rectified, when attention is directed to them, and are so extremely rare, that they are not [to] be regarded as belonging to the general course of the mathematician's work. You might as well say that his success in answering how much seven times eight is, is due to his not being struck by lightening before he has given his answer.

A bold sailor contrives to sail round the world alone in a small boat. As he lands at the end of his achievement, a person at the landing says to him, "Do you know that your success was due to whales?" "How so?" "Why if every whale in the ocean had not abstained from butting your boat you never would have brought it to port." The sailor, I think, might properly reply, "Oh, do have a little good sense!" A sound logician would certainly say that the sailor was right, as we shall see, in due time.

The second circumstance that has to be borne in mind at the same time with the first, is that the mathematician, as such, never knew that he had any such faculties as imagination and intellect. Consequently, he cannot be said to have relied upon them. The fact that they did not go wrong was simply one of the million matters of course the absence of which, if they had been replaced by something almost miraculous, might have resulted in his making a mistake in some few sporadic cases. If you do not agree with me you ought to find fault with a physician who reports yellow fever as the cause of a death instead of attributing it in part to the deceased not having lived in Judaea at the time of Jesus Christ and his apostles and consequently not having been healed by one of them. If you reply that the mathematician's imagination and intellect are the "direct cause" of his success, I shall rejoin that that is metaphysics too stale for me to deal with at this date.

The mathematician does not "rely" upon anything. He simply states what is *evident*, and notes the circumstances which make it evident. When a fact is evident, and nobody does or can doubt it, what could "reliance" upon anything effect?

If Euclid's geometry is regarded, as he no doubt regarded it, as the science of real, physical space, it is humanly speaking certain to a closer degree of approximation than any other special science. It is, in that case a branch of idioscopy; for the most refined methods of observation have to be employed in order to determine, for example, how great the area of a triangle must be in order that the sum of its angles should depart from 180° by as much as one thousandth or one ten-thousandth of the angular diameter of the moon as seen from the earth. But it does not, in that case, approach, — literally, *does not approach*, — the degree of precision of pure mathematics. For that purpose, the postulates, after having been thoroughly revised, must be considered as purely hypothetical; and the question that the geometer is answering must be understood to be, 'What *would be* the properties of spatial figures, in a space having the properties embodied in the postulates?'

To answer this question, he needs no external observation. Strictly speaking, he does not need even to look at his own diagrams. Indeed, a mathematician of good habits ordinarily only puts pen to paper to record his results, except in intricate work of a low order. The objects he studies are the creations of his own conscious volition, so that no part of their nature can be hidden from him.

There is thus nothing to surprise one in the (theoretical) infallibility of mathematics. The puzzle is how it can present any difficulties. Anybody who should take up that question, and give a thorough-going answer to it, would write an exceedingly helpful book. It would require a book, since the causes of difficulty are various. I can only indicate in the briefest terms a few of them. One is that much intellectual strength is required at each step of one's progress to put the really pertinent, pressing, and readily answered question. Then, although we say that the mathematician is dealing only with the creatures of his own conscious volition, yet, after all, who is this self? His whole discourse of reason is a dialogue between a past self and a future self; and much that was the creation of one in his skin is now strange to him and expressed in foreign language. He has gone on complicating his ideas to an extent of which nobody but logicians, and few of them have any adequate idea. To grasp his entire hypothesis clearly at one time has passed beyond his power. Moreover, his two selves are unavoidably talking to each other in two different languages, the one a language of images, necessary for tracing out results, the other a language of abstractions, necessary for the generalization.

Those items will serve well enough such ends as a bald list, even were it more nearly complete, could alone serve. What I am aiming at is to deduce from the mathematician's purposes and circumstances those features of his general method of procedure which it concerns the student of logic to understand before taking up the study of those parts of mathematics that are indispensable for any true comprehension of the nature of the different kinds of reasoning.

It is obvious, to begin with, that precision is everything in the kind of inquiry in which he is engaged, whether he is dealing with metrical branches of his science or not.[2] Rough statements are sometimes useful; such as Simpson's Rule, Poncelet's Theorem, and some methods of trisection of angles; but their utility lies mostly outside of pure mathe-

---

[2] Peirce has written on the page opposite, "I do not think that this precision is by any means so evident. The mathematician is tracing out necessary consequences. Now that which is necessary must be definite. But the only way will be to rewrite a part here on sheet that I shall insert." No such sheets are in the MS. today.

matics, and they generally must be mere abridgments of exact knowledge. Precise statement being thus the very staff of life to the mathematician, he is forced to use abstract statements of all his results. For in no other terms could anybody dream of precisely enunciating very general propositions, as those of the mathematician always are of their very essence.

But one can never trace out consequences by the help of general terms. That is a proposition which very few logicians have seen. John Stuart Mill partly saw it; I cannot say more. Friedrich Albert Lange fully saw it; but he restricted the means of tracing consequences too narrowly, by far. As for such writers as Boole, DeMorgan, Schröder, together with another class of whom the Abbé Gratry is a typical example, I am astonished that they should have failed to grasp a truth so near to their own constant practice. They must have been deceived by the old syllogistic, where it is true that deductive conclusions, — very puerile ones, though they be, — are obtained by no other apparent machinery than that general language. But in those cases, the conclusion comes so instantaneously that direct observation of how we come by it is impossible. It is not the question what is the process by which I am led to conclude that Socrates was mortal, or Andree who was lost in the Arctic regions, or that I myself shall die; because those are not pure deductions. But to say how, from the pure *hypothesis* that every three-legged animal has two heads and that every inhabitant of Mars is a three-legged animal, I trace out the consequence that every inhabitant of Mars has two heads, that is a question to which direct observation can furnish no even tolerably certain reply. The conclusion comes too quickly. It is so even with the least obvious of the old syllogisms, Frisesomorum, of which I will give an example. Now watch your process of thought, reader, and see if you can catch a glimpse of it, as it flashes by. Now! On the hypothesis that

> Some minstrels know the Basque Language and that
> Among the fireworshippers there is no minstrel

How are you led to be certain that

Somebody who knows Basque is not a fireworshipper. I will venture to guess that, as well as you can make out, you found out the conclusion by imagining a minstrel who knew Basque, and considering that the absence of minstrels among fireworshippers would preclude this Basque-speaker from being a minstrel. If so, you resorted to an image and did not perform your reasoning in general terms, as syllogistic prescribes. But in order to make sure of how necessary conclusions are drawn, you must take one which is more difficult. Watch again, if you please! Suppose we admit as true the testimony of two witnesses one of whom testifies that in the course of a single calendar year there was not a single monk in a certain monastery against whom at least one formal criminal accusation had not been laid by some monk of the same monastery; while against some there had been several such accusations. The other witness testifies, however, that no one monk had ever made accusations against more than one monk.

Then I ask, does it follow or not, that every monk during that year made an accusation of crime against a monk?

You may say, offhand, that it does not follow. But think again, and see how else it could be. There were only just so many monks. Every mother's son of them was accused. You might think from the first witness's testimony, that perhaps some one malicious and half-cracked fellow had accused all his fellows and finally, publicly confusing, had accused himself. But no; the other witness avers that no monk accused more than one. Then each one accused had his own distinct accuser; and the accusers must have been as many as the accused. Since then the accused were all, the accusers were as many as all. Could that be if any were not accusers? How now, reader who said it did not follow. You see, you depended on the general terms.

However, it is true that the conclusion does not necessarily follow from the premises, as you have doubtless remarked; because all that follows is, that every monk must have had among the monks a distinct and separate accuser. But since we are considering a pure hypothesis, and the evidence of experience is wholly discarded, there might have been, for aught we know, an infinite number of monks that year. They might have been numbered from 1 up indefinitely as far as whole numbers go, which is beyond all limit. Then each monk may have been accused by the monk whose number was double his own. Thus all monks would have been accused, while only those who bore even numbers would have been accusers. This shows how circumspect one must be in necessary reasoning.

The abstract expressions and the images are all that the mathematician deals with. There is no third object that they both represent. The images, being the creations of an intelligent man, conform to some purpose; and a purpose, being general, can only be thought in abstract or general terms. Thus, in one way, the images represent, or translate, the abstract language; while in another way, it is the latter which represents the former. The mathematicians certainly always have considered the images an external world, not supposed to be experienced, and sometimes (of late years always) acknowledged to be imaginary, as of course a universe not experienced must be. Nearly every image is regarded as standing, not for a definite individual object in that universe but for an individual object, yet as truly any one of a class as any other. Thus, when the geometer draws a straight line upon his diagram, he understands that, within what restrictions there may be, it represents any straight line. The imaginary universe to the parts of which all the images and propositions refer is always thought of as a single individual whole of individual parts. No particular purpose is subserved by this individualization of the mathematician's imaginary universe; and the explanation of it must be awaited in another part of this book. Yet it may be remarked here that nothing would be gained by making the universe general.

The mathematician always describes his universe, usually in some individual features of it and always, at any rate, as to certain regularities which he supposes to be without exception in it in a list of postulates or otherwise. This description always leaves it indeterminate what color the imagined objects are. To specify anything of that sort would be considered grotesque. Why? Because so long as those which might have been imagined blue, or any other color, are represented to be in some way distinguished from others, the distinguishing color chosen could make no difference as to the forms of necessary conclusions that might be drawn. Now the mathematician's whole interest is in these forms of necessary conclusions; and whatever does not concern them is regarded by him as foreign to mathematics. It is the same in regard to relations. Only a few relations between the individuals of his imaginary universe are noticed at all by the mathematician, and as to those few, what he cares for is the presence (or absence) of an unbroken rule as to the identity of objects in different sets of objects between which the relation subsists. If there is no such rule, which might serve as the means of drawing some necessary conclusion, he will regard the relation as having no mathematical interest. The obvious reason is that dealing with creatures of the imagination exclusively, nothing that he can say will be of any consequence except that if certain propositions are true certain others are infallibly true; and this would be a necessary consequence.

Of course, the unfailing universality of the mathematician's conclusions are due to every conclusive step being evidently nothing but the application of a plain rule to a manifest instance under that rule. In that sense, those logicians are quite right who say that the mathematician's whole inferential proceeding is of the type called *Barbara*, or the form of

Any M is P;
S is an M;
$\therefore$ S is P.

On the other hand, the very same general fact justifies those who say that all the mathematician's reasoning consists in observation; for whatever is "evident," or "plain", or "manifest" is so only to observation of it. Logicians of each of these classes are, however, apt to deny what the other class says, wherein they fall into opposite errors; while both kinds commit the same rash assumption that having found out one universal characteristic of a large variety of reasonings, they have thereby mastered the whole secret of the subject. That is just as if the first chemist who discovered that matter never alters its weight, —Sir Walter Raleigh assumed this according to a tradition, —should have supposed that chemistry had no further secrets for him.

Those hypotheses which form the matter of the postulates are of permanent and immutable states of things, —mostly imagined laws as to how changes would take place; but so far, the mathematician has not created any moveables to obey those laws. As long as nothing else is imagined, very few and trifling consequences, if any, can be drawn. For what he imagines simply remains as he imagined it. To bring to light any result that was not obvious at the outset, it is necessary that changes should take place. Now imaginary and unreal things have no force to move of themselves: it must be the mathematician's own will that changes them. To do this, he must have a purpose: he must be aiming to bring about some result. If, however, he could foresee what the result would be, he would be able to draw those consequences which he cannot draw. I will not undertake to say how he grows to asking what would happen if some change in the mutable objects which he has now imagined were to take place according to the immutable laws expressed in his postulates. He has to do something similar to Euclid's moving one triangle over into coincidence with another. But when he has supposed this change to take place he cannot be satisfied to trust to his imagination of what the result has been. He must appeal to the general rule contained in one of his postulates, and to the manifest fact that this comes under that rule, and to the evident result of the rule in that case.

Thus, hypotheses of objects imagined to be mutable furnish the mathematician with the data of his problems. It often happens that there are some questions about the relations between the parts of such an image which he can answer with certainty without making any changes in it. They are all obvious enough, Such make the chief substance of those *corollaries* which editors of Euclid have added to his text.* (*In Heiberg's text of the 13 books of the *Elements*, there are 132 definitions, 5 postulates, 9 axioms, 401 theorems, 64 problems, 17 lemmas, 2 scholia, or remarks, and 27 corollaries or πορίσματα. They are parenthetical inferences of an obvious kind. As to the genuineness of these 27 there are special reasons for doubt in many cases. The other corollaries, of which, for example, there are a score in the first book of Playfair's Euclid, they are all the work of editors, and are almost all trivial things that Euclid did not deem worth mention. It is supposed they were called corollaries because marked by a marginal wreath of triumph.) Hence, I propose to call this class of inferences *corollarial deductions*.

But all the consequences not readily seen, are only to be discovered or proved after making changes in the image, requiring more or less ingenuity. How the mathematician can guess in advance what changes to make is a mystery that deserves a life-time of study. Books have been written about it; but they are mere collections of empirical recipes for accomplishing what has substantially been done before, and are no help to any really great step, because they do not go into the rationale of the matter. The truth is that mathematicians have not at present any knowledge of how to plan an attack upon a problem altogether novel. This is shown by the fact that such a problem as that of proving how many colors will suffice to distinguish the regions upon any map on a given kind of surface has remained prominent for half a century without being solved.

11

AN APPRAISAL OF THE FACULTY OF REASONING

(616 and 617)

A query lately appeared in these columns that seems worthy of being followed out. It was whether, in case a given planet were known to be the habitation of a race of high psychical development, and that in the direction of knowledge, it would be safely presumable that that race was able to reason as Man does.

Next after the laws of inanimate nature and after sense-perception, nothing works so uniformly and smoothly as the Instinct of the lower animals. A downright blunder on the part of Instinct is extremely rare, to say the least, while our reasoning goes entirely wrong and reaches conclusions quite contrary to the truth and unwarranted by its premises with such distressing frequency that an incessant watch has to be maintained against these lapses. As for small divergences from strict logic, they are to be found in the majority of human inferences. We may as well acknowledge that Man's self-flattery about his "reason," though we all indulge in it, is prodigiously exaggerated. Reason itself winks satirically in its boasting, and broadly hints at its own mendacity; and yet were some divinity to offer to exchange any man's logical faculty for that "Intuition" that is usually attributed to women,—the Intuition promised reaching, however, the same pitch of perfection as the Instinct of bees and seals, or ever so much higher, how many, think ye, would close with the offer? Hastily to conclude that such an exchange would result in making its subject appear as a fool would be simply to furnish a new example of Reason's blundering. Far from that, it would surely enhance the transformed individual's reputation for sound judgment. If one had it in his power to collect a numerous sample of the men that are today known and honored for their intellect throughout the better informed classes of Europe and America,—only taking care not to draw too many from the ranks of exact science,—and, having withal authority to test their reasoning powers, were to set to each of them the task of reasonably proving or disproving a given promising scientific hypothesis, one would certainly find that a notable percentage of these justly respected minds would not know at all how to go to work. Not a few of them for example, would begin, as we have many a time seen just such men begin, by studying with promiscuous assiduity the facts upon which the hypothesis had been based, instead of beginning, as they ought, with the study, not of the facts, but of the hypothesis, in order to ascertain what observable consequences of the truth of the hypothesis, in case it were true, would contrast with the consequences of its falsity, in case it were false; and only thereafter turning to the facts, to examine them, not in a promiscuous way, but only in these pertinent respects. When one's purpose is to produce a reasonable hypothesis, it is right to begin by immersing oneself in the study of the facts until the mind is soaked through and through with the spirit of their interrelations; but when one has to estimate the ascertainable truth in a hypothesis already presented, it is the hypothesis that should take precedence in one's inquiry. The experiment supposed would not, however, be an altogether satisfactory test of a man's reasoning power, partly because the task set is too special and peculiar, and partly because a man who had been trained in testing hypotheses would at once set to work in the right way, even though he merely followed a rule of thumb, without at all knowing why that way should be the right way, and perhaps not even knowing there was any other; while the best reasoner in the world, if the problem were novel to him, might halt and stumble in his procedure owing to the very circumstance that he was taking his steps in the constraint and tight boots of too much reasoning instead of in the old slippers of habit.

A better test is the ability to follow a simple mathematical demonstration; because mathematical conceptions are all conceptions of visible objects, and involve no other difficulty than the extreme complexity of most of them which, however, does not affect the simple demonstrations; and because there is no element of mathematical reasoning which is not found in all reasoning unless forming a conjecture be called reasoning. Nevertheless, it is well-known that among those who have never been able to cross the *pons asinorum*, i.e. Euclid's demonstration that the angles at the base of an isosceles triangle are equal, there are men who are so far from being asses that their heads touch the very heavens of human intelligence, jurists whose opinions are authoritative the world over, diplomats renowned for their skill in unravelling the most tangled of human snarls, naturalists of the first order, and minds who shake the senate of philosophy. It is, no doubt, true that elementary mathematics is so abominably taught that a pretty bright mind may quite fail to apprehend the thought. He may, for example, suppose that mathematics deals with questions of fact, instead of with questions of whether the truth of an arbitrary hypothesis would involve the truth of another proposition or not. It is further true that the elliptical style of writing which mathematicians have inherited from the Greeks tends to veil the connexions between the different steps. But examples can be framed which are not open to these objections. At the end of this article will be found a little mathematical discussion which has been carefully designed to serve as a test of capacity for such mathematics as is not very intricate. The beginning of it is excessively easy; but the last part of it is not so; and who so clearly sees the cogency of the whole may rest assured that he labors under no mental defect in respect to mathematical reasoning. . . . [1]

---

[1]   MS. 617 tends to develop Peirce's thesis at this point and the material is edited so that it follows directly on MS. 616 as given here. However MS. 616 ends with the following statement.

"The difficulties of different minds may be different; so that it is desirable to enumerate the different kinds of mental processes that enter into mathematical discussions. In the ancient style, which is still much followed, a mathematical demonstration is prefaced by a statement of the proposition to be proved, expressed in the abstract terms of ordinary speech;—a form of expression relatively difficult of apprehension to the mathematical mind.

"Calling this the first step, the second will consist in translating the words which denote that which the proposition supposes, or takes for granted, into diagram-language. Thus, if the proposition was that the sum of the angles of a spherical triangle is greater than two right angles, the diagram should show a spherical triangle and two right angles. The word diagram is here used in the peculiar sense of a concrete, but possibly changing, mental image of such a thing as it represents. A drawing or model may be employed to aid the imagination; but the essential thing to be performed is the act of imagining. Mathematical diagrams are of two kinds; 1st, the geometrical, which are composed of lines (for even the image of a body having a curved surface without edges, what is mainly seen with the mind's eye as it is turned about, is its generating lines, such as its varying outline); and 2nd, the algebraical, which are arrays of letters and other characters whose interrelations are represented partly by their arrangement and partly by repetitions of them. If these change, it is by instantaneous metamorphoses.

"The diagram-language into which a proposition in mathematics is translated cannot possibly consist in nothing but a diagram, since no diagram, even if it be a changing one can present more than a single object, while the verbal expression of the proposition to be proved is necessarily general. To revert to our example, a proposition about any spherical triangle whatsoever, relates to something that no single image of a spherical triangle can cover. Accordingly, every diagram must be supplemented by certain general understandings or explicit rules, which shall warrant the substitution for one diagram of any other conforming to certain rules. These will be rules of permissible substitution, partly limited to the special proposition, partly extending to an entire class of diagrams to which this one belongs."

This is a most singular psychological phenomenon; because there is no element of mathematical reasoning that is not found in all reasoning whatever, excepting only in logical analysis and in the formation of conjectures. If the incapacity of high judicial and other intellects were limited to the majority of mathematical reasonings, we might attribute it to the difficulty of grasping the very intricate relations to which mathematical propositions usually refer. But since this incapacity extends to the *pons asinorum* and other propositions which present no such intricacy, we have to look elsewhere for the secret of it. Let us see, then, precisely what logical operations are involved in a very simple piece of mathematical reasoning. Suppose the question is how many rays, or unlimited straight lines, can at most each cut four given rays that cannot all be cut by each of an indefinitely great multitude of rays. For the sake of brevity, we will suppose that, of the four given rays, which we will distinguish as A, B, C, and D, no two intersect, whence no two lie in one plane. Consider any plane containing A. The rays B and C each cut this plane in a point; and the ray through these two points is the only ray in that plane that cuts all three of the rays A, B, and C. Considering any second plane containing A, the same will be true. But the ray in this second plane that cuts B and C cannot cut A in the same point where the ray in the first plane cut[s]A, since it is easy to see that if it did B and C would lie in one plane, contrary to our assumption that no two of the four lie in one plane. Consequently if we imagine, —for the mathematician much more than the poet must be "of imagination all compact," —that the plane containing A turns round on A as an axis, then the ray in that plane that cuts B and C will corkscrew round A; and its "wake" will form a surface, something like that of a dice-box that has been compressed so as to render its throat elliptical. This, to be sure, is not quite evident. But it is clear, —it would be easy to render the whole argument rigidly demonstrative, if we were not afraid of striking panic into the reader's heart, —that the surface is curved, and that D may be so situated as nowhere to come near this surface. Also that D may cut the surface at two points; and therefore intermediate between these two possible positions for D, there must be one in which it would touch the surface in single point. Now we ask is it possible for three points of D to be on that surface without all the points of D lying on that surface? If so, through each of these points there passes that ray in one of its positions that corkscrews round A. Let us call these three positions of that ray, X, Y, and Z. Then no two of the three rays X, Y, Z lie in one plane or have any common point. Thus, everything is true of X, Y, and Z that we assumed to be true of A, B, and C; and consequently all that we found to be true of A, B, and C is equally true of X, Y, and Z. But each of the rays X, Y, Z, cuts each of the rays A, B, and C; and since we found that all of the rays that cut A, B, and C, lie wholly in, and indeed make up, that supposedly dice box shaped surface on which lies every point of A, B, and C, it follows that every ray that cuts X, Y, and Z equally lies wholly in that surface. Hence, D lies wholly in that surface, and through each point of it there passes one of those positions of the ray that cork-screws round A and cuts A, B, and C. Thus, not more than two rays can each cut A, B, C, and D, unless there is an infinite number of such rays making up a surface. Such is the outline of the mathematician's reasoning concerning this problem. The reader will find it easy to fill out this outline so as to render every step demonstrative, if he likes; but for our purposes, the outline, as it stands, will suffice.

Now let us see what the general nature of the essential parts of this course of reasoning was. In the first place, a problem was expressed in general terms, — in this case, in ordinary language. In the second place, the mathematician had to translate this general language into a concrete diagram which he had to create in his imagination, with or without the aid of a copy of it upon paper. The diagram being a single object, it could not, by itself, be adequate to represent the full meaning of the general language. But the mathematician rendered it so by a supplementary understanding that it might be freely modified in certain respects while remaining unchanged in others. In the third place, the mathematician proceeded to experiment upon his diagram, especially by making addi-tions to it, in a way not unlike the way in which a chemist might experiment upon a specimen of ore that might be brought to him for examination, especially by adding this or that reagent to it. But in one particular, the mathematician's experiments are decidedly unlike those of the chemist. Namely, the latter are very costly. They cost labor, if nothing more. At the very outset his lump of hard ore has to be reduced, with much elbow grease, to an impalpable powder before he can do anything with it. On the other hand, the subject of the mathema-tician's experiments being his own creation each modification of it costs him considerably less effort than saying "Presto! Change!" would cost him. For this reason, if for no other (and perhaps there is another reason), he is soon enabled to say with positive certainty exactly what the result of a given kind of experimentation upon a given kind of diagram must be. But the mathematician does not make experiments at random, any more than the chemist does. Like the chemist he seeks to conquer the difficulties by dividing them. In our example, instead of undertaking to cope, at the outset with the four rays, A, B, C, and D, he begins by considering only three of them; and still further to divide the difficulty, he begins with a plane section through these three. But it is such a plane, that by rotating it round one of the rays, as an axis, it can be made to pass through every part of space. There is an immense variety of such devices. Indeed, it is so vast that though volume after volume has been written about them, nobody has, as yet, ever pretended to draw up a complete classification of them. It is a department of logic that still awaits a master. Finally, the mathematician, having completed his experimentation, has to retranslate his result into ordinary language.

Now all these processes are requisite in any exact reasoning whatsoever, unless it may be in logical analysis and in the forming of conjectural explanations. Turning then to the study of those minds which are incapable of comprehending the simplest mathematical demonstrations, we ask what one of the processes of mathematical reasoning is it that they are unable to perform. We note, in the first place, what we should naturally expect, that they never think with logical precision, and exhibit a dislike of precise thought, amounting to positive disgust and abhorrence; and naturally they have no confidence in a kind of thought that they [are] unable themselves to follow.

If we examine the kind of reasoning,—to dignify it by that title,—which these people themselves practise and trust to, we find that it is wholly of the kind that is called working by a "rule of thumb." That is to say, they trust to modes of inference which produce upon them the impression of being similar to inferences which they have found by experience to work well. The similarity is indefinite and may have nothing to do with the success of those similar instances that they have known to turn out well. This is plainly shown by the fact that the majority of such minds are entrapped by the ridiculous catch of Achilles and the tortoise, which has so little essential resemblance to any valid argument that it is only with difficulty, if at all, that a critic, by questioning its victim, can make out in what way he imagines it to be probative. A man of honour,—of which there was understood to have been definite proof,—who was certainly not in default in respect to mathematical imagination, since he was a noted player of blindfold chess, was nevertheless so incapable of logical analysis that he seriously assured the writer of this that he could not see why the following was not a logically valid argument:

> It either rains or it does not rain,
> Now it rains;
> Hence, it does not rain.

An almost incredible phenomenon! Evidently, he did not translate the words into a diagram. The jingle sounded to him like a syllogism. Such men make poor work of any positive investigation, especially in regard to theoretical questions; but in the majority of practical affairs their judgments are cautious and conservative; and it is by those qualities, often joined to penetrating observation, that they frequently gain great reputations for wisdom, for which the only real foundation is their inability to reason logically. For it is just this that causes them to be so prudent. It is the specious glitter of apparent inerrancy in their logic that betrays the mass of mankind to foolish hopes and causes them to take undue risks.

There are, it is needless to say, many men who cannot reason about mathematics and never gain any reputation for good sense. Those cases offer no enigma and are not here considered. But the writer has made careful studies of the ways of thinking of a number of men who either believe themselves to be incapable of mathematical reasoning or else are apt to commit fallacies in such reasoning. The latter class is composed of minds whose logical acumen is not sufficient to enable them always accurately to translate abstract statements into diagrams or the reverse. One will find them confusing quite different concepts, such as 'every quantity' and 'every assignable quantity' or 'every physically possible flying-machine' with 'every practically constructible flying-machine.' In the former class, the writer has not infrequently found reason to think that the supposed incapacity was a delusion due to bad instruction in mathematics (which is almost universal) acting on a mind that is not stimulated to activity by the meagre contents of a diagram. The writer is, of course, debarred from making public what he may have ascertained as to the mental weaknesses of intellects which he sincerely honors in common with the rest of the word. He can only say that he has uniformly found that where real inability to follow even a simple mathematical demonstration is conjoined with distinguished good sense, there has been a fine observation for circumstances significant of human facts, and a remarkable "intuition," or power of guessing rightly in human matters, with a grave defect of accuracy of thought.

An ancient famous king is said to have made it a rule to consider every important matter, first when sober, and afterward when drunk; though it is more likely that he reversed this order. The writer's rule is to consider whether a question is one for exact reasoning and is quite beyond the jurisdiction of good sense; or whether accuracy of thought ought to give way to sound instinct and wholesome feeling; or whether, finally, it ought first to be carefully reasoned out, the reasoning then being submitted to the review of common sense.

12

[NECESSARY REASONING] (760)

What is the nature of necessary reasoning? The true answer to this question is very different from any that the logical treatises give; very different too from any that the sensational philosophers give. It appears to be afforded, beyond all rational dispute, by the logic of relatives. To bring out this answer, I propose to submit some instances of necessary reasoning to analysis. I will not, however, choose any of the examples that the treatises afford: they are too difficult. Not that it is too difficult to infer their conclusions from their premises; but that the reasoning in them is so slight and instantaneous that acute minds have been unable to perceive that any reasoning lies therein, and a powerful logical microscope is required to detect in them the rudiments of the features which are characteristic of necessary reasoning. This is true even in those cases in which the multitude of terms and the repetition of some of them in three or more premises serve to conceal the conclusion a little.

The proper method of study is to commence with pieces of reasoning which are strongly marked with the characteristics of necessary inference, especially with that of decidedly advancing the comprehension of the subjects to which they relate. It is in mathematics chiefly in which such demonstrations are to be found; in other subjects demonstrations are either ineffectual or fallacious or seem to be mere applications of mathematics, or imitations of the procedures of mathematics. Taking then the demonstrations of a few mathematical theorems,—not mere corollaries,—we should analyze them with the utmost minuteness and accuracy, and note carefully what their effectiveness is due to. From these, we can pass on to other reasonings too slight to be easy of dissection.

This study is not a genial one. In fact, it is so excessively dry and crabbed that I have long hesitated to invite any readers to follow me in it; yet the results are of so great importance, and so unattainable in any other way, that my scruples have at length been overcome.

Each branch of mathematics is the study of the relations involved in some ideal system. I might say an ideal *infinite* system; for finite systems, such as the icosahedron, form only subordinate subjects of study. Of infinite systems there are two kinds, the discrete, or *number*, and the continuous, or *space*.

Of systems of number the chief is that of the *positive integers*.

13

OF THE PLACE AMONG THE SCIENCES OF PHILOSOPHY
AND OF EACH BRANCH OF IT (from 328)

Science consists in the business of a group of men organized together and specially equipped, mentally, physiologically, tactically, and materially, for the thorough survey of a province of truth, and going about it with devoted energy, with the most systematic thoroughness, and with the highest, broadest, and most detailed intelligence. All this is requisite to constituting a research scientific. But there are only two conditions required to bring it about, a single hearted desire to learn and great energy. Give these to any one man and he will animate others with the same spark and all the remaining conditions will soon be fulfilled.

The different sciences help one another, and that in multiform ways. No rules can be laid down as to where a science shall seek help; far less as to where it shall not. Yet in a general way the sciences are related like the rungs of a ladder. That is to say, some sciences are broader than others, look over a wider range of facts, but look less into details. The general rule is that the broader science furnishes the narrower science with principles by which to interpret its observations while the narrower science furnishes the broader science with instances and suggestions. Thus chemistry is a broader science than astronomy; and chemical principles enable the astronomer to use his observations to determine the relative ages, or stages of evolution, of stars. On the other hand Lockyer has almost conclusively proved that astronomy can exhibit to the chemist decompositions of elements that he has never been able to effect in his laboratory. Here is a collection of facts, but not a principle, which the narrower science furnishes the broader one. Astronomy also furnishes suggestions to the chemist. A solar system or a star-cluster suggests in a general way what the structure of a molecule may be like. But all such suggestions are very hazardous, in great contrast to the confidence with which principles of chemistry and of laboratory physics may be applied to astronomy.

A good classification is a diagram usefully expressive of significant interrelations of the objects classified. The best classification of sciences is a ladder-like scheme where each rung is itself a ladder of rungs; so that the whole is more like a succession of waves each of which carries other waves, and so on, until we should come to single investigations. Thus, all science has these three rungs:

A. Science that discovers truth, merely for the sake of knowing what is so discovered,

B. Science that sets forth and arranges discovered truth; for the sake of comprehending it as a whole, supplementing it where needful for this purpose,

C. Science that applies well-understood truth, supplementing it where needful, for the sake of other interests than that of the truth itself.

The first of these three rungs itself consists of three rungs, as follows:

I. *Mathematics*, which frames and studies the consequences of hypotheses without concerning itself about whether there is anything in nature analogous to its hypotheses or not.

II. *Philosophy*, which seeks such universal truth as can be discovered from everyman's hourly experience.

III. *Idioscopy*, or special science, which seeks such truth as can only be discovered from peculiar experiences sought out for the purpose.

Of these three kinds of study, that of mathematics ought logically to come first. It is [in] every way the easiest of the three, and need not accept any principles proved by the others; while both the philosopher and the idioscoper have to appeal to the mathematician every day. Historically, mathematics was scientifically pursued long before the others had ceased groping and tumbling in thick darkness. All modern physics was dependent for its existence upon the calculus; and the other divisions of idioscopy depended for a truly scientific development upon physics. Anatomy is, no doubt, much, very much; but without physiology, which is the application of physics to explain the functions of the organs, tissues, and cells that anatomy discovers, anatomy gropes in darkness. This will be still more manifest in the future than it is now; for, at present, only the molecular forces active in the organism, as in osmosis, have been successfully studied, while it is plain that the etherial forces, electricity and the like, have much concern in the matter.

Philosophy has always aped mathematics; and its want of success has been in large part due to philosophers having been too sluggish of mind to do more than ape the externals of mathematics, and that so crudely that they had better have left it alone. Mathematics has not been truly understood until the last generation even by mathematicians themselves. To this day, one will find metaphysicians repeating the phrase that mathematics is the science of quantity,—a phrase which is a reminiscence of a long past age when the three words "mathematics," "science," and "quantity" bore entirely different meanings from those now remembered. No mathematicians competent to discuss the fundamentals of their subject any longer suppose it to be limited to quantity. They know very well that it is not so.

At the same time quantity is a very important thing since, as I shall show presently, quantity is nothing but a standard linear arrangement by means of which other linear arrangements may be compared. Now, inasmuch as the relations of cause and effect and of reason and consequent are transitive, that is, linear, relations, and since all thought hinges upon these ideas, it follows that quantity is a most important adjunct to thought.

Inasmuch as metaphysicians, — ignorant and slothful tribe, — and even psychologists, are perpetually falling into grave errors due to their not understanding that nothing is expressed by numbers, as such (independent of accidental associations of other facts with them) but [as] an order of linear arrangement, it will be worth while to pause here long enough to make it quite clear that this is the case. In the first place, it is readily seen that what is called an imaginary quantity or a complex quantity is not purely quantity. Another element enters into it. For a complex quantity of any description, like imaginaries and quaternions, needs at least two numbers to define it; nor do these two numbers, taken by themselves, express such a "quantity." There is involved, besides, the relation between the two units of which those two, or more, numbers are the coefficients. That relation may be regarded as a quantity, if it be considered by itself. But it cannot be combined with the other numbers required to express the complex "quantity" without introducing a relation that is not quantitative. I mean that $x + yi$, for example, where $i^2 = -1$, is not defined by x and y without saying how the x and the y are related; and the relation $i^2 = -1$, being "absurd," transcends the conception of quantity. It is true that we may say that $\pi i = \log(-1)$ [to base e] and it must be admitted that the logarithms of real numbers, taken by themselves, constitute a scheme of quantity. But when we combine this i with the real numbers, whether in the form $x + yi$ or in the form $r e^{i\theta}$ the relation between those two numbers is not itself a quantitative relation. The mathematician would say that such a form merely represents a "group," like any "group." By a "group," mathematicians mean the system of all the relations that result from compounding certain relations which are fully defined in respect to how they are compounded. This must be illustrated by an example; otherwise I had better not have attempted to say what is meant by a "group." But first let me say that whatever else mathematics may treat of, all mathematicians will admit that it embraces the theory of all "groups." Now to say that every "group" is an affair of *quantity* would be a stretch of the extension of this term that would make it cover all sorts of relations. For a group is merely the whole of any system of relationship, regarded in a certain abstract way. But nobody would say that all relations are quantitative.

I will give an example of a group which nobody would say involves nothing but quantity. I will use certain capital letters to denote each a single object. I will put a colon after one such letter and before another, as $A:B$, to signify that relation in which A stands to B and to nothing else, and in which A alone stands to anything. We speak in logic of the *products* and *sums* of relations; but these words, as so used, have nothing to do with quantity. By the *sum* of two relations, we mean a relation in which one individual object stands to another if and only if, it stands in one of the two relations "summed" to that other. By the *product* of two relations we mean such a relation that any individual object, X, stands in this relation to an individual, Y, if, and only if, there is some individual which stands in the *multiplicand* relation to Y, while X stands in the *multiplier* relation to it. Accordingly if r is one relation, and s another, the product rs, where r is the multiplier and s the multiplicand, will generally be a different relation from sr.

Now since the i of imaginary quantity is such that i^4 or $iiii$ is unity, and since by the relation *unity* we mean *identity*, because whatever relation r may be, with this meaning $r \cdot 1 = r$ and $1 \cdot r = r$, it will be appropriate to choose as my example of a group one which arises from a relation, x, such that $xxxx = 1$. In order that there may be a relation of which this is true, although xxx is not 1, there must be at least 4 individuals in the universe considered. Let us suppose that there are only four; which we will denote by A, B, C, D. Then x may be

$$D:A + C:B + A:C + B:D$$

For xx will then be

$$B:A + A:B + D:C + C:D$$

and xxx will be

$$C:A + D:B + B:C + A:D$$

and $xxxx$ will be

$$A:A + B:B + C:C + D:D$$

or the relation in which everything in the universe stands to itself and to nothing else. The relation xxx, which we may write ξ, will be the converse of x. That is, if anything is x to anything, then, and then only, the latter is xxx to the former. But in this universe there will be two other relations with their converses which have the same property. They will be

$$B:A + C:B + D:C + A:D$$

which we may call y, with its converse, η, or

$$D:A + A:B + B:C + C:D,$$

and

$$C:A + A:B + D:C + B:D$$

which we may call z, with its converse, ζ, or

$$B:A + D:B + A:C + C:D$$

Now let us join to these relations all the relations that are "products" of them. There will be, in the first place, their squares or

$$x^2 = \xi^2 = B:A + A:B + D:C + C:D, \text{ which we will call } B$$
$$y^2 = \eta^2 = C:A + D:B + A:C + B:D = \Gamma$$
$$z^2 = \zeta^2 = D:A + C:B + B:C + A:D = \Delta$$

Then there will be relations, such as $\xi\eta$ etc., four in number with their converses, which we will denote by a, b, c, d, with the converses $\alpha, \beta, \gamma, \delta$. These will be such that $aaa = 1$, etc. Namely

$$a = \xi\eta = \eta\zeta = \zeta\xi = A:A + C:B + D:C + B:D$$
$$b = \xi y = yz = z\xi = D:A + B:B + A:C + C:D$$
$$c = x\eta = \eta z = zx = B:A + D:B + C:C + A:D$$
$$d = xy = y\zeta = \zeta x = C:A + A:B + B:C + D:D$$
$$\alpha = zy = yx = xz = A:A + D:B + B:C + C:D$$
$$\beta = \zeta\eta = \eta x = x\zeta = C:A + B:B + D:C + A:D$$
$$\gamma = \zeta y = y\xi = \xi\zeta = D:A + A:B + C:C + B:D$$
$$\delta = z\eta = \eta\xi = \xi z = B:A + C:B + A:C + D:D$$

There will be besides six products of x, y, z with B, Γ, Δ namely,

$$X = x\Gamma = \Gamma\xi = \xi\Delta = \Delta x = A:A + B:B + D:C + C:D$$
$$\Xi = \xi\Gamma = \Gamma x = x\Delta = \Delta\xi = A:B + B:A + C:C + D:D$$
$$Y = y\Delta = \Delta\eta = \eta B = By = A:A + D:B + C:C + B:D$$
$$\Upsilon = \eta\Delta = \Delta y = yB = B\eta = C:A + B:B + A:C + D:D$$
$$Z = zB = B\zeta = \zeta\Gamma = \Gamma z = A:A + C:B + B:C + D:D$$
$$\Sigma = \zeta B = Bz = z\Gamma = \Gamma\zeta = D:A + B:B + C:C + A:D$$

These relations together with the relation of identity,

$$1 = A:A + B:B + C:C + D:D$$

are 24 in number. If we were to write down A, B, C, D, in that order in a row, and were then to substitute for each of these letters, that one which is in one of the 24 relations to it, all the rows that could so result would be different, and since there are but 24 permutations of 4 objects, it follows that every permutation could be so produced. Now a permutation of a permutation is a permutation; and consequently, all possible "products" of this collection of 24 relations are among this collection of relations; and that is what is meant by calling the collection a group. In order to show that it is so, I here give the "multiplication table" of the group. You may enter any column, say the nth from the left, with the "multiplier" and enter the nth horizontal row from the top with the "multiplicand"; and at the junction of the row of the "multiplier" with the column headed by the "multiplicand," you find the "product." This table is just like a quantitative multiplication table, except that the "multipliers" are arranged differently from the "multiplicands," in that in the sequence of "multiplicands" the converse of each relation takes the place of the relation itself in the sequence of the "multipliers." All the relations denoted by capital letters are their own converses; and the converse of each relation denoted by a small Roman letter is the corresponding Greek letter and *vice versa*. Between certain of the rows and columns spaces have been left blank in the table in order to bring more prominently into view certain features of the products.[1]

On examining this multiplication table, it will be seen that not only do the entire four and twenty relations form a group, but that it is easy to pick out smaller collections of them that form groups. Such are

$1B\Gamma\Delta\,abcd\alpha\beta\gamma\delta$, $1B\Gamma\Delta\,X\Xi\,z\,\xi$, $1B\Gamma\Delta\,Y\,\Upsilon\,y$,
$1B\Gamma\Delta\,Z\Sigma z\,\zeta$, $1a\alpha XYZ$, $1b\beta x\Upsilon\,\Sigma$, $1c\gamma\Xi\,Y\Sigma$, $1d\delta\Xi\,\Upsilon Z$,
$1B\Gamma\Delta$, $1BX\Xi$, $1\Gamma Y\Upsilon$, $1\Delta Z\Sigma$, $1Bx\xi$, $1\Gamma y\eta$, $1\Delta z\zeta$,
$1a\alpha$, $1b\beta$, $1c\gamma$, $1B$, 1Γ, 1Δ, $1X$, 1Ξ, $1Y$, 1Υ, $1Z$, 1Σ.

To get a visual image of the group, let A, B, C, D be the four diagonal lines of a cube. Then each of the 6 relations x, ξ, y, η, z, ζ will be determined by one of the 6 possible quadrantal turns of the cube about an axis through centers of opposite faces, being the relation of any diagonal of the cube to the diagonal whose place that quandrantal turn would cause it to take. The three relations B, Γ, Δ will be the relation of any diagonal of the cube to that diagonal which a reversal about such an axis would cause it to displace. Each of the eight relations a, b, c, d, α, β, γ, δ will be determined by possible rotations through 120° about diagonals of the cube, being the relation of any diagonal after such a rotation to the diagonal that was in the same position before rotation. Each of the six

[1]   There is no such multiplication table in the manuscript although two blank pages following this statement would seem to indicate that he reserved space for such a table. Peirce says: "This should be the arrangement of all multiplication tables of relations, since the reversal of the factors of a product substituting for each its converse must give the converse of the product." Compare this presentation with the letter to J. M. P. (3, 18j) dated 13 February 1904.

relations X, Ξ, Y, Υ, Z, Σ^2 will be determined by a reversal about an axis through the middle points of opposite edges of the cube, being the relation of any diagonal of the cube after such reversal to the diagonal that was in the same position before the reversal. The following figures, where A_1, A_2, B_1, B_2, C_1, C_2, D_1, D_2 are the extremities of the four diagonals, show the effects of these displacements. [See Fig. 1 on the following page.]

---

[2]   Peirce wrote this as Σ in this place only, and as it appears on the figure. He seems to have overlooked his Σ notation.

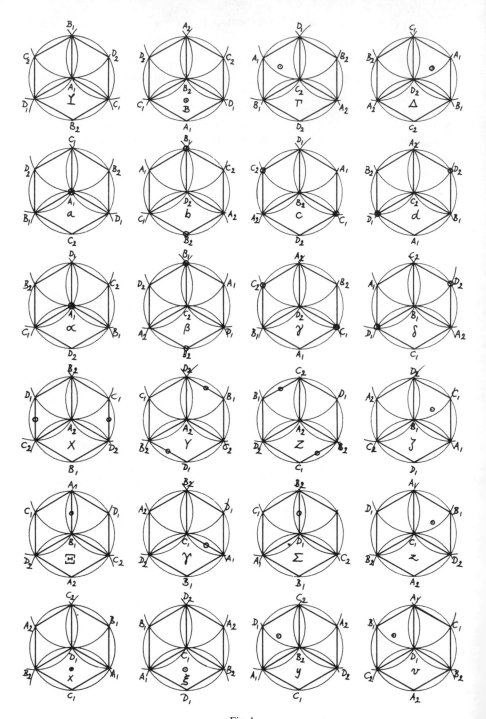

Fig. 1

14

$$K\alpha\iota\nu\grave{\alpha}\ \sigma\tau o\iota\chi\epsilon\hat{\iota}\alpha\ (517)$$

PREFACE

I deem it useful to say a few words about this piece. Some years ago I wrote a book entitled "New Elements of Mathematics." It was such a book as a man with considerable natural aptitude for logic and mathematics, who had devoted the best of his time for forty years to the study of the former and all that has been written about it, and had not neglected the latter, was able to write by devoting a year exclusively to it. If the author had been a German, he would have shared the loose ideas of logic that naturally are associated with subjectivism, and consequently could not have written the same book; but had he written it, it would have been in print long ago. As it was, he carried it to three publishers, one of whom had asked him to prepare the book. All of them were very modest men. They did not pretend to know much except about the elements of mathematics. One of them was at that time the publisher of a treatise on geometry which professed to show how to inscribe a regular polygon of any number of sides in a circle by the aid of rule and compasses only. Both the others published treatises on geometry of similar startling pretensions. None of them approved of my book, because it put perspective before metrical geometry, and topical geometry before either. This was the fault of the book; namely, that a publisher who was so well versed in the elements of mathematics was not convinced by it that this arrangement was logical, even though he took the book home with him and glanced at it during the evening. A writer on the logic of mathematics in America must meet American requirements.

Personally, I regret that that MS. has been lost; for it was the record of much close thought.[1] I can never reproduce it, because it was written in the strictest mathematical style, and with advancing years I have lost the power of writing about logic in mathematical style, although in my youth it was natural to me. In losing the power of writing this style, I

---

[1] The manuscript is 165, now in Volume 2 of this edition.

have equally lost my admiration of it. I beg permission to offer a criticism of the mathematical style in logic.

I say 'in logic,' because it is only with a view of presenting the logic of the subject that mathematicians employ the Euclidean style. When they are simply intent on the solution of difficult problems they forget all about it. I call it Euclidean because the first book of Euclid's elements is the earliest and the most perfect model of that style. Euclid follows it, in some measure, in his other writings; but it is only in the first book of the Elements that it is polished with endless labor and thought. It is easy to see that this style took its origin in the esthetic taste of the Greeks. Everything they did, in literature and in art, shows the predominance of their horror of the 'too much.' Perhaps this horror was due to the irrepressible activity of Greek minds, and their consequent impatience with useless considerations, together with the expensiveness, at that age in energy of every kind, of the mechanical processes of writing and reading. They took it for granted that the reader would actively think; and the writer's sentences were to serve merely as so many blazes to enable him to follow the track of that writer's thought. The modern book, which I only mention as a foil to the other, in order to be approved, must be approved by a densely stupid and unspeakably indolent young lady as she skims its pages while looking out of the window to be admired. In order to put an idea into such a shape that it cannot fail to be apprehended by her, the first requisite is that it shall fill a certain number of lines, and the second is that not the smallest step shall be left to her own intellectual activity.

The dominating idea of Euclid in writing his first book was plainly that the first elements of geometry can only be comprehended by understanding the logical structure of the doctrine. Yet, in his horror of the too much, he never says a single word about logic from beginning to end. He begins with a couple of dozen 'definitions,' which are followed by five 'postulates,' and these by several axioms, or 'notions of common sense'; yet he never tells us what a 'definition,' a 'postulate,' or a 'common notion' is supposed to be; and his meanings in all three cases have been seriously misapprehended. The forty-eight propositions of the book are set forth and arranged in a manner which betrays a profound understanding of their logical relations. Herein is the principal value of the work; today, its only value. Yet this knowledge is so concealed that it requires the same knowledge to detect it.

This profound work is put into the hands of boys who are not Greek, but overfed and logy. They meet with difficulties which they carry to a teacher who is far more incompetent than they; since he knows nothing of the logical structure which is the cryptic subject of the book, and long familiarity has rendered him incapable of perceiving the difficulties which his scholars can, at least, perceive. The old pedagogical method was to thrash the boys till they did understand; and that was tolerably efficient as an antidote to their over-doses of beef. Since that method has been abandoned it has been necessary to abandon the pedagogical use of the book.

II

It is extremely difficult to treat fully and clearly of the logic of mathematics in the Euclidean style, since this strictly requires that not a word should be said about logic. As an exact logician, however, I approve of addressing an actively intelligent reader in the ancient method by means of 1st, *definitions*; 2nd, *postulates*; 3rd, *axioms*; 4th, *corollaries*; 5th, *diagrams*; 6th, *letters*; 7th, *theorems*, 8th, *scholiums*. This distinction between a *general* proposition (which, if a postulate, is often erroneously called an axiom) and an *indefinite* proposition (to which, if indemonstrable, the word 'Postulat' is restricted in German) may also be maintained.

A *definition* is the logical analysis of a predicate in general terms. It has two branches, the one asserting that the definitum is applicable to whatever there may be to which the definition is applicable; the other (which ordinarily has several clauses), that the definition is applicable to whatever there may be to which the definitum is applicable. *A definition does not assert that anything exists.*

A *postulate* is an initial hypothesis in general terms. It may be arbitrarily assumed provided that (the definitions being accepted) it does not conflict with any principle of substantive possibility or with any already adopted postulate. By a principle of substantive possibility, I mean, for example, that it would not be admissible to postulate that there was no relation whatever between two points, or to lay down the proposition, that nothing whatever shall be true without exception. For though what this means involves no contradiction, it is in contradiction with the fact that it is itself asserted.

An *axiom* is a self-evident truth, the statement of which is superfluous to the conclusiveness of the reasoning, and which only serves to show a principle involved in the reasoning. It is generally a truth of observation; such as the assertion that something is true.

A *corollary*, as I shall use the word, is an inference drawn in general terms without the use of any construction.* (*At present, *corollary* is not a scientific term. The Latin word, meaning a gratuity, was applied to obvious deductions added by commentators to Euclid's propositions. Those his proofs compelled them to grant; and they added admissions of the corollaries without requiring proof. I propose to use the word in a definite sense as a term of logic.)

A *diagram* is an *icon* or schematic image embodying the meaning of a general predicate; and from the observation of this *icon* we are supposed to construct a new general predicate.

A *letter* is an arbitrary definite designation specially adopted in order to identify a single object of any kind.

A *theorem*, as I shall use the word, is an inference obtained by constructing a diagram according to a general precept, and after modifying it as ingenuity may dictate, observing in it certain relations, and showing that they must subsist in every case, retranslating the proposition into general terms.* (*This may exclude some propositions called theorems. But I do not think that mathematicians will object to that, in view of my making a sharp distinction between a corollary and a theorem, and thus furnishing the logic of mathematics with two exact and convenient technical terms in place of vague, unscientific words.) A theorem regularly begins with, 1st, the *general enunciation*. There follows, 2nd, a *precept* for a diagram, in which letters are employed. Then comes, 3rd, the *ecthesis*, which states what it will be sufficient to show must, in every case, be true concerning the diagram. The 4th article is the *subsidiary construction*, by which the diagram is modified in some manner already shown to be possible. The 5th article is the *demonstration*, which traces out the reasons why a certain relation must always subsist between the parts of the diagram. Finally, and 6thly, it is pointed out, by some such expression as Euclid's ὅπερ ἔδει δεῖξαι or by the usual Q.E.D., or otherwise, that it was all that it was required to show.

A *scholium* is a comment upon the logical structure of the doctrine. This preface is a scholium.

III

1. I now proceed to explain the difference between a theoretical and a practical proposition, together with the two important parallel distinctions between *definite* and *vague*, and *individual* and *general*, noting, at the same time, some other distinctions connected with these. A *sign* is connected with the 'Truth,' i.e. the entire Universe of being, or, as some say, the Absolute, in three distinct ways. In the first place, a sign is not a real thing. It is of such a nature as to exist in *replicas*. Look down a printed page, and every *the* you see is the same word, every *e* the same

letter. A real thing does not so exist in replica. The being of a sign is merely *being represented.* Now *really being* and *being represented* are very different. Giving to the word *sign* the full scope that reasonably belongs to it for logical purposes, a whole book is a sign; and a translation of it is a replica of the same sign. A whole literature is a sign. The sentence 'Roxana was the queen of Alexander' is a sign of Roxana and of Alexander, and though there is a grammatical emphasis on the former, logically the name 'Alexander' is as much *a subject* as is the name 'Roxana'; and the real persons Roxana and Alexander are *real objects* of the sign. Every sign that is sufficiently complete refers to sundry real objects. All these objects, even if we are talking of Hamlet's madness, are parts of one and the same Universe of being, the 'Truth.' But so far as the 'Truth' is merely the *object* of a sign, it is merely the Aristotelian *Matter* of it that is so. In addition however to *denoting* objects every sign sufficiently complete *signifies characters*, or qualities. We have a direct knowledge of real objects in every experiential reaction, whether of *Perception* or of *Exertion* (the one theoretical, the other practical). These are directly *hic et nunc.* But we extend the category, and speak of numberless real objects with which we are not in direct reaction. We have also direct knowledge of qualities in feeling, peripheral and visceral. But we extend this category to numberless characters of which we have no immediate consciousness. All these characters are elements of the 'Truth.' Every sign signifies the 'Truth.' But it is only the Aristotelian *Form* of the universe that it signifies. The logician is not concerned with any metaphysical theory; still less, if possible, is the mathematician. But it is highly convenient to express ourselves in terms of a metaphysical theory; and we no more bind ourselves to an acceptance of it than we do when we use substantives such as 'humanity,' 'variety,' etc. and speak of them as if they were substances, in the metaphysical sense. But, in the third place, every sign is intended to determine a sign of the same object with the same signification or *meaning.* Any sign, *B*, which a sign, *A*, is fitted so to determine, without violation of its, *A*'s, purpose, that is, in accordance with the 'Truth,' even though it, *B*, denotes but a part of the objects of the sign, *A*, and signifies but a part of its, *A*'s, characters, I call an *interpretant* of *A.* What we call a "fact" is something having the structure of a proposition, but supposed to be an element of the very universe itself. The purpose of every sign is to express "fact," and by being joined with other signs, to approach as nearly as possible to determining an interpretant which would be the *perfect Truth*, the absolute Truth, and as such (at least, we may use this language) would be the very Universe. Aristotle gropes for a conception of perfection, or *entelechy*, which he never succeeds in making clear. We may adopt the word to mean the very fact, that is, the ideal sign which should be quite perfect, and so identical, — in such identity as a sign may have, — with the very matter denoted united with the very form signified by it. The entelechy of the Universe of being, then, the Universe *quâ* fact, will be that Universe in its aspect as a sign, the 'Truth' of being. The 'Truth,' the fact that is not abstracted but complete, is the ultimate interpretant of every sign.

2. Of the two great tasks of humanity, *Theory* and *Practice*, the former sets out from a sign of a real object with which it is *acquainted*, passing from this, as its *matter*, to successive interpretants embodying more and more fully its *form*, wishing ultimately to reach a direct *perception* of the entelechy; while the latter, setting out from a sign signifying a character of which it *has an idea*, passes from this, as its *form*, to successive interpretants realizing more and more precisely its *matter*, hoping ultimately to be able to make a direct *effort*, producing the entelechy. But of these two movements, logic very properly prefers to take that of Theory as the primary one. It speaks of an *antecedent* as that which being known something else, the *consequent* may *also* be known. In our vernacular, the latter is inaccurately called a *consequence*, a word that the precise terminology of logic reserves for the proposition expressing the relation of any consequent to its antecedent, or for the fact which this proposition expresses. The conception of the relation of antecedent and consequent amounts, therefore, to a confusion of thought between the reference of a sign to its *meaning*, the character which it attributes to its object, and its appeal to an interpretant. But it is the former of these which is the more essential. The knowledge that the sun has always risen about once in each 24 hours (sidereal time) is a sign whose object is the sun, and (rightly understood) a part of its signification is the rising of the sun tomorrow morning. The relation of an antecedent to its consequent, in its confusion of the signification with the interpretant, is nothing but a special case of what occurs in all action of one thing upon another, modified so as to be merely an affair of being represented instead of really being. It is the representative action of the sign upon its object. For whenever one thing acts upon another it determines in that other a quality that would not otherwise have been there. In the vernacular we often call an effect a "consequence," because that which really is may correctly be represented; but we should refuse to call a mere logical consequent an "effect," because that which is merely represented, however legitimately, cannot be said really to be. If we speak of an argumentation as "producing a great effect," it is not the interpretant itself, by any means, to which we refer, but only the particular replica of it which is made in the minds of those addressed.

If a sign, *B*, only signifies characters that are elements (or the whole) of the meaning of another sign, *A*, then *B* is said to be a *predicate* (*or essential part*) of *A*. If a sign, *A*, only denotes real objects that are a part or the whole of the objects denoted by another sign, *B*, then *A* is said to be a *subject* (or *substantial part*) of *B*. The totality of the predicates of a sign, and also the totality of the characters it signifies, are indifferently each called its logical *depth*. This is the oldest and most convenient term. Synonyms are the *comprehension* of the Port-Royalists, the *content* (*Inhalt*) of the Germans, the *force* of DeMorgan, the *connotation* of J. S. Mill. (The last is objectionable.) The totality of the subjects, and also, indifferently, the totality of the real objects of a sign is called the logical *breadth*. This is the oldest and most convenient term. Synonyms are the *extension* of the Post-Royalists (ill-called *extent* by some modern French logicians), the *sphere* (*Umfang*) of translators from the German, the *scope* of DeMorgan, the *denotation* of J. S. Mill.

Besides the logical depth and breadth, I have proposed (in 1867) the terms *information* and *area* to denote the total of fact (true or false) that in a given state of knowledge a sign embodies.

3. Other distinctions depend upon those we have drawn. I have spoken of real relations as reactions. It may be asked how far I mean to say that all real relations are reactions. It is seldom that one falls upon so fascinating a subject for a train of thought [as] the analysis of that problem in all its ramifications, mathematical, physical, biological, sociological, psychological, logical, and so round to the mathematical again. The answer cannot be satisfactorily given in a few words; but it lies hidden beneath the obvious truth that any exact necessity is expressible by a general equation; and nothing can be added to one side of a general equation without an equal addition to the other. Logical necessity is the necessity that a sign should be true to a *real* object; and therefore there is *logical* reaction in every real dyadic relation. If *A* is in a real relation to *B*, *B* stands in a logically contrary relation to *A*, that is, in a relation at once converse to and inconsistent with the direct relation. For here we speak [not] of a vague sign of the relation but of the relation between two individuals, *A* and *B*. This very relation is one in which *A* alone stands to any individual, and it to *B* only. There are, however, *degenerate* dyadic relations, —*degenerate* in the sense in which two coplanar lines form a *degenerate* conic,—where this is not true. Namely, they are individual relations of identity, such as the relation of *A* to *A*. All mere resemblances and relations of reason are of this sort.

Of signs there are two different degenerate forms. But though I give them this disparaging name, they are of the greatest utility, and serve purposes that genuine signs could not. The more degenerate of the two forms (as I look upon it) is the *icon*. This is defined as a sign of which the character that fits it to become a sign of the sort that it is, is simply inherent in it as a quality of it. For example, a geometrical figure drawn on paper may be an *icon* of a triangle or other geometrical form. If one meets a man whose language one does not know and resorts to imitative sounds and gestures, these approach the character of an icon. The reason they are not pure icons is that the purpose of them is emphasized. A pure icon is independent of any purpose. It serves as a sign solely and simply by exhibiting the quality it serves to signify. The relation to its object is a degenerate relation. It asserts nothing. If it conveys information, it is only in the sense in which the object that it is used to represent may be said to convey information. An *icon* can only be a fragment of a completer sign.

The other form of degenerate sign is to be termed an *index*. It is defined as a sign which is fit to serve as such by virtue of being in a real reaction with its object. For example, a weather-cock is such a sign. It is fit to be taken as an index of the wind for the reason that it is physically connected with the wind. A weather-cock conveys information; but this it does because in facing the very quarter from which the wind blows, it resembles the wind in this respect, and thus has an icon connected with it. In this respect it is not a pure index. A pure index simply forces attention to the object with which it reacts and puts the interpreter into mediate reaction with that object, but conveys no information. As an example, take an exclamation "Oh!" The letters attached to a geometrical figure are another case. Absolutely unexceptionable examples of degenerate forms must not be expected. All that is possible is to give examples which tend sufficiently in towards those forms to make the mean suggest what is meant. It is remarkable that while neither a pure icon nor a pure index can assert anything, an index which forces something to be an *icon*, as a weather-cock does, or which forces us to regard it as an *icon*, as the legend under a portrait does, does make an assertion, and forms a *proposition*. This suggests the true definition of a proposition, which is a question in much dispute at this moment. A proposition is a sign which separately, or independently, indicates its object. No *index*, however, can be an *argumentation*. It may be what many writers call an *argument*; that is, a basis of argumentation; but an argument in the sense of a sign which separately shows what interpretant it is intended to determine it cannot be.

It will be observed that the icon is very perfect in respect to signification, bringing its interpreter face to face with the very character signified. For this reason, it is the mathematical sign *par excellence*. But in denotation it is wanting. It gives no assurance that any such object as it represents really exists. The index on the other hand does this most perfectly, actually bringing to the interpreter the experience of the very object denoted. But it is quite wanting in signification unless it involves an iconic part.

We now come to the genuine sign for which I propose the technical designation *symbol*, following a use of that word not infrequent among logicians including Aristotle. A symbol is defined as a sign which is fit to serve as such simply because it will be so interpreted. To recapitulate

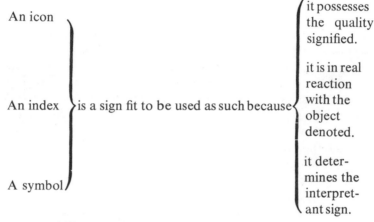

An icon ⎫
            ⎬ it possesses the quality signified.

An index ⎬ is a sign fit to be used as such because ⎨ it is in real reaction with the object denoted.

A symbol ⎭ it determines the interpretant sign.

Language and all abstracted thinking, such as belongs to minds who think in words, [are] of the symbolic nature. Many words, though strictly symbols, are so far iconic that they are apt to determine iconic interpretants, or as we say, to call up lively images. Such, for example, are those that have a fancied resemblance to sounds associated with their objects; that are *onomatopoetic*, as they say. There are words, which although symbols, act very much like indices. Such are personal demonstrative, and relative pronouns, for which *A*, *B*, *C*, etc. are often substituted. A *Proper Name*, also, which denotes a single individual well known to exist by the utterer and interpreter, differs from an index only in that it is a conventional sign. Other words refer indirectly to indices. Such is "*yard*" which refers to a certain bar in Westminster, and has no meaning unless the interpreter is, directly or indirectly, in physical reaction with that bar. Symbols are particularly remote from the Truth itself. They are abstracted. They neither exhibit the very characters signified as icons do, nor assure us of the reality of their objects, as indices do. Many proverbial sayings express a sense of this weakness; as "Words prove nothing," and the like. Nevertheless, they have a great power of which the degenerate signs are quite destitute. They alone express laws. Nor are they limited to this theoretical use. They serve to bring about reasonableness and law. The words *justice* and *truth*, amid a world that habitually neglects these things and utterly derides the words, are nevertheless among the very greatest powers the world contains. They create defenders and animate them with strength. This is not rhetoric or metaphor: it is a great and solid fact of which it behooves a logician to take account.

A symbol is the only kind of sign which can be an argumentation.*
(*I commonly call this an argument; for nothing is more false historically than to say that this word has not at all times been used in this sense. Still, the longer word is a little more definite.)

4. I have already defined an argument as a sign which separately monstrates what its intended interpretant is, and a proposition as a sign which separately indicates [what] its object is, and we have seen that the icon alone cannot be a proposition while the symbol alone can be an argument. That a sign cannot be an argument without being a proposition is shown by attempting to form such an argument. "Tully, c'est à dire a Roman," evidently asserts that Tully is a Roman. Why this is so is plain. The interpretant is a sign which denotes that which the sign of which it is interpretant denotes. But, being a symbol, or genuine sign, it has a signification and therefore it represents the object of the principal sign as possessing the characters that it, the interpretant, signifies. It will be observed that an argument is a symbol which separately monstrates (in any way) its *purposed* interpretant. Owing to a symbol being essential [1y] a sign only by virtue of its being interpretable as such, the idea of a purpose is not entirely separable from it. The symbol, by the very definition of it, has an interpretant in view. Its very meaning is intended. Indeed, a purpose is precisely the interpretant of a symbol. But the conclusion of an argument is a specially monstrated interpretant, singled out from among the possible interpretants. It is, therefore, of its nature single, although not necessarily simple. If we erase from an argument every monstration of its special purpose, it becomes a proposition; usually a copulate proposition, composed of several members whose mode of conjunction is of the kind expressed by 'and,' which the grammarians call a 'copulative conjunction.' If from a propositional symbol we erase one or more of the parts which separately denote its objects, the remainder is what is called a *rhema*; but I shall take the liberty of calling it a *term*. Thus, from the proposition 'Every man is mortal,' we erase 'Every man,' which is shown to be denotative of an object by the circumstance that if it be replaced by an indexical symbol, such as 'That' or 'Socrates,' the symbol is reconverted into a proposition, we get the *rhema* or *term*

'_____ is mortal.'

Most logicians will say that this is not a term. The term, they will say, is 'mortal,' while I have left the copula 'is' standing with it. Now while it is true that one of Aristotle's memoirs dissects a proposition into subject, predicate, and verb, yet as long as Greek was the language which logicians had in view, no importance was attached to the substantive verb, 'is,' because the Greek permits it to be omitted. It was not until the time of Abelard, when Greek was forgotten, and logicians had Latin in mind, that the copula was recognized as a constituent part of the logical proposition. I do not, for my part, regard the usages of language as forming a satisfactory basis for logical doctrine. Logic, for me, is the study of the essential conditions to which signs must conform in order to function as such. How the constitution of the human mind may compel men to think is not the question; and the appeal to language appears to me to be no better than an unsatisfactory method of ascertaining psychological facts that are of no relevancy to logic. But if such appeal is to be made (and logicians generally do make it; in particular their doctrine of the copula appears to rest solely upon this), it would seem that they ought to survey human languages generally and not confine themselves to the small and extremely peculiar group of Aryan speech. Without pretending, myself, to an extensive acquaintance with languages, I am confident that the majority of non-Aryan languages do not ordinarily employ any substantive verb equivalent to is. Some place a demonstrative or relative pronoun; as if one should say

"_____ is a man *that* is translated."

for "A man is translated." Others have a word, syllable, or letter, to show that an assertion is intended. I have been led to believe that in very few languages outside the Aryan group is the common noun a well-developed and independent part of speech. Even in the Shemitic languages, which are remarkably similar to the Aryan, common nouns are treated as verbal forms and are quite separated from proper names. The ordinary view of a term, however, supposes it to be a common noun in the fullest sense of the term. It is rather odd that of all the languages which I have examined in a search for some support of this ordinary view, so outlandish a speech as the Basque is the only one I have found that seems to be constructed thoroughly in the manner in which the logicians teach us that every rational being must think.*

(*While I am on the subject of languages I may take occasion to remark with reference to my treatment of the direct and indirect "objects" of a verb as so many subjects of the proposition, that about nine out of every ten languages regularly emphasize one of the subjects, and make it the principal one, by putting it in a special nominative case, or by some equivalent device. The ordinary logicians seem to think that this, too, is a necessity of thought, although one of the living Aryan languages of Europe habitually puts that subject in the genetive[2] which the Latin puts in the nominative. The practice was very likely borrowed from a language similar to the Basque spoken by some progenitors of the Gaels. Some languages employ what is, in effect, an ablative for this purpose. It no doubt is a rhetorical enrichment of a language to have a form "*B* is loved by *A*" in addition to "*A* loves *B*." The language will be still richer if it has a third form in which *A* and *B* are treated as equally the subjects of what is said. But logically, the three are identical.)

What is the difference between '_____ is a man' and 'man?' The logicians hold that the essence of the latter lies in a definition describing its characters; which doctrine virtually makes 'man' equivalent to 'what is a man.' It thus differs from '_____ is a man' by the addition is an index of that object; and the two taken together form a proposition. In respect to being fragmentary, therefore, the two signs are alike. It may be said that 'Socrates wise' does not make a sentence in the language at present used in logic, although in Greek it would. But it is important not to forget that no more do 'Socrates' and 'is wise' make a proposition unless there is something to indicate that they are to be taken as signs of the same object. On the whole, it appears to me that the only difference between my rhema and the 'term' of other logicians is that the latter contains no explicit recognition of its own fragmentary nature. But this is as much as to say that logically their meaning is the same; and it is for that reason that I venture to use the old, familiar word 'term' to denote the rhema.

It may be asked what is the nature of the sign which joins 'Socrates' to '_____ is wise,' so as to make the proposition 'Socrates is wise.' I reply that it is an index. But, it may be objected, an index has for its object a thing *hic et nunc*, while a sign is not such a thing. This is true, if under 'thing' we include singular events, which are the only things that are strictly *hic et nunc*. But it is not the two signs 'Socrates' and 'wise' that are connected, but the *replicas* of them used in the sentence. We do not say that '_____ is wise,' as a general sign, is connected specially with Socrates, but only that it is so as here used. The two replicas of the

---

²   This is probably a deliberate Peircean spelling out of the Latin form.

words 'Socrates' and 'wise' are *hic et nunc*, and their junction is a part of their occurrence *hic et nunc*. They form a pair of reacting things which the index of connection denotes in their present reaction, and not in a general way; although it is possible to generalize the mode of this reaction like any other. There will be no objection to a generalization which shall call the mark of junction a *copula*, provided it be recognized that, in itself, it is not general, but is an *index*. No other kind of sign would answer the purpose; no general verb 'is' can express it. For something would have to bring the general sense of that general verb down to the case in hand. An index alone can do this. But how is this index to signify the connection? In the only way in which any index can ever signify anything; by involving an *icon*. The sign itself is a connection. I shall be asked how this applies to Latin, where the parts of the sentence are arranged solely with a view to rhetorical effect. I reply that, nevertheless, it is obvious that in Latin, as in every language, it is the juxtaposition which connects words. Otherwise they might be left in their places in the dictionary. Inflexion does a little; but the main work of construction, the whole work of connexion, is performed by putting the words together. In Latin much is left to the good sense of the interpreter. That is to say, the common stock of knowledge of utterer and interpreter, called to mind by the words, is a part of the sign. That is more or less the case in all conversation, oral and scriptal. It is, thus, clear that the vital spark of every proposition, the peculiar propositional element of the proposition, is an indexical proposition; an index involving an icon. The rhema, say '_____ loves _____,' has blanks which suggest filling; and a concrete actual connection of a subject with each blank monstrates the connection of ideas.

It is the Proposition which forms the main subject of this whole scholium; for the distinctions of *vague* and *distinct*, *general* and *individual* are propositional distinctions. I have endeavored to restrain myself from long discussions of terminology. But here we reach a point where a very common terminology overlaps an erroneous conception. Namely those logicians who follow the lead of Germans, instead of treating of propositions, speak of "judgments" (*Urtheile*). They regard a proposition as merely an expression in speech or writing of a judgment. More than one error is involved in this practice. In the first place, a judgment, as they very correctly teach, is a subject of psychology. Since psychologists now-a-days, not only renounce all pretension to knowledge of the *soul*, but also take pains to avoid talking of the *mind*, the latter is, at present not a scientific term, at all; and therefore I am not prepared to say that logic does not, as such, treat of the mind. I should like to take mind in such a sense that this could be affirmed; but in any sense in which psychology,—the scientific psychology now recognized,—treats of mind, logic, I maintain, has no concern with it. Without stopping here to discuss this large question, I will say that psychology is a science which makes special observations: and its whole business is to make the phenomena so observed (along with familiar facts allied to those things), definite and comprehensible. Logic is a science little removed from pure mathematics. It cannot be said to make any positive phenomena known, although it takes account and rests upon phenomena of daily and hourly experience, which it so analyzes as to bring out recondite truths about them. One might think that a pure mathematician might assume these things as an initial hypothesis and deduce logic from these; but this turns out, upon trial, not to be the case. The logician has to be recurring to reexamination of the phenomena all along the course of his investigations. But logic is all but as far remote from psychology as is pure mathematics. Logic is the study of the essential nature of signs. A sign is something that exists in replicas. Whether the sign "it is raining," or "all pairs of particles of matter have component accelerations toward one another inversely proportional to the square of the distance" happens to have a replica in writing, in oral speech, or in silent thought, is a distinction of the very minutest interest to logic, which is a study, not of replicas, but of signs. But this is not the only, nor the most serious error involved in making logic treat of "judgments" in place of propositions. It involves confounding two things which must be distinguished if a real comprehension of logic is to be attained. A *proposition*, as I have just intimated, is not to be understood as the lingual expression of a judgment. It is, on the contrary, that sign of which the judgment is one replica and the lingual expression another. But a judgment is distinctly *more* than the mere mental replica of a proposition. It not merely *expresses* the proposition, but it goes further and *accepts* it. I grant that the normal use of a proposition is to affirm it; and its chief logical properties relate to what would result in reference to its affirmation. It is, therefore, convenient in logic to express propositions in most cases in the indicative mood. But the proposition in the sentence, 'Socrates est sapiens,' strictly expressed, is 'Socratem sapientem esse.' The defence of this position is that in this way we distinguish between a proposition and the assertion of it; and without such distinction it is impossible to get a distinct notion of the nature of the proposition. One and the same proposition may be affirmed, denied, judged, doubted, inwardly inquired into, put as a question, wished, asked for, effectively commanded, taught, or merely expressed, and does not thereby become a different proposition. What is the nature of these operations? The only one that need detain us is affirmation, including judgment, or affirmation to oneself.

As an aid in dissecting the constitution of affirmation I shall employ a certain logical magnifying-glass that I have often found efficient in such business. Imagine, then, that I write a proposition on a piece of paper, perhaps a number of times, simply as a calligraphic exercise. It is not likely to prove a dangerous amusement. But suppose I afterward carry the paper before a notary public and make affidavit to its contents. That may prove to be a horse of another color. The reason is that this affidavit may be used to determine an assent to the proposition it contains in the minds of judge and jury; — an effect that the paper would not have had if I had not sworn to it. For certain penalties here and hereafter are attached to swearing to a false proposition; and consequently the fact that I have sworn to it will be taken as a negative index that it is not false. This assent in judge and jury's minds may effect in the minds of sheriff and posse a determination to an act of force to the detriment of some innocent man's liberty or property. Now certain ideas of justice and good order are so powerful that the ultimate result may be very bad for me. This is the way that affirmation looks under the microscope; for the only difference between swearing to a proposition and an ordinary affirmation of it, such as logic contemplates, is that in the latter case the penalties are less and even less certain than those of the law. The reason there are any penalties are, as before, that the affirmation may determine a judgment to the same effect in the mind of the interpreter to his cost. It cannot be that the sole cause of his believing it is that there are such penalties, since two events cannot cause one another, unless they are simultaneous. There must have been, and we well know that there is, a sort of hypnotic disposition to believe what one is told with an air [of] command. It is Grimes's credenciveness, which is the essence of hypnotism. This disposition produced belief; belief produced the penalties; and the knowledge of these strengthens the disposition to believe.

I have discussed the nature of belief in the *Popular Science Monthly* for November 1877. On the whole, we may set down the following definitions:

A *belief* in a proposition is a controlled and contented habit of acting in ways that will be productive of desired results only if the proposition is true;

An *affirmation* is an act of an utterer of a proposition to an interpreter, and consists, in the first place, in the deliberate exercise, in uttering the proposition, of a force tending to determine a belief in it in the mind of the interpreter. Perhaps that is a sufficient definition of it; but it involves also a voluntary self-subjection to penalties in the event of the interpreter's mind (and still more the general mind of society) subsequently becoming decidedly determined to the belief at once in the falsity of the

proposition and in the additional proposition that the utterer believed
the proposition to be false at the time he uttered it.

A *judgment* is a mental act deliberately exercising a force tending to
determine in the mind of the agent a belief in the proposition: to which
should perhaps be added that the agent must be aware of his being liable
to inconvenience in the event of the proposition's proving false in any
practical aspect.

In order fully to understand the distinction between a proposition
and an argument, it will be found important to class these acts, affirmation, etc. and ascertain their precise nature. The question is a purely logical one; but it happens that a false metaphysics is generally current,
especially among men who are influenced by physics but yet are not
physicists enough fully to comprehend physics, which metaphysics
would disincline those who believe in it from readily accepting the
purely logical statement of the nature of affirmation. I shall therefore
be forced to touch upon metaphysics. Yet I refuse to enter here upon a
metaphysical discussion: I shall merely hint at what ground it is necessary to take in opposition to a common doctrine of that kind. Affirmation is of the nature of a symbol. It will be thought that this cannot be
the case since an affirmation, as the above analysis shows, produces
real effects, physical effects. No sign, however, is a real thing. It has
no real being, but only being represented. I might more easily persuade
readers to think that affirmation was an index, since an index is, perhaps, a real thing. Its replica, at any rate, is in real reaction with its
object, and it forces a reference to that object upon the mind. But a symbol, a word, certainly exists only in replica, contrary to the nature of a
real thing; and indeed the symbol only becomes a sign because its
interpreter happens to be prepared to represent it as such. Hence, I
must and do admit that a symbol cannot exert any real force. Still, I
maintain that every sufficiently complete symbol governs things, and
that symbols alone do this. I mean that though it is not a force, it is a
law. Now those who regard the false metaphysics of which I speak as
the only clear opinion on its subject are in the habit of calling laws
"uniformities," meaning that what we call laws are, in fact, nothing
but common characters of classes of events. It is true that they hold
that they are symbols, as I shall endeavor to show that they are;
but this is to their minds equivalent to saying that they are common
characters of events; for they entertain a very different conception
of the nature of a symbol from mine. I begin, then, by showing that
a law is not a mere common character of events. Suppose that a man
throwing a pair of dice, which were all that honest dice are supposed to
be, were to throw sixes a hundred times running. Every mathematician

will admit that that would be no ground for expecting the next throw to turn up sixes. It is true that in any actual case in which we should see sixes thrown a hundred times running we should very rightly be confident that the next throw would turn up sixes likewise. But why should we do so? Can anybody sincerely deny that it would be because we should think the throwing of a hundred successive sixes was an almost infallible indication of there being some real connection between those throws, so that the series [was] not merely a uniformity in the common character of turning up sixes, but something more, a result of a real circumstance about the dice connecting the throws? This example illustrates the logical principle that mere community of character between the members of a collection is no argument, however slender, tending to show that the same character belongs to another object not a member of that collection and not (as far as we have any reason to think) having any real connection with it, unless perchance it be in having the character in question. For the usual supposition that we make about honest dice is that there will be no real connection (or none of the least significance) between their different throws. I know that writer has copied writer in the feeble analysis of chance as consisting in our ignorance. But the calculus of probabilities is pure nonsense unless it affords assurance in the long run. Now what assurance could there be concerning a long run of throws of a pair of dice, if, instead of knowing they were honest dice, we merely did not know whether they were or not, or if, instead of knowing that there would be no important connection between the throws, we merely did not know that there would be. That certain objects A, B, C, etc. are known to have a certain character is not the slightest reason for supposing that another object Ξ, quite unconnected with the others so far as we know, has that character. Nor has this self evident proposition ever been denied. A "law," however, is taken very rightly by everybody to be a reason for predicting that an event will have a certain character although the events known to have that character have no other real connection with it than the law. This shows that the law is not a mere uniformity but involves a real connection. It is true that those metaphysicians say that if A, B, C, etc. are known to have two common characters and Ξ is known to have one of these, this is a reason for believing that it has the other. But this is quite untenable. Merely having a common character does not constitute a real connection; and those very writers virtually acknowledge this, in reducing law to uniformity, that is, to the possession of a common character, as a way of denying that "law" implies any real connection. What is a law, then? It is a formula to which real events truly conform. By "conform," I mean that, taking the formula as a general principle, if experience shows that the formula applies to a given event, then the result will be confirmed by experience. But that such a general formula is a symbol, and more particularly, an asserted symbolical proposition is evident. Whether or not this symbol is a reality, even if not recognized by you or me or any generations of men, and whether, if so, it implies an Utterer, are metaphysical questions into which I will not now enter. One distinguished writer seems to hold that, although events conform to the formula, or rather, although it conforms to the Truth of facts, yet it does not influence the facts. This comes perilously near to being pure verbiage; for, seeing that nobody pretends that the formula exerts a compulsive force on the events, what definite meaning can attach to this emphatic denial of the law's "influencing" the facts? The law had such mode of being as it ever has before all the facts had come into existence, for it might already be experientially known; and then the law existing, when the facts happen there is agreement between them and the law. What is it, then, that this writer has in mind? If it were not for the extraordinary misconception of the word "cause" by Mill, I should say that the idea of metaphysical sequence implied in that word, in "influence," and in other similar words was perfectly clear. Mill's singularity is that he speaks of the cause of a singular event. Everybody else speaks of the cause of a "fact," which is an element of the event. But, with Mill, it is the event in its entirety which is caused. The consequence is that Mill is obliged to define the cause as the totality of all the circumstances attending the event. This is, strictly speaking, the Universe of being in its totality. But any event, just as it exists, in its entirety, is nothing else but the same Universe of being in its totality. It strictly follows, therefore, from Mill's use of the words, that the only *causatum* is the entire Universe of being and that its only cause is itself. He thus deprives the word of all utility. As everybody else but Mill and his school more or less clearly understands the word, it is a highly useful one. That which is caused, the *causatum*, is, not the entire event, but such abstracted element of an event as is expressible in a proposition, or what we call a 'fact'. The cause is another 'fact.' Namely, it is, in the first place, a fact which could, within the range of possibility, have its being without the being of the *causatum*; but, secondly, it could not be a real fact while a certain third complementary fact, expressed or understood, was realized, without the being of the causatum; and thirdly, although the actually realized causatum might perhaps be realized by other causes or by accident, yet the existence of the entire possible causatum could not be realized without the cause in question. It may be added that a part of a cause, if a part in that respect in which the cause is a cause, is also called a *cause*. In other respects, too, the scope of the word will be somewhat widened in the sequel. If the cause so defined is a part of the causatum, in the sense that the causatum could not logically be without the cause, it is called an *internal cause*; otherwise, it is called an *external cause*. If the cause is of the nature of an individual thing or fact, and the other factor requisite to the necessitation of the *causatum* is a general principle, I would call the cause a *minor*, or *individuating* or perhaps a *physical cause*. If, on the other hand, it is the general principle which is regarded as the cause and the individual fact to which it is applied is taken as the understood factor, I would call the cause a *major*, or *defining*, or perhaps a *psychical cause*. The individuating internal cause is called the *material cause*. Thus the integrant parts of a subject or fact form its *matter*, or material cause. The individuating external cause is called the *efficient*, or *efficient cause*; and the *causatum* is called the *effect*. The defining internal cause is called the *formal* cause, or *form*. All those facts which constitute the definition of a subject or fact make up its form. The defining external cause is called the *final cause*, or *end*. It is hoped that these statements will be found to hit a little more squarely than did those of Aristotle and the scholastics the same bull's eye at which they aimed. From scholasticism and the medieval universities, these conceptions passed in vaguer form into the common mind and vernacular of Western Europe, and especially so in England. Consequently by the aid of these definitions I think I can make out what it is that the writer mentioned has in mind in saying that it is not the law which influences, or is the final cause of, the facts, but the facts that make up the cause of the law. He means that the general fact which the law of gravitation expresses is composed of the special facts that this stone at such a time fell to the ground as soon as it was free to do so and its upward velocity was exhausted, that each other stone did the same, that each planet at each moment was describing an ellipse having the centre of mass of the solar system at a focus, etc. etc.; so that the individual facts are the material cause of the general fact expressed by the law; while the propositions expressing those facts are the efficient cause of the law itself. This is a possible meaning in harmony with the writer's sect of thought; and I believe it is his intended meaning. But this is easily seen not to be true. For the formula relates to all possible events of a given description; which is the same as to say that it relates to all possible events. Now no collection of actual individual events or other objects of any general description can amount to all possible events or objects of that description; for it is possible that an addition should be made to that collection. The individuals do not constitute the matter of a general: those who with Kant, or long before him, said that they do were wanting in the keen edge of thought requisite for such discussions. On the contrary, the truth of the formula, its really being a sign of the indicated object, is the defining cause of the agreement of the individual facts with it. Namely, this truth fulfills the first condition, which is that it might logically be although there were no such agreement. For it might be true, that is, contain no falsity, that whatever stone there might be on earth would have a real downward component acceleration even although no stone actually existed on earth. It fulfills the second condition, that as soon as the other factor (in this case the actual existence of each stone on earth) was present, the result of the formula, the real downward component of acceleration would exist. Finally, it fulfills the third condition, that while all existing stones might be accelerated downwards by other causes or by an accidental concurrence of circumstances, yet the downward acceleration of every possible stone would involve the truth of the formula.

It thus appears that the truth of the formula, that is, the law, is, in the strictest sense, the defining cause of the real individual facts. But the formula, if a symbol at all, is a symbol of that object which it indicates as its object. Its truth, therefore, consists in the formula being a symbol. Thus a symbol may be the cause of real individual events and things. It is easy to see that nothing but a symbol can be such a cause, since a cause is by its definition the premiss of an argument; and a symbol alone can be an argument. Every sufficiently complete symbol is a final cause of, and "influences," real events, in precisely the same sense in which my desire to have the window open, that is, the symbol in my mind of the agreeability of it, influences the physical facts of my rising from my chair, going to the window, and opening it. Who but a Millian or a lunatic will deny that that desire influences the opening of the window? Yet the sense in which it does so is none other than that in which every sufficiently complete and true symbol influences real facts.

A symbol is defined as a sign which becomes such by virtue of the fact that it is interpreted as such. The signification of a complex symbol is determined by certain rules of syntax which are part of its meaning. A simple symbol is interpreted to signify what it does from some accidental circumstance or series of circumstances, which the history of any word illustrates. For example, in the latter half of the XVth century, a certain model of vehicle came into use in the town of Kots (pronounced, *kotch*) in Hungary. It was copied in other towns, doubtless with some modifications, and was called a *kotsi szeker*, or Kots cart. Copied in still other towns, and always more or less modified, it came to be called, for short, a *cotch*. It thus came about that the *coach* was used, first, for a magnificent vehicle to be drawn by horses for carrying persons in state and in such comfort as that state required; then, for a large and pretentious vehicle to be drawn by four or more horses for conveying passengers from one town to another; and finally, to any large vehicle for conveying passengers at a fare by the seat from one town to another. In all ordinary cases, it is, and must be, an accidental circumstance which causes a symbol to signify just the characters that it does; for were there any necessary, or nearly necessary, reason for it, it would be this which would render the sign a sign, and not the mere fact that so it would be interpreted, as the definition of a symbol requires. It will be well here to interpose a remark as to the identity of a symbol. A sign has its being in its adaptation to fulfill a function. A symbol is adapted to fulfill the function of a sign simply by the fact that it does fulfill it; that is, that it is so understood. It is, therefore, what it is understood to be. Hence, if two symbols are used, without regard to any differences between them, they are replicas of the same symbol. If the difference is looked upon as merely grammatical (as with *he* and *him*), or as merely rhetorical (as with *money* and *spondesime*), or as otherwise insignificant, then logically they are replicas of one symbol. Hardly any symbol directly signifies the characters it signifies; for whatever it signifies it signifies by its power of determining another sign signifying the same character. If I write of the 'sound of sawing,' the reader will probably do little more than glance sufficiently at the words to assure himself that he could imagine the sound I referred to if he chose to do so. If, however, what [I] proceed to say about that sound instigates him to do more, a sort of auditory composite will arise in his imagination of different occasions when he has been near a saw; and this will serve as an icon of the signification of the phrase 'sound of a saw.' If I had used, instead of that phrase, the word 'buzz,' although this would have been less precise, yet, owing to the sound of the word being itself a sort of buzz, it would have more directly called up an iconic interpretation. Thus some symbols are far superior to others in point of directness of signification. This is true not only of outward symbols but also of general ideas. When a person remembers something, as for example in trying in a shop to select a ribbon whose color shall match that of an article left at home, he knows that his idea is a memory and not an imagination by a certain feeling of having had the idea before, which he will not be very unlikely to find has been somewhat deceptive. It is a sort of sense of similarity between the present and the past. Even if he had the two colors before his eyes, he could only know them to be similar by a peculiar feeling of similarity; because as two sensations they are different. But in the case supposed, it is not the mere general feeling of similarity which is required but that peculiar variety of it which arises when a present idea is pronounced to be similar to one not now in the mind at all, but formerly in the mind. It is clear that this feeling functions as a symbol. To call it an *icon* of the past idea would be preposterous. For instead of the present idea being serviceable as a substitute for the past idea by virtue of being similar to it, it is on the contrary only known to be similar to it by means of this feeling that it is so. Neither will it answer to call it an *index*. For it is the essence of an index to be in real connection with its object; so that it cannot be mendacious insofar as it is indexical; while this feeling is not infrequently deceptive; sometimes, in everybody's life, absolutely baseless. It is true that it is on the whole veracious; and veracity,—necessitated truth,—can belong to no sign except so far as it involves an index. But a symbol, if sufficiently complete always involves an index, just as an index sufficiently complete involves an icon. There is an infallible criterion for distinguishing between an index and an icon. Namely, although an index, like any other sign, only functions as a sign when it is interpreted, yet though it never happened to be interpreted, it remains equally fitted to be the very sign that would be if interpreted. A symbol, on the other hand, that should not be interpreted, would either not be a sign at all, or would only be a sign in an utterly different way. An inscription that nobody ever had interpreted or ever would interpret would be but a fanciful scrawl, an index that some being had been there, but not at all conveying or apt to convey its meaning. Now imagine the feeling which tells us that a present idea has been experienced before not to be interpreted as having that meaning, and what would it be? It would be like any other feeling. No study of it could ever discover that it had any connection with an idea past and gone. Even if this should be discovered, and if it should further be discovered that that connection was of such a nature as to afford assurance that the present idea with which the feeling is connected is similar to that former idea, still this would be by an additional discovery, not involved in the sign itself; a discovery, too, of the nature of a symbol, since it would be the discovery of a general law. The only way in which an index can be a proposition is by involving an *icon*. But what *icon* does this feeling present? Does it exhibit anything similar to similarity? To suppose that the feeling in question conveys its meaning by presenting a new idea a vague duplicate of the idea first present; gratuitous as this hypothesis would be, would not suffice to prove the feeling to be an index, since a symbol would be requisite to inform us that the first idea and the newly presented idea were similar; and even then there would be the element of preterity to be conveyed, which no icon and consequently no index could signify. It is quite certain therefore that in this feeling we have a definite instance of a symbol which, in a certain sense, *necessarily* signifies what it does. We have already seen that it can only be by an accident, and not be inherent necessity, that a symbol signifies what it does. The two results are reconciled by the consideration that the accident in this case is that we are so constituted that that feeling shall be so interpreted by us. A little psychological examination will vindicate the first assertion. For although it is not a very rare experience to have a strong feeling of having been in a present situation before, when in fact one never was in such a situation, yet everybody, unless he be a psychologist, is invariably deeply impressed by such an experience (or at least by his *first* experience of this sort), will not forget it for long years, and is importuned by the notion that he is here confronted with a phenomenon profoundly mysterious, if not supernatural. The psychologist may waive the matter aside, in his debonair satisfaction with the state of his science; but he seems to me to overlook the most instructive part of the phenomenon. This is that though internal feelings generally are testimonies proverbially requiring to be received with reserve and caution, yet when this particular one plays us false people feel as if the bottom had fallen out of the universe of being. Why should they take the matter so seriously? It is that if we try to analyze what is *meant* by saying that a present idea "resembles" one that is past and gone, we cannot find that anything else is *meant* but that there is this feeling connected with it. From which it seems to result that for this feeling to be mendacious would be a self contradictory state of things; in which case we might well say that the bottom had fallen out of Truth itself. Now a symbol which should by logical necessity signify what it does would obviously have nothing but its own application, or predication, for its own signification as a predicate; or, to express the matter in abbreviation, would signify itself alone. For anything else that might be supposed to be signified by it might, by logical possibility, not be so signified. That such a symbol should be false would indeed involve a contradiction. Thus, the feeling of amazement at the "sense of pre-existence," as it has been called, amounts virtually to nothing more than the natural confusion of that which is necessary by virtue of the constitution of the mind with that which is logically necessary. The feeling under discussion is necessary only in the latter sense. Consequently, its falsity is not absurd but only abnormal. Nor is it its own sole signification, but only the sole signification recognizable without transcendental thought; because to say that a present idea is really similar to an idea really experienced in the past means that a being sufficiently informed would know that the effect of the later idea would be a revivification of the effect of the earlier idea, in respect to its quality of feeling. Yet it must be remarked that the only effect of a quality of feeling is to produce a memory, itself a quality of feeling; and that to say that two of those are similar is, after all, only to say that the feeling which is the symbol of similarity will attach to them. Thus, the feeling of recognition of a present idea as having been experienced has for its signification the applicability of a part of itself. The general feeling of similarity, though less startling, is of the same kind. All the special occurrences of the feeling of similarity are recognized as themselves similar, by the application to them of the same symbol of similarity. It is Kant's "I think," which he considers to be an act of thought, that is, to be of the nature of a symbol. But his introduction of the *ego* into it was due to his confusion of this with another element.

This feeling is not peculiar among feelings in signifying itself alone; for the same is true of the feeling 'blue' or any other. It must not be forgotten that a feeling is not a psychosis, or state of mind, but is merely a quality of a psychosis, with which is associated a degree of vividness, or relative disturbance, or prominence, of the quality in the psychosis, as measured chiefly by after-effects. These psychoses are icons; and it is in being a symbol that the feeling of similarity is distinguished from other feelings. But the signification of the psychosis as a sign is that the percept to which it ultimately refers has the same quality, as determined by the symbol-feeling of similarity.

My principal object in drawing attention to this symbol of similarity is to show that the significations of symbols have various grades of directness up to the limit of being themselves their own significations. An icon is significant with absolute directness of a character which it embodies; and every symbol refers more or less indirectly to an icon.

An index is directly denotative of a real object with which it is in reaction. Every symbol refers more or less indirectly to a real object through an index. One goes into a shop and asks for a yard of silk. He wants a piece of silk, which has been placed (either by measurement or estimation depending on habits involving indices) into reactive comparison with a yard-stick, which itself by successive reactions has been put into reaction with a certain real bar in Westminster. The word "yard" is an example of a symbol which is denotative in a high degree of directness. When we consider the successive compari[sons] more scientifically, we have to admit that each is subject to a probable error. The more pains we take to make the reaction significant, the more we are forced to recognize that each single act of comparison gives its own result; and a general inference which we make from a large number of these represents what each one would be if it could be performed more accurately. It is by the intervention of the inferential symbol that we virtually obtain a more intimate reaction. When a biologist labels a specimen, he has performed a comparison, as truly reactive in its nature as that of two standards of length, with an original "type-specimen," as he calls the prototype. His label thus involves something of the nature of an index, although less prominently than does the word "yard." It would be more scientific if in place of a single prototype comparisons were made with 25 different bars of different material and kept under different conditions, and that were called a "yard" which agreed with the mean of them all; and so in biology, there ought to be 25 type-specimens, exhibiting the allowable range of variation as well as the normal mean character. This more scientific proceeding is that of common-sense in regard to ordinary names, except that instead of 25 instances, there are many more. I go into a furniture shop and say I want a "table." I rely upon my presumption that the shop-keeper and I have undergone reactional experiences which though different have been so connected by reactional experiences as to make them virtually the same, in consequence of which "table" suggests to him, as it does to me, a movable piece of furniture with a flat top of about such a height that one might conveniently sit down to work at it. This convenient height, although not measured, is of the same nature as the yard of silk already considered. It means convenient for men of ordinary stature; and his reactive experience presumably agrees with mine as to what the ordinary stature is. I go into a shop and ask for butter. I am shown something; and I ask "Is this butter, or is it oleomargarine, or something of the sort?" "Oh, I assure you that it is, in chemical strictness, butter." "In chemical strictness, eh? Well, you know what the breed of meat cattle is, as well as I do. It is an individual object, of which we have both seen parts. Now I want to know whether this substance has been churned from milk drawn from the breed of meat cattle." That breed is known to us only by real reactional experience. What is gold? It is an elementary substance having an atomic weight of about $197\frac{1}{4}$. In saying that it is elementary, we mean undecomposable in the present state of chemistry, which can only be recognized by real reactional experience. In saying that its atomic weight is $197\frac{1}{4}$, we mean that it is so compared with hydrogen. What, then, is hydrogen? It is an elementary gas $14\frac{1}{2}$ times as light as air. And what is air? Why, it is this with which we have reactional experience about us. The reader may try instances of his own until no doubt remains in regard to symbols of things experienced, that they are always denotative through indices; such proof will be far surer than any apodictic demonstration. As to symbols of things not experienced it is clear that these must describe their objects by means of their differences from things experienced. It is plain that in the directness of their denotation symbols vary through all degrees. It is, of course, quite possible for a symbol to represent itself, at least in the only sense in which a thing that has no *real being* but only *being represented*, and which exists in *replica*, can be said to be identical with a real and therefore individual object. A map may be a map of itself; that is to say one replica of it may be the object mapped. But this does not make the denotation extraordinarily direct. As an example of a symbol of that character, we may rather take the symbol which is expressed in words as "the truth," or "Universe of Being." Every symbol whatever must denote what this symbol denotes; so that any symbol considered as denoting the truth necessarily denotes that which it denotes; and in denoting it, it *is* that very thing, or a fragment of it taken for the whole. It is the whole taken so far as it need be taken for the purpose of denotation; for denotation essentially takes a part for its whole.

But the most characteristic aspect of a symbol is its aspect as related to its interpretant; because a symbol is distinguished as a sign which becomes such by virtue of determining its interpretant. An interpretant of a symbol is an outgrowth of the symbol. We have used the phrase, a symbol *determines* its interpretant. Determination implies a *determinandum*, a subject to be determined. What is that? We must suppose that there is something like a sheet of paper, blank or with a blank space upon it upon which an interpretant sign may be written. What is the nature of this blank? In affording room for the writing of a symbol, it is *ipso facto* itself a symbol, although a wholly vague one. In affording room for an interpretant of that particular symbol, it is already an interpretant of that symbol, although only a partial one. An *entire* interpretant should involve a replica of the original symbol. In fact, the interpretant symbol, so far as it is no more than an interpretant *is* the original symbol, although perhaps in a more developed state. But the interpretant symbol may be at the same time an interpretant of an independent symbol. A symbol is something which has the power of reproducing itself, and that essentially, since it is constituted a symbol only by the interpretation. This interpretation involves a power of the symbol to cause a real fact; and although I desire to avoid metaphysics, yet when a false metaphysics invades the province of logic, I am forced to say that nothing can be more futile than to attempt to form a conception of the universe which shall overlook the power of representations to cause real facts. What is the purpose of trying to form a conception of the universe if it is not to render things intelligible? But if this is to be done, we necessarily defeat ourselves if we insist upon reducing everything to a norm which renders everything that happens essentially and *ipso facto*, unintelligible. That, however, is what we do, if we do not admit the power of representations to cause real facts. If we are to explain the universe, we must assume that there was in the beginning a state of things in which there was nothing, no reaction and no quality, no matter, no consciousness, no space and no time, but just nothing at all. Not determinately nothing. For that which is determinately not *A* supposes the being of *A* in some mode. Utter indetermination. But a symbol alone is indeterminate. Therefore Nothing, the indeterminate of the absolute beginning is a symbol. That is the way in which the beginning of things can alone be understood. What logically follows? We are not to content ourselves with our instinctive sense of logicality. That is logical which comes from the essential nature of a symbol. Now it is of the essential nature of a symbol that it determines an interpretant, which is itself a symbol. A symbol, therefore, produces an endless series of interpretants. Does anybody suspect all this of being sheer nonsense. *Distinguo*. There can, it is true, be no positive information about what antedated the entire Universe of being; because, to begin with, there was nothing to have information about. But the universe is intelligible; and therefore it is possible to give a general account of it and its origin. This general account is a symbol; and from the nature of a symbol, it must begin with the formal assertion that there was an indeterminate nothing of the nature of a symbol. This would be false if it conveyed any information. But it is the correct and logical manner of beginning an account of the universe. As a symbol it produced its infinite series of interpretants, which in the beginning were absolutely vague like itself. But the direct interpretant of any symbol must in the first stage of it be merely the *tabula rasa* for an interpretant. Hence the immediate interpretant of this vague Nothing was not even determinately vague, but only vaguely hovering between determinacy and vagueness; and *its* immediate interpretant was vaguely hovering between vaguely hovering between vagueness and determinacy and determinate vagueness or determinacy, and so on, *ad infinitum*. But every endless series must logically have a limit.

Leaving that line of thought unfinished for the present owing to the feeling of insecurity it provokes, let us note, first, that it is of the nature of a symbol to create a *tabula rasa* and therefore an endless series of *tabulae rasae*, since such creation is merely representation, the *tabulae rasae* being entirely indeterminate except to be representative. Herein is a real effect; but a symbol could not be without that power of producing a real effect. The symbol represents itself to be represented; and that representedness is real being owing to its utter vagueness. For all that is represented must be thoroughly borne out.

For reality is compulsive. But the compulsiveness is absolutely *hic et nunc*. It is for an instant and it is gone. Let it be no more and it is absolutely nothing. The reality only exists as an element of the regularity. And the regularity is the symbol. Reality, therefore, can only be regarded as the limit of the endless series of symbols.

A symbol is essentially a purpose, that is to say, is a representation that seeks to make itself definite, or seeks to produce an interpretant more definite than itself. For its whole signification consists in its determining an interpretant; so that it is from its interpretant that it derives the actuality of its signification.

A *tabula rasa* having been determined as representative of the symbol that determines it, that *tabula rasa* tends to become determinate. The vague always tends to become determinate, simply because its vagueness does not determine it to be vague. (As the limit of an endless series.) In so far as the interpretant is the symbol, as it is in some measure, the determination agrees with that of the symbol. But in so far as it fails to be its better self it is liable to depart from the meaning of the symbol. Its purpose, however, is to represent the symbol in its representation of its object; and therefore, the determination is followed by a further development, in which it becomes corrected. It is of the nature of a sign to be an individual replica and to be in that replica a living general. By virtue of this, the interpretant is animated by the original replica, or by the sign it contains, with the power of representing the true character of the object. That the object has at all a character can only consist in a representation that it has so, — a representation having power to live down all opposition. In these two steps, of determination and of correction, the interpretant aims at the object more than at the original replica and may be truer and fuller than the latter. The very entelechy of being lies in being representable. A sign cannot even be false without being a sign and so far as it is a sign it must be true. A symbol is an embryonic reality endowed with power of growth into the very truth, the very entelechy of reality. This appears mystical and mysterious simply because we insist on remaining blind to what is plain, that there can be no reality which has not the life of a symbol.

How could such an idea as that of *red* arise? It can only have been by gradual determination from pure indeterminacy. A vagueness not determined to be vague, by its nature begins at once to determine itself. Apparently we can come no nearer than that to understanding the universe.

That is not necessarily logical which strikes me today as logical; still less, as mathematics amply exemplifies is nothing logical except what appears to me so. That is logical which it is necessary to admit in order to render the universe intelligible. And the first of all logical principles is that the indeterminate should determine itself as best it may.

A chaos of reactions utterly without any approach to law is absolutely nothing; and therefore pure nothing was such a chaos. Then pure indeterminacy having developed determinate possibilities, creation consisted in mediating between the lawless reactions and the general possibilities by the influx of a symbol. This symbol was the purpose of creation. Its object was the entelechy of being which is the ultimate representation.

We can now see what *judgment* and *assertion* are. The man is a symbol.

Different men, so far as they can have any ideas in common, are the same symbol. Judgment is the determination of the man-symbol to have whatever interpretant the judged proposition has. Assertion is the determination of the man-symbol to determining the interpreter, so far as he is interpreter, in the same way.

ON QUANTITY, WITH SPECIAL REFERENCE TO COLLECTIONAL AND MATHEMATICAL INFINITY (15)

§1. PURPOSE OF THE MEMOIR

Art. 1.   A mathematical investigation is acknowledged to be proper matter for a memoir of a scientific academy.[1] Now that logic has been rendered mathematically exact, why should not a logical inquiry be admitted to the same rank? The latter is necessarily more verbose, because it has to direct attention to observations which are open to all men; but every branch of science has some peculiarities which are not to the taste of those who do not pursue it.

The inquiry whose principal results are here set forth had for its aim to apply mathematically exact logic, particularly the logic of relatives, to the solution of the following problems:

1st, What is *mathematics*? That is to say, granted that the theory of numbers, algebra, topology, intersectional and metrical geometry, the differential calculus, the theory of functions, etc. are branches of mathematics, it is required to form a distinct and unitary conception of a study embracing what is characteristic of the aggregate of those doctrines, this conception being so limited as to have the greatest possible utility in the natural classification of the sciences, and to be intimately connected with a sufficient explanation of the part which mathematics has been found to play in the development of knowledge.

2nd, What is *quantity*? That is to say, granted that the system of whole numbers is a system of quantity, that the system of rational numbers is another, the system of real numbers is another, the system of imaginaries is another, and the system of quaternions is another, it is required to form a distinct and unitary conception applicable to those systems, so limited as to be of the greatest possible utility in the theory of cognition, bringing into view those features of the above systems which give them

---

[1]   Apparently this paper was prepared with the National Academy of Sciences in mind. Peirce read a paper there "On the Logic of Quantity" in April 1896; another on "Mathematical Infinity" in November 1896.

their importance for science, and connecting itself directly with an explanation of that importance. In connection with this question, I also consider what the chief systems of quantity are.

3rd, What are *multitude* and *collectional quantity*? This problem is found to divide itself into several parts, of which the chief are, What is the place of predications concerning the multitudes of collections in the natural logical classification of propositions? Is it true that of any two possible multitudes not equal one is less than the other? That is, is one of two collections always capable of being in one-to-one correspondence with the other? What are all the grades of multitudes, and how are they distinguished? Is there any maximum possible multitude? Are the grades of collectional quantity the same as those of multitude?

4th, What is *continuity*? That is to say, if we develop the natural common-sense idea of time and space into a logically exact conception, in what does the unbroken flow of time and the capacity of space for admitting unbroken motion consist, as so conceived? That answered, I very briefly consider what the evidence is that the natural common-sense idea of time and space is true to nature. I also inquire how far it can concern the mathematician.

5th, What is *mathematical infinity*? Is it possible to construct a scale of quantity containing a doubly endless series of orders of infinity, related as integral powers [of] one base, and if so how is the neglect of quantities of each order in comparison with those of the next higher order a mathematically exact proceeding? In this connection, I ask some questions of secondary importance, as follows: Is it a consistent attitude freely to admit of imaginaries while rejecting infinitesimals as inconceivable? Are there any particular logical difficulties in the literal acceptance of infinitesimals? Is the doctrine of literal infinitesimal particularly abstruse and hard to apprehend? Is there any reason to believe that mathematically infinite quantities exist as a matter of fact? I then pass to the study of the logic of the method of limits. I inquire: What is the exact mathematical definition of a limit, and in what sense can it be said that a limit is never reached?

§2. THE NATURE OF MATHEMATICS

Art. 2.   As a general rule, the value of an exact philosophical definition of a term already in familiar use lies in its bringing out distinct conceptions of the function of objects of the kind defined. In particular, this is true of the definition of an extensive branch of science; and in order to assign the most useful boundaries for such a study, it is requisite to consider what part of the whole work of science has, from the nature of things, to be performed by those men who are to do that part of the work which unquestionably comes within the scope of that study; for it does not conduce to the clearness of a broad view of science to separate problems which have necessarily to be solved by the same men. Now a mathematician is a man whose services are called in when the physicist, or the engineer, or the underwriter, etc. finds himself confronted with an unusually complicated state of relations between facts and is in doubt whether or not this state of things necessarily involves a certain other relation between facts, or wishes to know what relation of a given kind is involved. He states the case to the mathematician. The latter is not at all responsible for the truth of those premises: that he is to accept. The first task before him is to substitute for the intricate, and often confused, mass of facts set before him, an imaginary state of things involving a comparatively orderly system of relations, which, while adhering as closely as possible or desirable to the given premises, shall be within his powers as a mathematician to deal with. This he terms his *hypothesis*. That work done, he proceeds to show that the relations explicitly affirmed in the hypothesis involve, as a part of any imaginary state of things in which they are embodied, certain other relations not explicitly stated.

Thus, the mathematician is not concerned with real truth, but only studies the substance of hypotheses. This distinguishes his science from every other. Logic and metaphysics make no special observations; but they rest upon observations which have been made by common men. Metaphysics rests upon observations of real objects, while logic rests upon observations of real facts about mental products, such as that, not merely according to some arbitrary hypothesis, but in every possible case, every proposition has a denial, that every proposition concerns some objects of common experience of the deliverer and the interpreter, that it applies to that some idea of familiar elements abstracted from the occasions of its excitation, and that it represents that an occult compulsion not within the deliverer's control unites that idea to those objects. All these are results of common observation, though they are put into scientific and uncommon groupings. But the mathematician observes nothing but the diagrams he himself constructs; and no occult compulsion governs his hypotheses except one from the depths of mind itself.

Thus, the distinguishing characteristic of mathematics is that it is the scientific study of hypotheses which it first frames and then traces to their consequences. Mathematics is either *applied* or *pure*. Applied mathematics treats of hypotheses in the forms in which they are first suggested by experience, involving more or less of features which have no bearing upon the forms of deduction of consequences from them. Pure mathematics is the result of afterthought by which these irrelevant features are eliminated.

It cannot be said that all framing of hypotheses is mathematics. For that would not distinguish between the mathematician and the poet. But the mathematician is only interested in hypotheses for the forms of inference from them. As for the poet, although much of the interest of a romance lies in tracing out consequences, yet these consequences themselves are more interesting in point of view of the resulting situations than in the way in which they are deducible. Thus, the poetical interest of a mental creation is in the creation itself, although as a part of this a mathematical interest may enter to a slight extent. Detective stories and the like have an unmistakable mathematical element. But a hypothesis, in so far as it is mathematical, is mere matter for deductive reasoning.

On the other hand, it is an error to make mathematics consist exclusively in the tracing out of necessary consequences. For the framing of the hypothesis of the two-way spread of imaginary quantity, and the hypothesis of Riemann surfaces were certainly mathematical achievements.

Mathematics is, therefore, the study of the substance of hypotheses, or mental creations, with a view to the drawing of necessary conclusions.

Art. 3.   Before the above analysis is definitively accepted, it ought to be compared with the principal attempts that have hitherto been made to define mathematics.

Aristotle's definition shows that its author's efforts were, in a general way, rightly directed; for it makes the characteristic of the science to lie in the peculiar quality and degree of abstractness of its objects. But in attempting to specify the character of that abstractness, Aristotle was led into error by his own general philosophy. He makes, too, the serious mistake of supposing metaphysics to be more abstract than mathematics. In that he was wrong, since the former aims at the truth about the real world, which the latter disregards.

The Roman schoolmasters defined the mathematical sciences as the sciences of *quanta*. This definition would not have been admitted by a Greek geometer; because the Greeks were aware that the more fundamental branch of geometry treated of the intersections of unlimited planes. Still less does it accord with our present notion that as geometrical metrics is but a special problem in geometrical graphics, so geometrical graphics is but a special problem in geometrical topics. The only defence the Romans offered of the definition was that the objects of the four mathematical sciences recognized by them, viz: arithmetic, geometry, astronomy, and music, are things possessing quantity. It does not seem to have occurred to them that the objects of grammar, logic, and rhetoric equally possess quantity, although this ought to have been obvious even to them.

Subsequently, a different meaning was applied to the phrase "mathematics is the science of quantity." It is certainly possible to enlarge the conception of quantity so as to make it include tridimensional space, as imaginary quantity is two-dimensional and quaternions are four-dimensional. In such a way, this definition may be made coextensive with mathematics; but, after all, it does not throw so much light upon the position of mathematics among the sciences as that which is given in the last article.

De Morgan and Sir William Rowan Hamilton, influenced indirectly, as it would seem, by Kantianism, defined mathematics as the science of Time and Space, algebra being supposed to deal with Time as geometry does with Space. Among the objections to this definition, the following seem to be each by itself conclusive.

1st, this definition makes mathematics a positive science, inquiring into matters of fact. For, even if Time and Space are of subjective origin, they are nevertheless objects of which one thing is true and another false.

The science of space is no more a branch of mathematics than is optics. That is to say, just as there are mathematical branches of optics, of which projective geometry is one, but yet optics as a whole is not mathematics, because it is in part an investigation into objective truth, so there is a mathematical branch of the science of space, but this has never been considered to include an inquiry into the true constitution and properties of space. Euclid terms statements of such properties *postulates*. Now by a postulate the early geometers understood, as a passage in Aristotle shows, notwithstanding a blunder which the Stagyrite here makes, as he often blunders about mathematics, a proposition which was open to doubt but of which no proof was to be attempted. This shows that inquiry into the properties of space was considered to lie outside the province of the mathematician. In the present state of knowledge, systematic inquiry into the true properties of space is called for. It must appeal to astronomical observations on the one hand, to determine the metrical properties of space, to chemical experiment on the other hand to determine the dimensionality of space, and the question of the artiad or perissid character of space, or of its possible topical singularities (suggested by Clifford) [which] remain as yet without any known methods for their investigation. All this may be called Physical Geometry.

2nd, this definition erroneously identifies algebra with the science of Time. For it is an essential character of time that its flow takes place in one sense and not in the reverse sense; while the two directions of real quantity are as precisely alike as the two directions along a line in space. It is true that $+1$ squared gives itself while -1 squared gives the negative of itself. But there is another operation precisely as simple which performed upon -1 gives itself, and performed upon $+1$ gives the negative of itself. Besides, the idea of time essentially involves the notion of reaction between the inward and outward worlds: the future is the domain over which the Will has some power, the past is the domain of the powers which have gone to make Experience. The future and past which are essential parts of the idea of time cannot be otherwise accurately defined. Yet algebra does not treat of Will and Experience.

3rd, this definition leaves no room for some of the chief branches of mathematics, such as the doctrine of N-dimensional space, the theory of imaginaries, the calculus of logic, including probabilities, branches which it would be doing great violence to the natural classification of the sciences to separate from algebra and geometry.

4th, this definition is absurd, because it confines number to the domain of time, when time and space are, according to its own doctrine, *two* forms of intuition, so that their existence supposes number. Kant, himself, was too good a logician to make number a character of time. It is true that the cognition of number supposes time; but so does the cognition of Colorado silver mining. It no more follows that the science of number is a part of the science of time than that the science of Colorado silver mining is a part of the science of time.

5th, this definition would exclude *Quantity* from among the subjects of mathematical study. For *quantity*, according to the Kantian doctrine which this definition follows does not belong to intuition but is one of the four branches of the categories of the understanding, having precisely the same relation to time and space that Reality, Active Agency, and Modality have,—matters which certainly lie quite out of the province of mathematics. It may be added that this definition is far from receiving any countenance from Kant, who makes mathematics relate to the categories of quantity and quality.

In short, this defintion is probably of all definitions of branches of science that have ever gained numerous adherents the very worst.

In 1870, Benjamin Peirce defined mathematics as "the science which draws necessary conclusions." Since it is impossible to draw necessary conclusions except from perfect knowledge, and no knowledge of the real world can be perfect, it follows that, according to this definition mathematics must exclusively relate to the substance of hypotheses. My father seems to have regarded the work of treating the statements of fact brought to the mathematician and of creating from their suggestions a hypothesis as a basis of mathematical deduction to be the work of a logician, not of a mathematician. It cannot be denied that the two tasks, of framing hypotheses for deduction and of drawing the deductive conclusions are of widely different characters; nor that the former is similar to much of the work of the logician. But it is a mistake to suppose that everybody who reasons skillfully makes an application of logic. Logic is the science which examines signs, ascertains what is essential to being signs and describes their fundamentally different varieties, inquires into the general conditions of their truth, and states these with formal accuracy, and investigates the law of the development of thought, accurately states it and enumerates its fundamentally different modes of working. In metaphysics, no skill in reasoning can avail, unless that reasoning is based upon the exact generalizations of the logician as premises; and this may truly be said to be an application of logic. But in framing mathematical hypotheses no logic is required, since it is indifferent from a mathematical point of view how far the hypothesis agrees with the observed facts. It is for the employer of the mathematician to decide that. The framing of a mathematical hypothesis does not, therefore, come within the province of the logician.

Perhaps the definition of Benjamin Peirce may be defended on the ground that the transformation of the suggestions of experience into exact mathematical hypotheses is effected by drawing necessary conclusions. The drawing of a necessary conclusion is by no means the simple act which it is commonly supposed to be; and among the acts of which it is made up there are some which would suffice, or nearly suffice, to transform the result of experience into a mathematical hypothesis. The reply is that the two parts of the mathematician's functions are markedly dissimilar and therefore require to be distinguished in the definition.

Mr. George Chrystal, in the *Encyclopedia Britannica* (9th Ed. Article, *Mathematics*), endeavors to define mathematics by describing the general characters essential to a mathematical hypothesis (or, in his language, a "mathematical conception"). This is an effort in the right direction. But the definition is not very clear. The principal feature insisted upon is that the mathematical hypothesis must be marked by a *finite number* of distinct specifications. This implies that by "specifications" is meant some act, or element of an act, of which it is possible to make an infinite number; and what this is remains unexplained.

Art. 4.   It may be objected to the definition of art. 2 that it places mathematics above logic as a more abstract science the different steps [of] which must precede those of logic; while on the contrary logic is requisite for the business of drawing necessary conclusions. But this I deny. An application [of] logical theory is only required by way of exception in reasoning, and not at all in mathematical deduction. In probable reasoning where the evidence is very insufficient, or almost entirely wanting, we often hear men appeal to the "burden of proof" and other supposed logical principles; and in metaphysics logical theory is the only guide. But in the perspicuous and absolutely cogent reasonings of mathematics such appeals are altogether unnecessary. Many teachers of geometry think that it is desirable that a course in logic should precede the study of the elements. But the reasoning of the elements of geometry is often bad, owing to the fact that the hypothesis is not fully stated. The matter being confused and not made interesting, the pupil's mind becomes confused; and the teacher, not knowing enough either of geometry or of psychology to know what the difficulty is or how to remedy it, turns to the apparatus of the traditional syllogistic, which, as he teaches it, will serve to throw dust in the eyes of pupil and of teacher and make them both fancy the difficulty conquered. Publishers are of opinion that teachers would not use books in which the hypotheses of geometry should be fully set forth, and in which by taking up topics, graphics, and metrics in their logical order, the reasoning should be rendered unimpeachable; but it is certain that even dull pupils find no difficulties when they are taught in that way, so that only correct mathematical reasonings fully developed are offered to them. Undoubtedly, for any mind there is a point of complexity at which that mind will become confused, from inability to hold so many threads; and this point is very different for different minds. But such a difficulty is in no degree lessened by any appeal to the generalizations of logic.

Art. 5. In order to suggest the place which mathematics would seem to take in the system of the sciences, according to the above analysis, I venture to propose the following scheme of classification of all the sciences, modified from that of Auguste Comte, and like his proceeding from the more abstract to the more concrete. Each science (except mathematics) rests upon fundamental principles drawn from the truths discovered by the science immediately preceding it in the list, while borrowing data and suggestions from the discoveries of those which follow it.

1. *Mathematics*, which observes only the creations of the mathematician himself. It borrows suggestions from all other sciences, from philosophy Mathematical Logic, from psychics Mathematical Economics, from physics Mathematical Optics, Metrics, etc.

2. *Philosophy*, which makes no special observations, but uses facts commonly known. In order to be exact, it must rest on mathematical principles. It divides into *Logic*, which studies the world of thought, and *Metaphysics*, which studies the world of being; and the latter must rest upon the principles of the former.

3. *The science of time* and the *science of space* are the most abstract of the special sciences. They must be based upon metaphysical *principles*, and geometry largely upon the science of time, though the one draws *data* from psychology, and the other from astronomy and chemistry.

The science of time is psychical and that of space physical; and from this point on Psychics and Physics are widely separated, and influence one another little.

PSYCHICS

PHYSICS

4. *Nomological*

4a. The general law of psychics, i.e. *Association*, etc. It is founded on the science of time, but draws suggestions generally from psychology.

4a. The general laws of physics, i.e. *Dynamics*. It is founded on geometry, but draws suggestions principally from molar physics.

4b. The laws of psychics in single individuals, i.e. *Psychology*. It must be directly based on principles drawn from the science of Association.

4b. The law of the interaction of single particles at sensible distances, i.e. molar physics, or *The Science of Gravitation*. It is directly based on dynamics.

4c. The laws of physics which do not clearly appear in the study of individuals, but only in that of great aggregations, i.e. *Nomological Sociology*. It must be based on principles drawn from psychology.

4c. The laws of interaction of particles which cannot be individually studied, but only in the statistical effects of great aggregations; i.e. Molecular Physics, or *Elaterics* and *Thermics*. It is based on the principles of molar physics.

4d. Some persons maintain that there are minds of an altogether extraordinary nature, namely *spirits*. Whether this is really so or not, there is an active *inquiry* into the matter in a scientific spirit. But as yet its true principles are unsettled.

4d. A certain kind of matter, the *ether*, has properties so different from ordinary matter as to give rise to a special branch of nomological physics, which in treating of a special kind of matter, approaches chemistry, namely the Physics of Ether, or *Optics* and *Electrics*. It is based on Elaterics.

5. *Classificatory*

5a. The description of different kinds of minds and their works. The science of character, or *anthropology*. But it will treat of lower animals, as well as of man.

5a. The description of the properties of different kinds of matter. This, I think, ought to be called *chemistry*, although this term is, at present, restricted to the study of changes in the constitution of molecules. But it ought to be united with the whole science of the properties and constants of material substances, such as their crystalline forms, etc. Chemistry must draw its principles from nomological physics.

5b. Among kinds of minds, those whose common character is that of belonging to the same *race* or *society*, are particularly interesting. This gives ethnological psychics. The works of such societies, which are almost individual entities, are so various, and their study so difficult, that this branch breaks up into many, treating of different kinds of psychical products, such as language, manners and customs, government, etc.

5b. Among the kinds of matter, there is one class, which instead of forming mere crystals, develops into excessively complex forms. These substances are the protoplasms. The science of protoplasm is so far removed in all its methods from general chemistry that it must be set apart. It is *biology*, embracing *botany* and *zoology*. Each class is here more than a group of similar objects. It is, by virtue of hereditary descent, almost an individual entity. Thus, this science approaches the nature of a descriptive one.

6. *Descriptive*

The narrative of events of all kinds, and also the description of individual objects, with a view to the explanation of their characters upon the principles of psychics.

The description of individual objects of whatever nature, with the inquiry into the explanation, upon principles drawn from dynamics and chemistry, of their presenting certain combinations of characters frequently, or rarely, or never. The order of succession of these studies is as follows:

6a. *General History*, or the narrative and explanation of what has happened to societies and races. Although called *general*, it may be limited to special kinds of events.

6a. *Astronomy*, or the description and explanation of the stars.

6b. *Biography*, including Genealogy, or the narrative and explanation of what has happened to families and to individuals.

6b. *Meteorology*, including Ocean currents and Tides.

6c. *History of intellectual products*.
  (α) The history of Science;
  (β) The history of Useful Arts;
  (γ) The history of Art and Literature.

6c. *Geology*, or the description and explanation of the masses composing the earth, including the Atmosphere and Ocean; but not the rapid movements of the fluids.

6d. *Geography*, or the description and explanation of the different parts of the habitable shell outside the earth, including the fauna and flora of the Sea, etc.

7. Psychics and Physics now unite to produce the *Practical Sciences*, which with a few doubtful exceptions, such as Religion (or the art of dealing with Deity), and Ethics, equally depend upon the principles of Psychics and of Physics.

§3. THE NATURE OF QUANTITY

Art. 6.   Modern exact logic shows that every operation of deductive reasoning consists of four steps as follows:

1st, a diagram, or visual image, whether composed of lines, like a geometrical figure, or an array of signs, like an algebraical formula, or of a mixed nature, like a graph, is constructed, so as to embody in iconic form, the state of things asserted in the premise (there will be but one premise, after all that is known and is pertinent is collected into one copulative proposition).

2nd, Upon scrutiny of this diagram, the mind is led to suspect that the sort of information sought may be discovered, by modifying the diagram in a certain way. This experiment is tried.

3rd. The results of the experiment are carefully *observed*. This is genuine experiential observation, even though the diagram exists only in the imagination; for after it has once been created, though the reasoner has power to change it, he has no power to make the creation already past and done different from what it is. It is, therefore, just as real an object as if drawn on paper. Included in this observation is the analysis of what is seen and the representation of it in general language. What is so observed is a new relation between the parts of the diagram not mentioned in the precept by which it was constructed.

4th, By repeating the experiment, or by the similarity of the experiment to many others which have often been repeated without varying the result, the reasoner infers inductively, with a degree of probability practically amounting to certainty, that every diagram constructed according to the same precept would present the same relation of parts which has been observed in the diagram experimented upon.

Art. 7.   But in order to comprehend the nature of quantity, it is needful to examine more in detail the different kinds of deductively inferential steps. These are most philosophically exhibited in the unsimplified form of a system of logical graphs which I call *negative logical graphs*. In order to make the statements of these forms of inference intelligible, I must define a term of logic introduced by Duns Scotus, the *consequentia simplex de inesse*. A conditional proposition of the form, "If one thing, *A*, is true, then another thing, *B*, is true," is understood by that class of logicians who follow the doctrine of Philo, a logician mentioned by Cicero and other ancient authors, to mean that under whatever possible circumstances, *A* is true, (without asserting that there are any such circumstances), under those circumstances *B* is true. Hence, according to the Philonians, if there are no possible circumstances under which *A* is true, the conditional proposition makes no assertion at all, and thus, escaping all conflict with facts, is true whatever *B* may be. In ordinary speech, we do not make an impossible supposition the antecedent of a conditional proposition, and usually mean to imply that the range of possibilities is to be taken so wide that the antecedent always will be true under some possible circumstances of the kind of possibility meant. But in logic it is inconvenient thus to allow a simple proposition to assert two things; and the best logicians adhere to the Philonian convention about the meanings of conditionals. Possibility is of various kinds, some wider, others narrower. The possible is in every case that which is not known to be false, under a certain state of information.

A *consequentia simplex de inesse* is a Philonian conditional proposition which refers to the possible in the sense of that which is not known to omniscience to be false. In other words, in such a conditional, the *possible* is merely that which is true. That is to say, supposing the proposition "If *A* is true then *B* is true" to be understood as a *consequentia simplex de inesse*, it merely means, Either *A* is not true or *B* is true. This proposition is true if *B* is true, and also if *A* is false; it is false only if *A* is true while *B* is false.

The *consequentia de inesse* is, according to the analysis which I have long defended, by far the most important of logical forms, because upon the one hand it is the essential element of the illative relation, out of which logic springs, while upon the other hand every other form of proposition is expressible as a combination of *consequentiae simplices de inesse.** (*I wish this position might be vigorously assailed by some powerful and perspicuous logician of the exact school; for I feel that there is another side of the question to which justice has not been done; and until that side has been examined, a shadow of doubt rests upon the doctrine in my mind.) The antecedent and consequent of a *consequentia de inesse* may be any propositions whatever. They may, therefore, themselves be consequences *de inesse*. In fact, every proposition may be regarded as a consequence *de inesse*, provided we admit the possibility that the consequent may be such as to have no effect. Suppose the consequence *de inesse*, "If *A* is true, then *B* is true," have such a consequent *B*, that this branch of the consequence has no effect. Then, *B* must be a proposition known to be false. The whole consequence will, in that case, amount to a denial that *A* is true. For to assert, as a consequence *de inesse*, that if a given opinion is true, then what is false is true, [is] no more than a way of saying that that opinion is false. Any ordinary Philonian consequence which should assert that if a given opinion were true, then something known to be false would be true, would assert that the given opinion had no place in the range of possibility. But the range of possibility of a consequence *de inesse* is merely the actual state of things; so that to deny the possibility of the truth of an opinion in this sense of the word possibility is merely to deny the truth of the opinion. Thus, by imagining a consequence *de inesse* whose consequent has no effect, we have imagined in such a consequent, a proposition known to be false; and by taking this proposition as the consequent of a consequence *de inesse*, with any proposition we please as antecedent, we express the denial of that proposition in the form of a consequence *de inesse*. Let a proposition known to be false be termed a *zero proposition*. All such propositions are logically equivalent; for two propositions are logically equivalent if neither can possibly be true while the other is false. Hence, in logic all *zero* propositions may be regarded as the same proposition which may be termed the *zero* proposition; and we may simply write *zero* in place of the absurd assertion "the zero proposition is true."

Next suppose a consequence *de inesse* of which the antecedent is any proposition whatever, A, while its consequent is a consequence *de inesse* having the zero proposition for *its* consequent. Thus, the whole proposition will be of the form "If A is true, then if B is true, zero." This proposition is true if A is false and also if B is false; but it is false if both A and B are true. Let this whole proposition be made the antecedent of a consequence *de inesse* of which the consequent is zero, and the result is equivalent to the assertion that both A and B are true. We thus have the means of asserting two or more propositions simultaneously by means of the single logical form of the *consequentia de inesse*.

An ordinary Philonian conditional proposition, having a range of possibility and asserting that—"If A is true, B is true," that is, that in every possible state of things in which A is true, B is likewise true, may be regarded as a simultaneous assertion of a multitude of propositions each asserting that if a single state of things in which A is true, B is true, each therefore being a *consequentia de inesse*.

A *verb*, being understood in a generalized sense, may be defined as something logically equivalent to a word or combination of words, either making a complete proposition, or having certain *blanks*, or quasi-omissions, which being filled each with a proper name, will make the verb a complete proposition. According to the number of blanks the verb may be termed a *medad* ($\mu\eta\delta\epsilon\nu\acute{o}\varsigma$ + -$\acute{\alpha}\delta\alpha$), *monad, dyad, triad*, etc.

Every proposition may be regarded as a Philonian hypothetical concerning the filling of the blanks of verb. For example, the proposition, "No woman is adored by all Catholics" may be expressed as follows: If the second blank of the verb "—adores—" be filled with the name of a woman, then the first blank may afterward be filled with the name of a Catholic, so as to make the proposition false.

Thus, every proposition or system of propositions may be regarded as a simple or complex consequence *de inesse*; and consequently the forms of logical deduction from consequences *de inesse* are all the forms of logical deduction possible.

Now, it can be strictly demonstrated that the only forms of deduction from simple or complex consequences *de inesse* are included in the enumeration which here follows.

In order to describe these forms it will be useful to define certain *orders of antecedentals*. An *antecedental of the Nth order* belonging to any consequence is a proposition forming a part of that consequence and such that to describe its exact place in the consequence it is necessary to use the word antecedent N times. Thus, the consequent, the consequent of the consequent, the consequent of the consequent of the consequent, etc. are antecedentals of the 0 order. The antecedent, the consequent of the antecedent, the antecedent of the consequent, the consequent of the antecedent of the consequent, etc. are antecedentals of the 1st order. The antecedent of the antecedent, the consequent of the antecedent of the consequent of the consequent of the antecedent, etc. are antecedentals of the 2nd order. And so on.

It will also be useful to term the antecedent of a consequence *de inesse* together with the antecedent of its consequent, the antecedent of the consequent of its consequent, the antecedent of the consequent of the consequent of its consequent, and so on, as far as the analysis is carried, and finally together with that antecedental of zero order whose parts are not separately considered, the *first constituents*, or the *constituents of the first degree* of the whole consequence. The first constituents of the first constituents may be termed the *second constituents*, or the *constituents of the second degree*, and so forth. The whole consequence will be its own *constituent of the zero degree*.

It will also be useful to write "If A is true, then B is true," considered as a consequence *de inesse*, thus:

$$A \curlyvee B;$$

"If it be true that if A is true B is true, then it is true that if C is true D is true," where all the "ifs" signify consequences *de inesse* will be written

$$(A \curlyvee B) \curlyvee (C) \curlyvee (D).$$

The rule will be to enclose all constituents of the same degree in parentheses, brackets, braces or other enclosing marks of the same shape.

Art. 8. The different modes of deductive inference may now be enumerated.

The first mode is *inference by omission*. The rule is that in any consequence *de inesse* any constituent of even degree can be illatively omitted. If the omitted constituent be a consequent its place must be formally filled by the zero proposition. For example, from

$$(A \curlyvee B) \curlyvee (C),$$

we may infer by omission either

$$B \curlyvee C \qquad \text{or} \qquad (A \curlyvee 0) \curlyvee C$$

The second mode of necessary deduction is *inference by insertion*. The rule is that in any consequence *de inesse*, in place of any antecedental of even order may illatively be substituted any consequence *de inesse* that has that antecedental for its consequent. For example, from $A \;⤙\; B$, we may infer either $(C) \;⤙\; (A \;⤙\; B)$ or $(A) \;⤙\; (C \;⤙\; B)$.

The third mode of necessary deduction is *inference by iteration*. The rule is that in any odd constituent, x, of any consequence *de inesse*, substitution may illatively be made, the place of any even constituent, y, of x being taken by a consequence *de inesse* composed of y and of any first constituent, z, of x in such a manner that the antecedentalities of x and of y in the conclusion shall agree in being odd or even with their antecedentalities in the premise and that their combination shall be an odd constituent of the whole proposition. Namely, the forms of combination are as follows

Antecedentality of z	of y	Combination
Even	Odd	$z \;⤙\; y$
Odd	Odd	$(z \;⤙\; 0) \;⤙\; (y)$
Odd	Even	$(y \;⤙\; z) \;⤙\; (0)$
Even	Even	$[(y) \;⤙\; (z \;⤙\; 0)] \;⤙\; [0]$

For example, suppose the following consequence *de inesse* to have been accepted: $(\{[(A \;⤙\; B) \;⤙\; (C)] \;⤙\; [D]\} \;⤙\; \{[E] \;⤙\; [F]\}) \;⤙\; (G)$. The antecedent of this is a constituent of odd order. Call it x. The antecedent of the antecedent of the antecedent of x is $A \;⤙\; B$. Hence, A and B are even constituents of x. A is of odd antecedentality in the whole proposition, B of even. The first constituents of x besides its antecedent are E and F, the former of even, the latter of odd antecedentality in the whole proposition. Hence, $A \;⤙\; B$ may be replaced, either by $(E \;⤙\; A) \;⤙\; (B)$, or by $[(F \;⤙\; 0) \;⤙\; (A)] \;⤙\; [B]$, or by $[A] \;⤙\; [(B \;⤙\; F) \;⤙\; (0)]$, or by $\{A\} \;⤙\; \{[(B) \;⤙\; (E \;⤙\; 0)] \;⤙\; [0]\}$, **etc.**

If a verb is thus iterated, its blanks must in the iteration be filled by signs of the same individuals as in the place of its occurrence in the premise.

The fourth mode of necessary deduction is *inference by deiteration*. The rule is that if in a consequence *de inesse* any constituent, x, of an even constituent, y, is exactly like a first constituent, z, of y, (and if it is a verb, has its blanks filled by signs of the same objects) and if the orders of antecedentality of x and z agree in being both even or both odd, then x can be illatively omitted. If the term omitted is a consequent, its place is to be formally filled by zero. For example, suppose it admitted that (A) ᴄ $(B$ ᴄ $C)$. Then, by insertion, we infer $[B]$ ᴄ $[(A)$ ᴄ $(B$ ᴄ $C)]$. Here the two Bs are first constituents of the zero constituent, and both are first antecedentals. Hence, the second can be illatively omitted by deiteration, and we finally infer $[B]$ ᴄ $[(A)$ ᴄ $(C)]$. Since (A) ᴄ $(B$ ᴄ $C)$ and (B) ᴄ $(A$ ᴄ $C)$ are thus logically equivalent, we may write A ᴄ B ᴄ C, the enclosure of the consequent being misleading. But $(A$ ᴄ $B)$ ᴄ (C) is different. Again, suppose it to be admitted that $[(A$ ᴄ $B)$ ᴄ $(C)]$ ᴄ $[D]$. Then, by insertion, we can infer that $[A]$ ᴄ $[(A$ ᴄ $B)$ ᴄ $(C)]$ ᴄ $[D]$. Here, the first A is a first constituent of the zero constituent, and is a first antecedental, while the other A is a third antecedental. Hence, the latter may be illatively omitted by deiteration, and we infer $[A]$ ᴄ $[(B)$ ᴄ $(C)]$ ᴄ $[D]$. This example illustrates one of the principal illative elements of syllogism.

The fifth mode of necessary deduction is *inference from contradiction*. The rule is that if in a consequence *de inesse* any constituent, x, of an odd constituent, y, is exactly like a first constituent, z, of y (and if it be a verb, has its blanks filled by signs of the same individual objects) and if, of x and z, one is an even and the other an odd antecedental, then the whole of that constituent of highest odd degree of which x forms a part (or the whole) can be illatively omitted. But if that constituent is a consequent, its place must be formally filled by the zero proposition. Suppose, for example, that we have $\{[(A$ ᴄ $A)$ ᴄ $(B)]$ ᴄ $[C]\}$ ᴄ $\{D\}$. Then we can infer $\{[B]$ ᴄ $[C]\}$ ᴄ $\{D\}$.

The sixth mode of necessary deduction is *inference by excluded middle*. The rule is that if two first constituents of an even constituent of a consequence *de inesse* are precisely alike (and if verbs, have their blanks filled by signs of the same individual objects), but if one is an even and the other an odd antecedental, then that whole even constituent can be illatively omitted. But if it is a consequent, its place must be formally filled with *zero*. For example, given

$$\{[(A)\ ᴄ\ (B\ ᴄ\ A)]\ ᴄ\ [C]\}\ ᴄ\ \{D\},$$

the inference is valid to

$$\{C\}\ ᴄ\ \{D\}.$$

The remaining modes of necessary deduction are modes of compounding relations. They are divisible into two kinds, those which concern relations which necessarily enter into the forms of propositions and those which concern other relations which are still such that no information has to be sought out in order to ascertain the effects of compounding them. The former are *logical relations*, the latter *mathematical relations*. The above six modes of inference depend upon the compounding of the relation of antecedent and consequent, which need not be further considered by itself. But since it has been found necessary to introduce the zero proposition as a consequent, thus producing denial, it is necessary to take account of combinations of the relation of contradiction.

This gives the seventh mode of necessary deduction, which is *inference by commutation*. The rule is that if any constituent of a consequence *de inesse* has zero for the consequent of its antecedent, its consequent and the antecedent of its antecedent may be illatively made to change places. For example, if it be admitted that $(A \mathrel{⤙} 0) \mathrel{⤙} B$, then the inference is valid that $(B \mathrel{⤙} 0) \mathrel{⤙} A$. I will now give an example which illustrates almost all the above modes of inference. Suppose we admit the two propositions $A \mathrel{⤙} B$ and $B \mathrel{⤙} C$. Colligating these we get $[(A \mathrel{⤙} B) \mathrel{⤙} (B \mathrel{⤙} C) \mathrel{⤙} (0)] \mathrel{⤙} [0]$. Thence, we can infer by insertion $[A] \mathrel{⤙} [(A \mathrel{⤙} B) \mathrel{⤙} (B \mathrel{⤙} C) \mathrel{⤙} (0)] \mathrel{⤙} [0]$; for this is a consequence having the premise (which is a zero antecedental of itself) as consequent. Thence, we can, secondly, infer by deiteration $[A] \mathrel{⤙} [(B) \mathrel{⤙} (B \mathrel{⤙} C) \mathrel{⤙} (0)] \mathrel{⤙} [0]$; for the first A in the premise is a first constituent of the zero constituent of that premise and is a first antecedental, while the second A is also a constituent of the zero constituent and is a third antecedental (being the antecedent of the antecedent of the antecedent of the consequent of the consequence), and the conclusion results from omitting this second A. Thence, we can, thirdly, infer by iteration $\{A\} \mathrel{⤙} \{[B] \mathrel{⤙} [(B \mathrel{⤙} B) \mathrel{⤙} (C)] \mathrel{⤙} [0]\} \mathrel{⤙} \{0\}$; for taking as the x of the rule, the first constituent $(B) \mathrel{⤙} (B \mathrel{⤙} C) \mathrel{⤙} (0)$ of the consequence, as the y of the rule, the second constituent of x, the second B, and as the z of the rule the first B, which is a first constituent of x, z is a second antecedental, being the antecedent of the antecedent of the consequent of the consequence, while y is a third antecedental, being the antecedent of the antecedent of the consequent of the antecedent of the consequent of the consequence. Hence, by the rule $B \mathrel{⤙} B$ can be illatively substituted for the second B, which produces the conclusion. Thence, we can, fourthly, infer by omission $\{A\} \mathrel{⤙} \{[(B \mathrel{⤙} B) \mathrel{⤙} (C)] \mathrel{⤙} [0]\} \mathrel{⤙} \{0\}$; for the first B is a second constituent of the consequence, and its omission gives the conclusion. Thence, we can, fifthly, from contradiction, infer $\{A\} \mathrel{⤙} \{[C] \mathrel{⤙} [0]\} \mathrel{⤙} \{0\}$; for we may take as the y of the rule, $B \mathrel{⤙} B$, which is a third constituent of the consequence, and as x and z, the two Bs, which are both first constituents of y, one being a third and the other a fourth antecedental of the consequence, being the antecedent and consequent of the antecedent of the antecedent of the antecedent of the consequent of the consequence. Then the constituent of highest odd degree of which x forms a part is $B \mathrel{⤙} B$, and consequently this can be omitted. This omission produces the conclusion. Thence, sixthly, by commutation, we can infer $\{A\}$ ⦚ $\{[0]$ ⦚ $[0]\}$ ⦚ $\{C\}$; for C is the A of the rule and the second zero is the B of the rule. Thence, seventhly, from contradiction, we can infer $\{A\}$ ⦚ $\{C\}$. Thus, the syllogism is analyzed into seven virtual steps.

There are also inferences depending upon identifying subjects of different verbs or upon abstracting from identifications. Namely, the eighth mode of necessary deduction. . . .

SKETCH OF DICHOTOMIC MATHEMATICS (4)

Scholium. The purposes of this sketch are

1st, to put the reader into a position to understand what dichotomic mathematics is, why it is so called, and that it is the simplest possible mathematics and the foundation of all other mathematics;

2nd, to develop its main propositions and methods;

3rd, to exhibit in mathematical style the analysis of its foundations into their simplest formal elements, making clear the different categories of conceptions involved, and indicating the logical doctrine with which the subject is connected as well as can be done while strictly avoiding all direct logical discussion.

I shall prefix to each article a descriptive heading, inventing terms for the purpose when necessary, but as far as possible availing myself of those found in editions of Euclid's Elements; viz., Definition, Postulate, Axiom, Theorem and Demonstration, Problem and Solution, Corollary, and Scholium. The first three of these are technical terms whose signification could not be changed without a violation of the ethics of terminology. To the others, I consider myself at liberty to attach definite meanings not in violation of their present vague meanings.

A *Definition* is either *Nominal* or *Real*. A nominal definition merely explains the meaning of a term which is adopted for convenience. I shall not make separate articles for such definitions nor state them formally. For they do not affect the course of development of the thought. A *Real Definition* analyzes a conception. As Aristotle well says (and his authority is well-nigh absolute upon a question of logical terminology), a definition asserts the existence of nothing. A definition would consist of two members, of which the first should declare that any object to which the *definitum*, or defined term, should be applicable would possess the characters involved in the definition; while the second should declare that to any object which should possess those characters the definitum would be applicable. And any proposition consisting of two members of this description and really contributing to the development of the thought would be a Real Definition.

The term *Postulate* is carefully defined by Aristotle, whose acquaintance with the language of mathematics is guaranteed by the fact that he was long in Plato's school at the Academy, taken in connection with his known intellectual character; and Heiberg's recension of the Elements renders it manifest that Euclid used the term in the same sense. This meaning has become established in the English language, which has been peculiarly fortunate in inheriting much of the medieval scholastic precision, and is by far the best of the modern languages for the purposes of logic. To this meaning, therefore, the ethics of terminology peremptorily commands us to adhere. Long after Euclid a quite different sense came to be attached by some writers to the word. Namely, they took it to mean an indemonstrable practical proposition. The influence of Christian Wolff has caused the word *Postulat*, in German, to be generally used to signify the mental act of adhesion to an indemonstrable practical proposition. This usage is very abusive. Generally, the German treatment of logical terms is bad. Little of medieval precision of logic is traditional in Germany. I have read a great many German nineteenth-century treatises on logic;—probably fifty. But I have never yet found a single one, not even Schroeder's, which did not contain some unquestionable logical fallacy. It would be difficult to find a single instance of such a phenomenon in an English book by a sane author. German logical terminology was early vitiated by the national tendency toward subjectivism. It was further corrupted by the nominalism of Leibniz. With Hegel came chaos, and all the restraints of terminological ethics seemed to give way.

A *Postulate* would be a proposition necessary as a premiss for a course of deductive reasoning and predicating a contingent character of the hypothetical subject of that course of reasoning; and any proposition of that description would be a Postulate.

It seems, at first sight, as if it must be easy to distinguish between a fundamental assumption concerning the subject of a deductive reasoning, and a fundamental convention respecting the general signs employed in that reasoning. But the primary subject of a course of deduction is itself of the nature of a general sign, since there can be no necessary reasoning about real things except so far as it is assumed that certain general signs are applicable to them. It therefore becomes a delicate matter to draw a just line between the subject of the reasoning and the machinery of the reasoning. The subject of geometry will afford an illustration. Be it observed, in the first place, that it does not fall within the province of the mathematician to determine whether it is a fact that a plane has a line at infinity, as projective geometry assumes, or a point at infinity, as the theory of functions assumes, or whether or not any postulate of geometry is precisely true in fact. Consequently, his whole course of reasoning relates to an idea, or representation, or symbol. But everybody would agree that the description of a system of coördinates was a convention and not a postulate concerning pure space. A system of measurement, however, is of the same nature as a system of coordinates. For to the same space may be applied at will elliptical, hyperbolical, or parabolic measurement. If, therefore, the description of a system of coordinates is a convention and not a postulate, which is unquestionable, the same must be said of every proposition determining the system of measurement employed. Yet this is the nature of Euclid's fourth and fifth postulates. The truth is that the only true postulates of geometry are the topical postulates. I shall head every article describing the machinery to be employed in the discussion as a *Convention*.

A *Convention* would be a proposition concerning a subject which we imagine to exist as an aid in drawing conclusions concerning our main subject; and any such proposition would be a Convention.

The *Axioms* are called by Euclid *common notions*, κοιναὶ ἔννοιαι, and Aristotle occasionally applies the same designation to them. A little later than Euclid, the term common notion, κοινὴ ἔννοια, became a technical term with the stoics; and they undoubtedly meant by the appellation, "common," κοινή that they were common to all men, a part of the "common sense" from which the human mind cannot escape. It was the same idea, though less distinctly apprehended, perhaps, which Aristotle had in view. An axiom was a proposition which any learner would recognize as true, and which, since no doubt was entertained about it, could not but be assumed to be really true. But considering that pure mathematics or any pure deduction can deal only with a purely hypothetical state of things which is described, so far as it is supposed at all, in the postulates, how can we know anything about it, or what would it mean to say that anything was true of it, beyond what can be deduced from the postulates? What sort of a proposition could an axiom be? My reply to this question may be embodied in a definition, as follows:

Any *Axiom* would be a proposition not deducible from anything asserted in the definitions and postulates, but immediately evident in view of the facts that have been laid down and these are true.

In order to make the meaning of this clear, suppose that I make a judgment, that is, that I say to myself '*S* is *P*.' Some German logicians tell us that this is logically identical with saying to myself 'I know that *S* is *P*.' But this is not so. In the former proposition nothing whatever is said about me or my knowledge. And if the two propositions were identical, their denials would be identical; but the denial of one is, '*S* is not *P*,' while that of the other is the very different proposition 'I do not know that S is P.' But the truth is that, granting that S is P, then by the observed fact that I have said to myself that S is P, I perceive that it is true that I know that S is P. Thus, the enunciation of the definitions and postulates puts us into position to observe other facts than those which they assert; and I suppose that an axiom is of that nature, because there is no other nature that it can have. Besides, this is confirmed by a rather striking remark of Aristotle to the effect that it is not necessary in a demonstration to refer to an axiom. Why should it not be necessary to refer to a premiss? My definition affords the explanation. It is because at every step of a demonstration it is necessary to make similar observations in order to apprehend the force of the reasoning. A stupid person may admit almost in one breath the two premisses of a syllogism, and think them, too, and yet not see the truth of the conclusion until it is pointed out. To do that it is necessary to observe that that very character, M, which belongs to S, is the same character that carries with it P. The introduction of an axiom, therefore, spares the reader no difficulty, since to see the application of it is quite as difficult as to see the consequence for which the axiom was supposed to afford a stepping-stone. Axioms are, in truth, of no other use than that of showing what the foundations of theorems are: they render them neither more certain nor more clear. In a backhanded way, indeed, a good logician will make use of them; for where an axiom is cited there is a likely place to find a fallacy. Euclid in his first book uses his axiom that the whole is greater than its part seven times. In the 6th and 26th propositions it was not needed; and these propositions are strictly true. The 7th, 16th, 18th, 20th, and 24th propositions, where it is used, are, in consequence of this, untrue for triangles whose sides pass through infinity, or are untrue for elliptic space, or both. About as true as the axiom about the whole and part are those about equals subtracted from equals giving equals, etc., as if infinite lines cut from infinite lines would necessarily give equal finite remainders. Yet I have known a professed modern mathematician to cite this axiom to prove that infinitismals are strictly zero. Since he wrote this to me, a logician, he seems not to have dreamed that the argument was open to any objection. He voluntarily gave me permission to make any use of his letter; but I have no taste for seeing respectable people held up to ridicule.

Any *Corollary* (as I shall use the term) would be a proposition deduced directly from propositions already established without the use of any other construction than one necessarily suggested in apprehending the enunciation of the proposition; and any such proposition would be a Corollary.

The proof of a corollary should not only make it evident, but should show clearly upon what it depends. The proof should, therefore, never cite another corollary as premiss but should be drawn from postulates and definitions, as far as this can be done without a special construction.

Any *Theorem* (as I shall use this term) would be a proposition pronouncing, in effect, that were a general condition which it describes fulfilled, a certain result which it describes in a general way, except so far as it may refer to some object or set of objects supposed in the condition, will be impossible, this proposition being capable of demonstration from propositions previously established, but not without imagining something more than what the condition supposes to exist; and any such proposition would be a Theorem.

For example, the *Pons Asinorum* may be proved by first proving that a rigid triangle may be exactly superposed on the isosceles triangle, and that it may be turned over and reapplied to the same triangle. But since the enunciation of the *Pons* says nothing about such a thing; and since the *Pons* cannot be demonstrated without some such hypothesis, and since, moreover, the *Pons* does not pronounce anything to exist or to be possible, but only pronounces the inequality of the basal angles of an isosceles triangle to be *impossible*, it is a theorem. Perhaps, however, spatial *equality* could not be better defined than by saying that were space to be filled with a body called "rigid," which should be capable of continuous displacement, freely interpenetrating all other bodies, its rigidity consisting of the facts, first, that every film (or bounding *part* between *portions*) of it which at any instant (or absolutely determinate state of its displacement) should fully occupy any one of the surfaces of a certain continuous unlimited family of fixed and topically non-singular surfaces of which any three should have one, and only one, point in common, would at *any* instant fully occupy one of those surfaces, and second, that there should be one of those surfaces that should at all instants be fully occupied by the same film, then any two places, or parts of space, which could be fully occupied at different instants by one and the same part of that rigid body would be *equal*; and any two parts of space that would be equal would be capable of being so occupied, provided the rigid body had all the freedom of continuous displacement that was consistent with its rigidity. Were such a definition admitted, we may admit that the idea of a rigid triangle being turned over and reapplied to the fixed isosceles triangle would be so nearly suggested in the enunciation that the proposition might well be called a corollary. Perhaps when any branch of mathematics is worked up into its most perfect form all its theorems will be converted into corollaries. But it seems to be the business of mathematicians to discover new theorems, leaving the grinding of them down into corollaries to the logician.

There are propositions whose proofs are accomplished by means of constructions so well understood to be called for in proving propositions of the classes to which those propositions belong, that it is a delicate matter to determine whether they are best called corollaries or theorems. But though this were inherently impossible in some cases, my distinction between a corollary and a theorem would not thereby be proved ill-founded.

Sundry advantages not to be despised recommend the adoption of a canonical form of statement for all demonstrations of theorems, the conformity to the canon being subject in each case to such limitation as good sense may impose. The following is the canon of demonstration which I shall, in that sense, adopt:

1. The theorem having been enunciated in that form which is most convenient for carrying it in mind may, in the first place, be put by logical transformation, into such form as is most convenient for the purpose of demonstration.

2. Supposing the enunciation to be still in general terms, an icon, or diagram must next be created, representing the condition of the theorem, in the statement to which it has been brought in 1. At the same time, indices (usually, letters) must be attached to those parts of the icon which are to be made objects of attention in the demonstration, for the purpose of identifying them.

3. Next we must state the *ecthesis*, which is that proposition, which, in order to prove the theorem, it will manifestly be necessary and sufficient to show will be true of the icon created and of every equivalent icon.

4. Such additions to the icon as may be needful must now be created and supplied with indices.

5. It must be proved that these additions are possible. If the postulates (with the aid of previously solved problems) expressly render them so, this part of the demonstration will involve no peculiar difficulty. Even when this is not the case the possibility may be axiomatically evident. Otherwise it can only be proved by means of analogy with some experience; and to render the sufficiency of this analogy evident will be difficult. The logic of substantive possibility will be of assistance here.

6. It will be convenient at this point to make all the applications of previous propositions that are to be made to parts of the augmented icon, [the] several immediate consequences being set down and numbered, or otherwise indicated. Of course, the interpreter of the demonstration will have to observe for himself that the previous propositions do apply; but unobvious cases should be elucidated by the demonstrator.

7. The demonstrator will now call attention to logical relations between certain of the numbered propositions which lead to new propositions which will be numbered and added to the list, until the ecthesis is reproduced.

8. When this proposition is reached, attention will be drawn to its identity with the ecthesis by means of the letters Q.E.D. or something else equivalent to Euclid's phrase, ὅπερ ἔδει δεῖξαι. Euclid frequently employs the same phrase when he reaches the general enunciation itself without any ecthesis. But it would be more correct in such a case to vary the phrase, since the logical situation is not the same. I shall say "which is our theorem," or, since a repetition of the enunciation is needless and fatiguing, "whence, or from such and such propositions, our theorem directly follows."

If I wish to employ the *reductio ad absurdum*, the first article of the demonstration will consist in throwing the theorem into the form, "To suppose so and so leads to a contradiction," and my ecthesis will state definitely what contradiction it is that I shall find it convenient to bring out. Since a hypothesis which involves contradiction may be shown to be contradictory in any feature that may be chosen, the ecthesis in this case is determined by convenience.

Any *Problem* (as I shall use the term) would be a proposition pronouncing, in effect, that under circumstances known to exist or to be possible, a certain sort of result, described partly, at least, in a general way, exists or is possible, this vague proposition being usually, though not essentially capable of definition, showing under what general conditions it always will be realized, and at any rate, being capable of demonstration from propositions previously established, but only by means of a hypothesis not immediately suggested in the vague enunciation; and any proposition of that description would be a Problem.

This definition certainly ventures to break somewhat with ordinary usage (for I do not think the word 'problem,' in its mathematico-logical sense, can be called a term of art), and appears to do so more than it does. It is true that as commonly used, it is supposed to describe a thing to be *done*. Every proposition has its practical aspect. If it means anything it will, on some possible occasion, determine the conduct of the person who accepts it. Without speaking of its acceptance, every proposition whatsoever, although it has no real existence but only a *being represented*, causes practical, even physical, facts. All that is made evident by the study which I call *speculative rhetoric*. But I do not think that the practical aspect of propositions is pertinent to the designation of classes of mathematical propositions. For example, Euclid's first proposition ought to read that upon any terminated right line as base some equilateral triangle is possible. He first makes this definite by describing precisely how the triangle is defined. There can be a circle of any centre and radius. Then there can be two circles with centres at the two ends of the line and with the length of the line as radius. Had Euclid looked at his proposition in this light, he could not well have failed to perceive the necessity for proving that the two circumferences (which, as usual, he calles circles) will intersect. He would have proved this by the definitions of *figure* and of *circle*, and by his third postulate.

So far, the difference between the ordinary acception of the term *problem* and that which I propose is trifling. But it frequently happens that [there is] a problem, or the vague statement that an object of a certain description exists, such as, a real root of the equation of an algebraic polynomial of odd degree involving but one unknown, while it still remains impossible to give an exact general description applicable to every such thing and to nothing else; which would constitute, as I should say, a *solution* of the problem. The distinction between such a proposition affirming existence or possibility, and a theorem, or proposition denying existence or possibility, is the most important of all logical distinctions between propositions; while the distinction between *knowing* and *ability to do* is a distinction quite irrelevant to logic. Considering further that the limitation of the word *problem* to practical matters is confined to geometry, while in every sense of the word the idea of vagueness attaches to it, encourages me to believe that I may venture, without blame, to use the word in the sense defined. . . .

THE FIRST THEORY

Definition 1.   Any *Blank* is a symbol which could not be vaguer than it is (although it may be so connected with a definite symbol as to form with it, a part of another partially definite symbol), yet which has a purpose.

Axiom 1.   It is the nature of every symbol to be blank in part.

Illustration.   The Library of the British Museum is a pretty full symbol; yet it is a blank as to what I had for breakfast this morning, or whether I had any.

Definition 2.   Any *Sheet* would be that element of an entire symbol which is the subject of whatever definiteness it may have; and any such element of an entire symbol would be a Sheet.

Remark.   I might, with more scientific logicality, have said that the *sheet* is the *matter* of a symbol, using "matter," by no means in the Kantian, but in the Aristotelian sense. But that would have brought with it the necessity of explaining what Matter is,—a difficult conception, at best, and complicated when we speak of the matter of a symbol. Nor could matter have been explained without first explaining what *Form* is. Form is quality, suchness,—red, for example. Even a color-blind person, who cannot see it, may be convinced that there is a suchness, *red.* The resolute metaphysician will alone deny it. Wherein is *red* what it is? Not in being felt. That would be an answer to a different question. The peculiar suchness of the feeling, wherein is that? It is wholly in itself. The quality or form is whatever it is in itself, irrespective of anything else. No embodiment of it in this or that object or feeling in any degree modifies the suchness. It is something positive in itself. The inclination to deny that there is any such thing arises from a feeling that to apply a noun to it, as if it were a thing, is to misrepresent it. This is true, in a sense. Not that there is any falsity in doing so; but it is looking at the category under a form belonging to another category. The suchness does not *exist*, but it is something definite. Neither does it consist in being represented. The being represented is one thing; the being represented such as *red* is represented, is another definite thing. It is general. It is an element of existing things; but it is not and has nothing to do with the element of existence. The suchness of *red* is such as it is in its own suchness, and in nothing else.

Matter, that something which is the subject of a fact, is, in every respect the contrary of form, except that both are elements of [the real world] that are independent of how they are represented to be. Form is not an existent. Matter is precisely that which exists. (Remember, that whether corporeal, or physical matter is, or is not, the only matter is beyond my present scope.) Form is definite. Whatever *red* is, it is of its very essence, and is nothing else. Matter is an element of something definite. But it is in itself, as the subject of that determination, vague. If the same matter cannot at one time be heavy and at another time light, that is because it is subject to a law. There is nothing in nature, as matter, to prevent, since, as matter, it is merely the subject of any characters it may possess. Laws prevent matter from having at once high atomic weight and great elasticity, and forbid numberless combinations of characters. But matter, in itself, as matter, is as favorable to one determination as another. We vaguely call all those characters which law forbids a *red* matter to have, as long as it is red, the name "not-red," and say that nothing can be red and not-red at once. But this is only a vague expression of law. Matter, in itself, has no aversion to contradiction. This is self-evident; yet thought, however evident, is so fallible that confirmation is welcome; and such confirmation we find in this case in the fact that a point on the boundary between a red and a green surface is at once red and green. If this is only true of a point and not of a surface, this is because when we speak of a surface, we do not speak of pure matter, but of whatever man is within the limits of a certain general regularity, or law. It is true one might ask why we might not say that the point on the boundary is neither red nor green. But the answer is that as individual it must be determined and there are only these two determinations open to it. For while Form is general and without self-identity, having no existence as Form, a matter, or *quoddam*, on the contrary, is individual. For if two matters, or *quoddams*, agree, it is in respect to some character. True they may be other than one another, and indeed, must be so in order to agree. There may seem to be a confusion here; but if there is any, it arises from not distinguishing vagueness from generality, definiteness from individuality. Definiteness is self-determination. Individuality is determination with reference to other. Form, as we have seen, is all that it is in itself. Matter being the subject of fact, and being nothing but the subject of a fact, is all that it is in reference to something else than itself. A quoddam is something which is whatever it is simply because it is nothing else. Being nothing else whatever, it must completely determinate, while it equally inclined to being determined in one way as another, so long as it remains other than everything else. It has absolutely nothing which belongs to it *per se* except its determined opposition to others. It is its reactions against those others which we have in mind when we say it exists. If we call experience all the thought which is forced upon us, matter is a word which serves merely to express the element of reaction which we find in experience, which by the nature of it can have place only between the members of a pair, and of which we have direct experience when we exert ourselves to push against anything. We express this experience by saying that a *quoddam* pushes against us. Then, if we see a red patch moving about, and others see it, too, we say that a *quoddam* is red. This is as good a formulation of the experience as any philosopher could construct. The contrast of this red patch to what surrounds it leads us to think there is something of the nature of resistance to the will there; and experiment will confirm this, generally speaking.

Aristotle held that Matter and Form were the only elements of experience. But he had an obscure conception of what he calls *entelechy*, which I take to be a groping for the recognition of a third element which I find clearly in experience. Indeed it is by far the most overt of the three. It was this that caused Aristotle to overlook it. Attention cannot be concentrated on that which covers the entire field; and this element is so universal that it is difficult to find a point of view from which there shall be any unquestionable contrast in this respect. Philosophers do right in using their utmost endeavors to form a conception of the universe with the smallest number of categories possible, and those the simplest possible. There is a sect of nominalism which, as I understand it, can only be regarded as an attempt to express the universe in terms of Form alone. If it could succeed, it would certainly be the simplest possible system of metaphysics. It is of all philosophies the most inadequate, and perhaps the most superficial, one is tempted to say the silliest possible. But its adherents are thinkers of power, and their work has been eminently valuable; for they have made the futility of their own theory most manifest. In the concrete, it is, however, difficult to distinguish this sect from another variety of nominalism which endeavors to express the universe in terms of Matter alone. This is individualism. This is another useful but futile attempt. Aristotle, so far as he is a nominalist, and he may, I think, be described as a nominalist with vague intimations of realism, endeavors to express the universe in terms of Matter and Form alone. It is important to recognize how much may be adequately so expressed. So far as the real world, as distinguished from the figments more or less natural to men, consists of a mere heap of blind facts, it can be expressed in terms of Matter and Form. All nominalists emphasize a real and true world which they contrast with a lot of more or less deceptive notions. It may be remarked that if, as I hold, there are three categories, Form, Matter, and Entelechy, then there will naturally be seven schools of philosophy; that which recognizes Form alone, that which recognizes Form and Matter alone, that which recognizes Matter alone (these being the three kinds of nominalism); that which recognizes Matter and Entelechy alone; that which recognizes Entelechy alone (which seems to me what a perfectly consistent Hegelianism would be); that which recognizes Entelechy and Form alone (these last three being the kinds of imperfect realism); and finally the true philosophy which recognizes Form, Matter, and Entelechy. The imperfect forms of realism have this manifest superiority over nominalism, that they are not obliged to maintain the wonderful position that any hypothesis about the world can be false in the sense of involving elements the like of which does not exist or really be at all. But they are obliged to say that something can present itself, or seem to present itself to the mind, which is utterly unlike anything real and true, not by containing a positive element that the truth lacks, but by not containing something that the truth contains. The philosopher who recognizes all three can alone take the same position that nothing can be fundamentally and of its nature unlike the truth.

This *Entelechy*, the third element which it is requisite to acknowledge besides Matter and Form, is that which brings things together. It is the element which is prominent in such ideas as Plan, Cause, and Law. The philosopher who recognizes only Form, will do best to insist that Form fulfills this uniting function by virtue of its generality. But it is not so; since Form remains entirely within its own self. Moreover, Plan, Cause, and Law all suppose actual events. Now there must be something besides Form in an event; since Form is immutable, because it is all that it is in its own nature. Now it is precisely in the event that the Bringing Together takes place. If, however, we take into account Matter as well as Form, we have undoubtedly connexion. For Matter consists in reaction; and an agreement or connection is nothing but a special mode of reaction, or opposition. It is of the nature of Matter to be determinable by Form; and a single fact is nothing but Matter determined by Form. No third element of connexion is there; for Matter is nothing but the connected,—that which what it is in reference to another. Nor does the idea of a single event involve anything but the determination of Matter by Form and the reaction of Matter with Matter. A succession of events, then, will involve nothing further. No doubt, the ideas of Plan, Cause, and Law involve connection which cannot be reduced to mere dualistic relations. But the nominalist will simply say that these ideas are illusory; that there are no such things in the real world. What we call Law really consists merely in the circumstance that, taking any two facts of a certain description, they are alike in a certain respect; and Causation in like manner is nothing but the succession of a second upon a first, invariably. There are in reality no other connections than these dualistic connections. The rest is metaphysical nonsense. If the realist objects that ideas are something, and that they cannot take aspects that have no being, the nominalist ought to deny that there are even any such ideas as Law and Cause, otherwise than in the sense of Uniformity and Antecedent. That the semblance of an idea is necessarily an idea he must deny. But if one is going to get rid of modes of connexion by calling them unreal, why not at once say that all connexion is unreal, and say that the world is really made up of individual facts consisting of pure Matter, in the sense of that which is individual, and that these are in themselves isolated, it being merely our way of looking at them which, as it were, projects one upon another? Kant bears traces of having passed through this phase of opinion; and indeed he never altogether emancipated himself from it; although his "deductions" constitute a quite needlessly indirect refutation of it. Any logically healthy mind must instinctively feel that this position is unsound; as a scientific logic will show it to be without resort to Kant's subjective method. Some Kantians, by the way, hold that Kant's argument is not subjective; and perhaps they are right in a certain sense; but Kant plainly understood it to be so.

My principal object in glancing at those three forms of nominalism has been to bring out the truth that there are two sorts of connection which do not involve anything but Matter and Form; namely, the determination of Matter by Form, and the blind reaction of Matter with Matter.

There are, however, forms of connexion of which this is not true. Such is the action of a sign in bringing its interpreter into relation with its object. Indeed, if we fully set before ourselves all that is involved in this action, we shall see that *signification*, meaning the action of a sign, covers all connexions of this description. The nominalists are fond of insisting upon the distinction between words and things, between signs and realities. Now it is very true that a word is not a thing, and there is a sense in which a sign is not a reality; although in another sense the very entelechy of reality is of the nature of a sign. One can hardly glance down a printed page without seeing a number of things, or individual objects, determined like this: the. These "replicas," as I shall call them, embody one and the same word. This one word is not an individual object. No more is it a thought, if by a "thought" be meant an individual act of the mind. Not being individual, it is not *Matter*. Nor is it, properly speaking, Form. For instead of being what it is of itself, and remaining altogether such as it is even if not connected with matter, the sign's mode of being is, on the contrary, such that it consists in the existence of replicas destined to bring its interpreter into relation to some object. A Form is a quality or character. The sign is usually related to a character as a part of its object: it plainly is not itself that character. The sign has its essential significant character, which is the character of causing the interpretation of its object. But the quality of an act and the act itself are utterly different natures. It is in causing the interpretation that the being of the sign consists. Let us examine this operation of interpretation. I shall show below that we ought not to think that what are signs to us are the only signs; but we have to judge signs in general by these. What then is it for us to interpret a sign? Let us take first the example of a very perfect sign, say a demonstration of Euclid; for an entire demonstration is a sign. What is it to interpret this? Probably the first attempt at an answer will be that it is the immediate feeling of the force of the demonstration. But no; we ought to take the whole effect of the sign, or at least the whole of the essential part of it, to be the interpretation, so long as we have used the word interpretation to mean that effect the causing of which constitutes the being of the sign. Now it is not in any feeling or even in any particular act of thought that that effect consists but in the *belief*, with all that the belief essentially effects. Belief does not principally consist in any particular act of thought, but in a *habit* of thought and a conduct. A man does not necessarily believe what he thinks he believes. He only believes what he deliberately adopts and is ready to make a habit of conduct. It would thus be plain enough that a sign, as a sign, produces physical effects, even if there were any other way than that for one mind to communicate with another, and even if the action of the will were not the most important fact in the world. It is not surprising that the legitimate induction of physics that every acceleration is immediately caused by physical circumstances should be supposed to contradict the action of signs upon matter, although it is not really so. For if the air feels close in my study; and I conclude that the window should be open; but remark that if the window is to be opened, I must open it; and if I am to open the window I must cross the room; and if I am to cross the room, I must rise from my chair; and if I am to rise from my chair, I must first bend forward; I soon lose sight of what takes place in my mind, and when I come to myself the action is taking place. Now it is quite possible that in that short interval there was an endless series of conclusions followed by a beginningless series of physical events. For instance, it may be that matter consists of atoms, each a vortex in a fluid, itself consisting of atoms each a vortex in a subtler fluid and so on *ad infinitum*, and it may be that effects in all these preceded the molar accelerations. No matter how baseless this hypothesis may be, it still serves to show that that which seems to involve a contradiction does not truly do so. It is not surprising that this should be generally overlooked; but what is remarkable, not to say comical, is that the dynamical theory, which, however well founded, is only a theory, should be allowed to override the very plainest and most important fact in the world. As for the doctrine now in vogue of psycho-physical parallelism, even if, instead of the vague flotsam it is, it were definitely shown to be adaptable to the phenomena, and though there were facts that seem specially to point to it, instead of its resting mainly on the absence of any better theory, still it would remain unacceptable for the reason that while the only possible justification of any hypothesis is that it renders facts intelligible that without it could not be made so, this hypothesis is marked by the peculiarity of misrepresenting the only thing that is *per se* intelligible, that is to say reasoning and signification in general, in such a way as to make its intelligibility to appear illusory and to suppose it to be such as to be forever hopelessly and absolutely unintelligible.

There are certain Forms of which we are immediately aware, as Red, Beauty, Pitifulness, etc. No metaphysician can explain away this fact by any account of how they come about, however true it may be; for he will leave their peculiar suchnesses untouched; and suchnesses are all that they are. Having these indubitable examples of suchness we suppose a thousand others of which we have no direct consciousness. We find that a diamond cannot be scratched. We ask ourselves,

Why not? There must be some *reason*. But a reason is founded on some general condition. It cannot have for its condition that it is *this* thing and not *that*. If there is any reason for the diamond's being hard, it must be that it has a peculiar quality which, being of a general nature, may be the condition of the reason. If on the other hand, there is no reason, it is either arbitrary chance that this diamond never submits to being scratched, or else it possesses hardness as its quality. It is good reasoning to say that it is not arbitrary chance, and that the diamond has a quality. But if we ask ourselves what do we mean by a quality; we can only answer that it is something without us like *red* within us, in that it is what it is in itself, without reference to anything else. True, the *possession* of the quality is supposed, *by virtue of the law*, or otherwise, to cause certain reactions between the diamond and any stone used to compare with it. But the quality itself does not consist in that reaction, since it continues though the reaction ceases. Nay the *quality* continues though every diamond be burned up. One must needs be as reason-proof as the stiffest metaphysician to deny the being as suchnesses of other qualities than those that we immediately feel.

Of Matter, too, we have direct experience in pushing against an obstacle in the dark, and in many other ways. As with Form, we presume the existence of matter which we do not directly experience. Many writers seem to think that they eliminate Matter by saying that if a ball, for example, is red, round, rough, rigid, resilient, etc., the subject of these qualities is merely the reunion of them. But they do not at all get rid of Matter in that way; they only make the dualistic nature of it a little plainer. Qualities, suchnesses, *per se*, cannot be reunited. It is only the occasion, the Matter, which is determined by them which may inaccurately be said to reunite the qualities, although, in fact, it produces no effect upon them whatever. But they are quite right in their implication that the existence of the Matter consists in its reference to the Forms and to other Matter.

It is the same with Signs, or *Entelechies*. Some address themselves to us, so that we fully apprehend them. But it is a paralyzed reason that does not acknowledge others that are not directly addressed to us, and that does not suppose still others of which we know nothing definitely. We see that by the action of reason and will, that is, by the action of a sign, matter becomes determined to a Form; and we infer that wherever Matter becomes determined to a Form it is through a sign. Much that happens certainly happens according to Natural Law; and what is this Law but something whose being consists in its determining Matter to Form in a certain way? Many metaphysicians will answer that Law does not *make* Matter to become determined to Form but only *recognizes* in a general way, that which happens quite independently. But do these men mean to say that it is merely by chance that all stones allowed to drop have hitherto fallen? If so, there is no reason to suppose that it will be so with the next stone we may let loose. To say that would be to paralyze reason. But if it is not mere chance then evidently it has some cause or reason. To say that this is a sign is merely to say that it has its being in producing that union of Form and Matter. Why suppose it has any further being, especially since in order to do so, you must evolve a conception that the human mind has never possessed? You might talk of such a thing, but think it you could not. Nor does anybody propose that. Those who hate so to admit that anything of the nature of a sign can act upon matter imagine that they can express the phenomena with less, and do not dream of insisting upon more. Less and more are equally impossible.

I have not explained, however, why I call the sign the *Entelechy*, or perfectionment, of reality. Though approximate explanation will be easy: the exacter examination must be postponed a little longer. We have seen that a law is a sign. Take away from nature all conformity to law, all regularity and what have we? A chaos? No matter would possess any property, nor even an accident for any duration however short. Indeed there would be no time. Form and Matter would be almost disunited. Properly speaking there would be no reality at all. The true and perfect reality, the very thing, is the thing as it might be truly represented, as it would be truly represented were thought carried to its last perfection. As a perissid curve passing further and further toward the positive side traversed infinity and appears coming nearer and nearer from the negative side, so thought passing always from object to interpretation at its extremest point reaches the absolute reality of objectivity. The real and true thing is the thing as it might be known to be.

It is to be observed that a sign has its being in the *power* to bring about a determination of a Matter to a Form, not in an *act* of bringing it about. These are several good arguments to show that this is the case. Perhaps none of them is more conclusive than the circumstance that there is no such act. For an act has a Matter as its subject. It is the union of Matter and Form. But a sign is not Matter. An act is individual. The sign only exists in replicas. The sign has its being in being represented; and it is absurd to say that that which has such a mode of being immediately affects real existence, which is essentially independent of how it is represented to be.

ON THE NUMBER OF DICHOTOMOUS DIVISIONS:
A PROBLEM IN PERMUTATIONS (74)

In the calculus of logic, a proposition is separated by its copula, $-\!\!\prec$, into two parts, as $A-\!\!\prec B$. But these parts may again be separated in like manner, as $(A-\!\!\prec B)-\!\!\prec C$ and $A-\!\!\prec (B-\!\!\prec C)$, and so on indefinitely. It becomes pertinent to inquire how many such propositional forms with a given number of copulas there are. The same problem presents itself in general algebra, where $-\!\!\prec$ is replaced by any non-associative sign of operation; and, indeed, the question not unfrequently arises; but I do not know that the solution has been given.

We may consider a row of letters, A, B, C, etc. which we may call the ABC, separated into two parts by a punctuation mark, and each part (not consisting of a single letter) into two parts by a subordinate punctuation mark, and so on until all the letters are separated. I shall call the resulting form an *ABC-separation*. The following are examples

$A:B.C;D,E:F$
$A.B;C:D;E,F$

Let n be the number of punctuations; then, the number of letters will be $n + 1$. Let Fn be the number of ABC-separations with n punctuations, or say of n-point separations. Then, if i be the number of letters to the left of the highest punctuation, so that $n + 1 - i$ is the number to the right, the number of ABC-separations of the row to the left is $F(i-1)$ and the number of ABC-separations of the row to the right is $F(n-1)$, and the number of those ABC-separations of the total row in which the first punctuation has i letters to the left of it is $F(i-1) \cdot F(n-i)$. And the total number of n-point separations is

$$Fn = \sum_{1}^{n}{}_i F(i-1) \cdot F(n-i)$$

$$= F0 \cdot F(n-1) + F1 \cdot F(n-2) + \ldots + F(n-2) \cdot F1 +$$
$$+ F(n-1)F0.$$

And it is evident that $F0 = 1$; so that this formula gives

$$F1 = F0 \cdot F0 = 1 \cdot 1 = 1$$
$$F2 = F0 \cdot F1 + F1 \cdot F0 = 1 \cdot 1 + 1 \cdot 1 = 2$$
$$F3 = F0 \cdot F2 + F1 \cdot F1 + F2 \cdot F0 = 1 \cdot 2 + 1 \cdot 1 + 2 \cdot 1 = 5$$
$$F4 = F0 \cdot F3 + F1 \cdot F2 \cdot + F2 \cdot F1 + F3 \cdot F0 =$$
$$1 \cdot 5 + 1 \cdot 2 + 2 \cdot 1 + 5 \cdot 1 = 14$$

And so we should find $F5 = 42$, $F6 = 132$, $F7 = 429$, etc.

Taking the ratio of successive Fs, we find

$$\frac{F1}{F0} = 1, \frac{F2}{F1} = 2, \frac{F3}{F2} = \frac{5}{2}, \frac{F4}{F3} = \frac{14}{5}, \frac{F5}{F4} = 3, \frac{F6}{F5} = \frac{22}{7}.$$

Now, noticing that the last of these has its denominator equal to 7 and that next but one before it its denominator 5, we are naturally led to express the intermediate one as a fraction with 6 for its denominator. The three numerators are then 14, 18, 22, which are in arithmetical progression. If this holds good we should have $\dfrac{F3}{F2} = \dfrac{10}{4}, \dfrac{F2}{F1} = \dfrac{6}{3}, \dfrac{F1}{F0} = \dfrac{2}{2}$, and all these values are verified. We next predict that $\dfrac{F7}{F6} = \dfrac{26}{8}$, and

finding it to be so, we entertain a high degree of confidence that this is the general rule. For unless the true formula is of a degree of complexity which the simple nature of the problem prevents us from expecting, it can hardly agree with that we have found for the first seven ratios of the series without according completely. We have, then, as we may be morally sure,

$$Fn = \frac{(4n-2)(4n-6)(4n-10) \ldots 2}{(n+1)\,n\,(n-1) \ldots 2}$$
$$= 2^n \frac{(2n-1)(2n-3)(2n-5) \ldots 1}{(n+1)\,n\,(n-1) \ldots 1}$$
$$= \frac{2^n}{(n+1)\,n\,(n-1) \ldots 1} \cdot \frac{2n(2n-1)(2n-2)(2n-3) \ldots 1}{2n\,(2n-2) \ldots 2}$$
$$= \frac{1}{(n+1)\,n\,(n-1) \ldots 1} \cdot \frac{2n(2n-1)(2n-2) \ldots 1}{n(n-1)(n-2) \ldots 1}$$
$$= \frac{(2n)!}{n!\,(n+1)!}$$

It remains to demonstrate this mathematically.

Suppose that in writing the ABC-separation, before we insert any letters we set down all those punctuations that are superordinate to the first letter, that is, all which help to separate in regular order from left to right to the telltale row; when a letter is inserted in the former we will add a j (for jot) to the latter, and when a punctuation is added to the former we will add a *t* (for *tittle*) to the latter. Thus in writing the *ABC*-separation

$A:B.C;D,E:F$

so as to get successively

$A:.$
$A:B.$
$A:B.C;$:
$A:B.C;D,$:
$A:B.C;D,E:$
$A:B.C;D,E:F$

the telltale row would be

ttjjttjtjjj

and in writing $A.B;C:D;E,F$
the telltale row would be

tjttjjtjtjj

Such a telltale row shows us how to make the corresponding *ABC*-separation without any doubt or ambiguity. For the rule will simply be that in writing the *ABC*-separation, we must insert each punctuation and letter in the first vacant space from the left, every letter filling up the space up to the next punctuation to the right, and each punctuation dividing the space in which it is inserted into two. In regard to the grade of the punctuations, the rule simply is that each point inserted must be subordinate in grade to the point which produced the space in which it is inserted by dividing a larger space.

Not every row of *j*s and *t*s can be taken as representing an *ABC*-separation; but it is plain that the necessary and sufficient condition of its being able to do so is that, as it is written letter after letter, as soon as the number of *j*s exceeds that of the *t*s, the row must end. For then, plainly, all the spaces of the *ABC*-separation will be filled up. It is important to remark that this is the same as to say that if we take a row beginning with a *t* and having *j*s and *t*s alternately except that there are two *j*s at the end, as *tjtjtjtjj*. . . .

. . . . For it is clear that in bringing them into their last order, no *t* was ever carried across the pair of *j*s last spoken of, for that would have been contrary to rule. But we have seen that every arrangement of the *j*s and *t*s produced from the alternate order terminated with two *j*s by moving *t*s to the left is a possible telltale. Thus the first proposition is demonstrated.

To prove the second, take any possible telltale row. The properties of this row are, first that it has one more *j* than it has of *t*s, and second that taking any place in it between two letters, there are to the left of that place at least as many ts as js. There are consequently more js than ts to the right of any place between two letters. If therefore the tail of such a row is joined to its head to form a circle, and this circle is cut at another place, from this new beginning forward to the old beginning there are more js than ts, so that the row cannot be a telltale.

If, then we take all the arrangements of n ts and $n + 1$ js (which we know by a familiar proposition in permutations to be $\dfrac{(2n + 1)!}{n!(n + 1)!}$ in number), they may be distributed into classes each containing $2n + 1$ arrangements convertible into one another by joining to make a circle and cutting again, or say, cyclically identical. For the same arrangement cannot occur twice or more in one cycle. For if it did, all the arrangements of the cycle must recur, and one or other of the last two propositions would be violated. Now each of these classes contains just one possible telltale row and consequently the number of these, which is the number of ABC-separations, must be

$$\frac{(2n)!}{n!(n + 1)!}$$

It can readily be shown that this number contains any prime number. p, as many times as a factor as there are powers of p which being taken as divisors of $n + 1$ leave remainders at least one more than $\frac{1}{2}p$.

Thus, take $n = 13, n + 1 = 14$:

$14 \div 2 = 7$ exact		This yields no factor.
$14 \div 2^2 = 3\frac{2}{4}$	$2 < 1 + \frac{1}{2}\cdot 4$	This yields no factor.
$14 \div 2^3 = 1\frac{6}{8}$	$6 > 1 + \frac{1}{2}\cdot 8$	Makes 2 a factor.
$14 \div 2^4 = \frac{14}{16}$	$14 > 1 + \frac{1}{2}\cdot 16$	Makes 2 a factor.

Hence $F13$ has 2^2 as factor.

$14 \div 3 = 4\frac{2}{3}$	$2 < 1 + \frac{1}{2}\cdot 3$	This yields no factor.
$14 \div 3^2 = 1\frac{5}{9}$	$5 < 1 + \frac{1}{2}\cdot 9$	This yields no factor.
$14 \div 3^3 = \frac{14}{27}$	$14 < 1 + \frac{1}{2}\cdot 27$	This yields no factor.

Hence $F13$ does not contain 3 as a factor.

$14 \div 5 = 2\frac{4}{5}$	$4 > 1 + \frac{1}{2}\cdot 5$	Makes 5 a factor.
$14 \div 5^2 = \frac{14}{25}$	$14 > 1 + \frac{1}{2}\cdot 25$	Makes 5 a factor.

Hence $F13$ contains 5^2 as factor.

$14 \div 7 = 2$ exact		
$14 \div 7^2 = \frac{14}{49}$	$14 < 1 + \frac{1}{2}\cdot 49$	

Hence $F13$ does not contain 7 as factor.

$14 \div 11 = 1\frac{3}{11}$	$3 < 1 + \frac{1}{2}\cdot 11$	This yields no factor.

F13 does not contain 11 as factor.

$14 \div 13 = 1\frac{1}{13}$ $1 < 1 + \frac{1}{3} \cdot 13$ F13 does not contain 13 as factor.

$14 \div 17 = \frac{14}{17}$ $14 > 1 + \frac{1}{2} \cdot 17$ F13 contains 17 as factor.

$14 \div 19 = \frac{14}{19}$ $14 > 1 + \frac{1}{2} \cdot 19$ F13 contains 19 as factor.

$14 \div 23 = \frac{14}{23}$ $14 > 1 + \frac{1}{2} \cdot 23$ F13 contains 23 as factor and no higher prime factor.

Hence $F13 = 2^2 \cdot 5^2 \cdot 17 \cdot 19 \cdot 23.$[1]

---

[1] MS. 73 is related to this material but is too incomplete for publication.

17

THE CATEGORIES (717)

Art. 1. A triad is something more than a congeries of pairs. For example, *A* gives *B* to *C*. Here are three pairs: *A* parts with *B*, *C* receives *B*, *A* enriches *C*. But these three dual facts taken together do not make up the triple fact, which consist [s] in this that *A* parts with *B*, *C* receives *B*, *A* enriches *C*, *all in one act*. Take another illustration. There is a two-way mode of freedom of a particle on a line from *A* to *B*. But if there is a furcation of the line, so that it leads from *A* to *B* and *C* and from *B* to *A* and *C*, there is an essentially different feature. [See Fig. 1.] Thus, in triads we must expect to find peculiarities of which pairs give no hint.

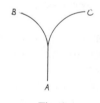

Fig. 1

Art. 2. Systems of more than three objects may be analyzed into congeries of triads. Thus, if a line branches from *O* five ways, so as to run separately to *A*, *B*, *C*, *D*, *E*; this result may be attained by supposing that at *P* there is a three-way node, two of whose branches run to *A* and *B*, while the third, infinitely short, runs to *Q*. [See Fig. 2.] At *Q* there

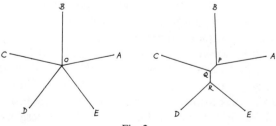

Fig. 2

is a three-way node, one of whose branches runs to *C*, while the other two, infinitely short, run to *P* and *R*. At *R* there is a three-way node, two of whose branches run to *D* and *E*, while the third, infinitely short, runs to *Q*. Thus, when the idea of branching three-ways is attained, we have all that is necessary to produce all higher orders of branching.

Art. 3.   The two principles contained in the last two articles explain a very important fact which has been remarked in making philosophical analyses, namely, that a given sort of phenomenon of any very general kind usually appears under three essentially different modifications according as it belongs to single objects, pairs, or triads, but presents nothing essentially new when carried to higher systems.

It will, at any rate, be found a most helpful maxim, in making philosophical analyses to consider, first, single objects, then pairs, last triads.

We have already applied this maxim in Article 1, where Cunning is that skill that resides only in the single persons, Wisdom is that which can be stated to others, Theory is that which can be fortified by means (observed that a *means*, or *medium*, is a *third*) of a reason.

Art. 4.   The above maxim crystallizes itself in the statement that there are three grand elementary formal ideas, as follows:

I. The *First*, or Original, expressed by the root AR. The plough goes first.

II. The *Second*, or Opponent, expressed by the root AN, as in Latin *in*, our *other*, and also more strongly, but with an idea of *success* in opposition, in AP, whence *ob*, apt, *opus*, *opes*, *optimus*, copy.

III. The *Third*, or Branching, or Mediation expressed by such roots as PAR, TAR, MA. These three ideas may be called the *Categories*.

Art. 5.   It is very useful in philosophizing to be already familiar with some of the principal disguises of the categories. We will glance at a few of them.

The idea of First is predominant in the ideas of freshness, life, freedom. The free is that which has not another behind it, determining its actions; but so far as the idea of the negation of another enters, the idea of another enters; and such negative idea must be put in the background, or else we cannot say that the Firstness is predominant. Freedom can only manifest itself in unlimited and uncontrolled variety and multiplicity; and thus the First becomes predominant in the ideas of measureless variety and multiplicity. It is the leading idea of Kant's "manifold of sense." But in Kant's synthetic unity, the idea of thirdness is predominant. It is an attained unity; and would better have been called Totality; for that is the one of his categories in which it finds a home. In the idea of Being, Firstness is predominant, not necessarily on account of the abstractness of that idea, but on account of its self-containedness.

It is not in being separated from qualities that firstness is most predominant, but in being something peculiar and idiosyncratic. The First is predominant in Feeling, as distinct from objective perception, will, and thought.

The idea of Second is predominant in the ideas of causation and of statical force. For cause and effect are two; and statical forces always occur between pairs. Constraint is a Secondness. In the flow of time in the mind, the past appears to act directly upon the future, its effect being called Memory, while the future only acts upon the past through the medium of Thirds. Phenomena of this sort in the outward world shall be considered below. In Sense and Will there are reactions of Secondness between the *ego* and the *non-ego* (which non-ego may be an object of direct consciousness). In will, the events leading up to the act are internal, and we say that we are agents more than patients. In sense, the antecedent events are not within us; and besides, the object of which we form a perception (though not that which immediately acts upon the nerves) remains unaffected. Consequently, we say that we are Patients, not Agents. In the idea of Reality, Secondness is predominant; for the real is that which insists upon forcing its way to recognition as something *other* than the mind's creation. (Remember that before the French word, *second*, was adopted into our language, *other* was merely the ordinal numeral corresponding to *two*.) The real is active: we acknowledge it, in calling it the actual. (This word is due to Aristotle's use of ἐνέργεια, action, to mean existence, as opposed to a mere germinal state.) Again, the kind of thought of those dualistic philosophers who are fond of laying down propositions as if there were only two alternatives, and no gradual shading off between them, as when they say that in trying to find a law in a phenomenon, I commit myself to the proposition, that law bears absolute sway in nature, such thought is marked by Secondness.

The ideas in which Thirdness is predominant are, as might be expected, more complicated, and mostly require careful analysis to be clearly apprehended; for ordinary, unenergetic thought slurs over this element as too difficult. There is all the more need of examining some of these ideas.

The easiest of those which are of philosophical interest is the idea of a sign, or representation. A sign stands *for* something *to* the idea which it produces, or modifies. Or, it is a vehicle conveying into the mind something from without. That for which it stands is called its *Object*; that which it conveys, its *Meaning*; and the idea to which it gives rise, its *Interpretant*. The object of a representation can be nothing but a representation of which the first representation is the interpretant. But an endless series of representations each representing the one behind it may be conceived to have an absolute object at its limit. The meaning of a representation can be nothing but a representation. In fact, it is nothing but the representation itself conceived as stripped of irrelevant clothing. But this clothing never can be completely stripped off; it is only changed for something more diaphanous. So there is an infinite regression here, too. Finally, the interpretant is nothing but another representation to which the torch of truth is handed along; and as representation, it has its interpretant again. [Lo] another infinite series.

Some of the ideas of prominent Thirdness which require closer study, preliminary to philosophy, are Continuity, Diffusion, Growth, and Intelligence.

Art. 6.   Let us examine the idea of continuity, the most difficult, the most important, the most worth study of all philosophical ideas. We say that time flows on in an unbroken stream; in what does the unbroken-ness of time and space consist? Let us confine ourselves to one dimen-sion; for the continuity of a line is complete, and other dimensions only render the idea of continuity more difficult to extricate.

In time, or in the description of a line, there is a succession. But succession might exist without continuity. There is a succession in the rows of poplar trees on a French road. Each poplar-tree has others after it. *Before* and *after* are evidently dyadic relations. But the idea of *next after* refers in a negative way to a third; it denies that there is any third after the first and before the second. Moreover, when we say that after *every* object of the series there is an object, we give Thirdness a different kind of prominence from what it has when we say that *some* object that comes after another has a third after it. The latter kind of thirdness is degenerate, it is a mere compound of *two pairs*; but the former kind permeates the whole series with thirdness, and denies any Last, or Second extremity. Thus, the idea of an endless row of discrete objects, which is the image of the system of whole numbers, contains the idea of Thirdness in considerable prominence.

But there is no instant of time *next after* another instant. Between any two instants there is a Third. The great philosophical analyst Kant took this to be a satisfactory analysis of continuity. But it is not so. As an instance of such a series, we may take all the quantities capable of exact expression in decimals between 0 and 1, *exclusive.** (*We often hear the words *inclusive* and *exclusive*, used as here, pronounced like the adjec-tives. They are Latin adverbs, and should be pronounced inkleusei vee and ekskleusei vee.) These numbers are supposed to be arranged in the order of their values, the larger after the smaller. The series is without a first and without a last; and between any two there are others.

But if we reverse the order of the figures which express each of these numbers, and take away the decimal point, so that, for example .461 becomes 164, we get for each of these numbers a distinct whole number; and arranging them in the order of these whole numbers, each one has another *next after* it. This cannot be done with a continuous series.

But the character which most manifestly distinguishes such a series from a continuous series, is that it does not contain the limit of every endless series that forms a part of it. For example, take the following series of numbers,

$$.1$$
$$.11$$
$$.111$$
$$.1111$$
etc.

This is the series of numbers written entirely with ones, beginning in the tens-place and each continuing one place further to the right than the number which went next before. Take also the following series

$$.2$$
$$.12$$
$$.112$$
$$.1112$$
etc.

This is the series of numbers each written with one 2 in a decimal place and with all the preceding places of decimals filled with ones, the two in each being one place further to the right than in the number which next preceded. All of the numbers in these two series belong to the series of the last paragraph. The former of the two continually advances while the latter continually recedes in the great series. Every number of the latter series of the two comes in the great series after every number of the former series of the two. Yet there is no number of the great series after every one of the former and before every one of the latter.

In a continuous series, this would not be so. There would be the number $\frac{1}{9}$, intermediate between the two. Imagine that on a line two particles move toward one another. Whether they move by leaps or run along the line makes no difference. But suppose that as they approach they move slower and slower. Now if there be no points between them that neither will ever attain or pass by, then they certainly must pass one another, if they move on forever. This is manifest. Yet those two series of numbers one increasing, the other decreasing, have between them no number of the great series that one or the other will not pass, and yet they never come together. There is precisely wherein that series is not continuous.

If in place of numbers exactly expressible in decimals between 0 and 1 we consider all the numbers between those limits which differ from one another in some place of decimals or other, although they may have significant figures further to the right than any assignable point, then we find that, if they do not form a continuous series, they, at least, have the above property of continuity. This series, for example, would include the number which has a 1 in every place of decimals. This number would be greater than every one of the former of the two series and less than every one of the latter.

If the reader turns this supposed series over in his mind, he will find it full of logical pitfalls. . . .

18

(PAP) (293)

[PROLEGOMENA FOR AN APOLOGY TO PRAGMATISM]

A variety of mob-madness to which we all seem to be more or less subject is manifested in taking up vague opinions about which our associates and companions seem strenuous. I am apt, in sane moments, to be wary of admitting doctrines of which no definition can be given. An eminent and admirable physiologist concludes a volume of great interest with this sentence: "The idea that mutation is working in a definite direction is a mere anthropomorphism, and like all anthropomorphisms is in contradiction with the facts." If I were to attach a definite meaning to "anthropomorphism," I should think it stood to reason that a man could not have any idea that was not anthropomorphic, and that it was simply to repeat the error of Kant to attempt to escape anthropomorphism. At the same time, I am confident a man can pretty well understand the thoughts of his horse, his jocose parrot, and his canary-bird, so full of *espièglerie*; and though his representation of those thoughts must, I suppose, be more or less falsified by anthropomorphism, yet that there is a good deal more truth than falsity in them, —and more than if he were to attempt the impossible task of eliminating the anthropomorphism, I am for the present sufficiently convinced.

I am led to these remarks from reflecting that a good many persons who tell themselves that they hold anthropomorphism in reprobation will nevertheless opine (though not in these terms) that I am not anthropomorphic enough in my account of logic as a science of signs and in describing signs without making any explicit allusion to the human mind.

A line of bricks stand on end upon a floor, each facing the next one of the line. An end one is tilted so as to fall over upon the next; and so they all successively fall. The mechanical statement of the phenomenon is that a portion of the sum of the energy of motion that each brick had at the instant its centre of gravity was directly over its supporting edge, added to the energy of its fall is transformed into an energy of motion of the next brick. Now I assert no more than this, but less, since I do not say whether it was mechanical energy, or what it was that was communicated, when, applying my definition of a Sign, I assert (as I do) that each brick is a Sign (namely, an Index), to the succeeding bricks of the line, of the original effect produced on the first brick. I freely concede that there is an anthropomorphic constituent in that statement; but there is none that is not equally present in the mechanical statement, since this asserts all that the other form asserts. Until you see this, you do not grasp the meaning that I attach to the word "sign."

I maintain that nothing but confusion can result from using in logic a more anthropomorphic conception than that. To ask how we *think* when we reason has no more to do with the security of the particular form of argumentation that may be under criticism than the histology of the cortex of the brain has to do with the same question of security. Here you may ask, "How does this gibe with your former statement that reasoning is 'self-controlled thought'? There is no difference between 'Thought and Thinking,' is there?" I reply, There is indeed. 'Thinking' is a fabled 'operation of the mind' by which an imaginary object is brought before one's gaze. If that object is a Sign upon which an argument may turn, we call it a Thought. All that we know of the 'Thinking' is that we afterwards remember that our attention was actively on the stretch, and that we seemed to be creating Objects or transformations of Objects while noting their analogy to something supposed to be real. We choose to call 'an operation of the mind'; and we are, of course, quite justified in doing so, provided it be well understood that its being so consists merely in our so regarding it, just as Alexander, Hannibal, Caesar, and Napoleon constitute a single quaternion, or plural of four, as long as we put them together in thought. The 'operation of the mind' is an *ens rationis*. That is my insufficient excuse for speaking of it as 'fabled'.

———

All necessary reasoning is diagrammatic; and the assurance furnished by all other reasoning must be based upon necessary reasoning. In this sense, all reasoning depends directly or indirectly upon diagrams. Only it is necessary to distinguish reasoning, properly so called, where the acceptance [of the] conclusion in the sense in which it is drawn, is seen *evidently* to be justified, from cases in which a rule of inference is followed because it has been found to work well, which I call following a rule of thumb, and accepting a conclusion without seeing why further than that the impulse to do so seems irresistible. In both those cases, there might be a sound argument to defend the acceptance of the conclusion; but to accept the conclusion without any criticism or supporting argument is not what I call reasoning. For example, a person having been accustomed to considering finite collections only might contract a habit of using the syllogism of transposed quantity, of which the following is an instance.

Every Hottentot kills a Hottentot
No Hottentot is killed by more than one Hottentot
Therefore every Hottentot is killed by a Hottentot.

Later forgetting why this necessarily follows for finite collections (if he ever did understand it), this person might by mere force of habit apply the same kind of reasoning to endless generations or other infinite [classes]; or he might apply it to a finite class, but with so little understanding that, only luck would prevent his applying it to infinite collections. Such a case is an application of a rule of thumb and is not reasoning. Many persons are deceived by the catch about Achilles and the Tortoise; and I knew one extremely bright man who could not, for the life of him, perceive any fault in this reasoning:

It either rains or it doesn't rain:
It rains;
Therefore, it doesn't rain.

Such people appear to mistake the rule of thumb for reasoning.

Descartes, in one of his letters, is quite explicit that his *Je pense, donc je suis* is not a syllogism with a suppressed premiss. I infer, then, that he thought it not impossible that an imaginary being should think (i.e. be conscious) albeit he had no real existence. Of course, there would be a fallacy here, but not one that Descartes might not easily fall into. In the same fallacious manner, I suppose he said, It would be quite possible antecedently that I had never existed. But when he tried to suppose, not of a being in general, who might be imaginary, but of himself, that *he* was conscious without existing, he found *that* quite impossible; while yet he had no reason or principle that could serve as major premiss in the argument. This confused inability to suppose his being false, as long as he *thought*, was not, in my terminology, Reasoning, because Reasoning[1] renders the truth of its conclusion plain and comprehensible, and

---

[1] Several pages, marked by Peirce "keep for reference," give an alternative passage as this point.

"Reasoning renders the truth of its conclusion plain and comprehensible, and does not like Descartes' plagiaristic formula stumble in the dark against a hard wall of inability to conceive something.

"In order to expound my proposition that all necessary reasoning is diagrammatic, it is requisite that I explain exactly what I mean by a Diagram, a word which I employ in a wider sense than is usual. A Diagram, in my sense, is in the first place a Token, or singular Object used as a Sign; for it is essential that it should be capable of being perceived and observed. It is, however, what is called a General sign; that is, it denotes a general Object. It is, indeed, constructed with that intention, and thus represents the Object of that intention. Now the Object

does not, like the plagiaristic formula of Descartes, stumble in the dark against an invisible wall of inability to conceive something.

In order to expound fully my proposition that all necessary reasoning is diagrammatic, I ought to explain exactly what I mean by a Diagram. But at present it would be extremely difficult to do quite that. At a later place in this paper I will endeavour to do so; but just now, I think it will better meet the reader's needs to give an exposition that shall cover the main points, and to leave the others, whose needfulness is only perceived after deep study, to follow when the need of them comes out.

To begin with, then, a Diagram is an Icon of a set of rationally related objects. By *rationally* related, I mean that there is between them, not merely one of those relations which we know by experience, but know not how to comprehend, but one of those relations which anybody who reasons at all must have an inward acquaintance with. This is not a sufficient definition, but just now I will go no further, except that I will say that the Diagram not only represents the related correlates,

---

of an intention, purpose, or desire is always General. The Diagram represents a definite Form of Relation. This Relation is usually one which actually exists, as in a map, or is intended to exist, as in a Plan. But this is so far from being essential to the Diagram as such, that if details are added to represented existential or experiential peculiarities, such additions are distinctly of an undiagrammatic nature. The pure Diagram is designed to represent and to render intelligible, the Form of Relation merely. Consequently, Diagrams are restricted to the representation of a certain class of relations; namely, those that are intelligible. We may make a diagram of the Battle of Gettysburgh, because in a certain [sense], it may thus be rendered comprehensible. But we do not make a diagram simply to represent the relation of killer to killed, though it would not be impossible to represent this relation in a Graph-Instance; and the reason we do not is that there is little or nothing in that relation that is rationally comprehensible. It is known as a fact, and that is all. I believe I may venture to affirm that an intelligible relation, that is, a relation of thought, is created only by the act of representing it. I do not mean to say that if we should some day find out the metaphysical nature of the relation of killing, that intelligible relation would thereby be created. For if such be the nature of killing, such it always was, from the date of a certain "difficulty" and consurrection in a harvest field. No; for the intelligible relation has been signified, though not read by man, since the first killing was done, if not long before. The thought of God,—if the anthropomorphism is too distasteful to you, you can say the thought in the universe,—had represented it. At any rate, a Diagram is clearly in every case a sign of an ordered Collection or Plural,—or, more accurately, of the ordered Plurality or Multitude, or of an Order in Plurality. Now a Plural,—say, for example, Alexander, Hannibal, Caesar, and Napoleon,—seems unquestionably to be an *ens rationis*, that is to be created by the very representation of it; and Order appears to be of the same nature; that is, to be an Aspect, or result of taking account of things in a certain way. But these are subtle points; and I should like to give the question maturer consideration before risking much on the correctness of my solution. No such doubt bedims our perception that it is as an Icon that the Diagram represents the definite Form of intelligible relation which constitutes its Object, that is, that it represents that Form by a more or less vague resemblance thereto. There is not usually much vagueness, but I use that word because the Diagram does not itself define just how far the likeness extends, and in some characteristic cases such definition would be impossible, although the Form of Relation is in itself Definite, since it is General.

but also, and much more definitely represents the relations between them, as so many objects of the Icon. Now necessary reasoning makes its conclusion *evident*. What is this "Evidence"? It consists in the fact that the truth of the conclusion is *perceived*, in all its generality, and in the generality the how and why of the truth is perceived. What sort of a Sign can communicate this Evidence? No index, surely, can it be; since it is by brute force that the Index thrusts its Object into the Field of Interpretation, the consciousness, as if disdaining gentle "evidence." No Symbol can do more than apply a "rule of thumb" resting as it does entirely on Habit (including under this term natural disposition); and a Habit is no evidence. I suppose it would be the general opinion of logicians, as it certainly was long mine, that the Syllogism is a Symbol, because of its Generality. But there is an inaccurate analysis and confusion of thought at the bottom of that view; for so understood it would fail to furnish Evidence. It is true that ordinary Icons, — the only class of Signs that remains for necessary inference, — merely suggest the possibility of that which they represent, being percepts *minus* the

---

It is, however, a very essential feature of the Diagram *per se* that while it is as a whole an Icon, it yet contains parts which are capable of being recognized and distinguished by the affixion to cach of a distinct Semantic Index (or Indicatory Seme, if you prefer this phrase). Letters of the alphabet commonly fulfil this office. How characteristic these Indices are of the Diagram is shown by the fact that though in one form or another they are indispensable in using the diagram, yet they are seldom wanted for the general enunciation of the proposition which the Diagram is used for demonstrating. That which is most of all requisitionable from a definition of an artificial contrivance such as a Diagram is, is that it should state what the Definition does and what it is for; so that these points must now be touched upon even at the risk that this Definition of a Diagram might be threatened with danger to its absolute preeminence over all others, of what sort soever, that ever have been or ever shall be given, in respect to the chief grace of definitions, that of Brevity. That which every sign does is to determine its Interpretant. The responsive Interpretant, or Signification, of one kind of signs is a vague presentation, of another kind is an Action, while of a third it is involved in a Habit and is General in its nature. It is to this third class that a Diagram belongs. It has to be interpreted according to Conventions embodied in Habits. One contemplates the Diagram, and one at once prescinds from the accidental characters that have no significance. They disappear altogether from one's understanding of the Diagram; and although they be of a sort which no visible thing be without (I am supposing the diagram to be of the visual kind), yet their [disappearance] is only an understood disappearance and does not prevent the features of the Diagram, now become a Schema, from being subjected to the scrutiny of observation. By what psychical apparatus this may get effected the logician does not inquire. It suffices for him, that one can contemplate the Diagram and perceive that it has certain features which would always belong to it however its insignificant features might be changed. What is true of the geometrical diagram drawn on paper would be equally true of the same Diagram when put on the blackboard. The assurance is the same as that of any description of what we see before our eyes. But the action of the Diagram does not stop here. It has the same percussive action on the Interpreter that any other Experience has. It does not stimulate any immediate counter-action, nor does it, in its function as a Diagram, contribute particularly to any expectation. As Diagram, it excites curiosity as to the effect of a transformation of it."

insistency and percussivity of percepts. In themselves, they are mere
Semes, predicating of nothing, not even so much as interrogatively. It
is, therefore, a very extraordinary feature of Diagrams that they *show*, —
as literally *show* as a Percept shows the Perceptual Judgment to be
true, — that a consequence does follow, and more marvellous yet, that it
would follow under all varieties of circumstances accompanying the
premisses. It is not, however, the statical Diagram-icon that directly
shows this; but the Diagram-icon having been constructed with an
Intention, involving a Symbol of which it is the Interpretant (as Euclid,
for example, first enounces in general terms the proposition he intends
to prove, and then proceeds to draw a diagram, usually a figure, to
exhibit the antecedent condition thereof) which Intention, like every
other, is General as to its Object, in the light of this Intention determines
an Initial Symbolic Interpretant. Meantime, the Diagram remains in the
field of perception or imagination; and so the Iconic Diagram and its
Initial Symbolic Interpretant taken together constitute what we shall
not too much wrench Kant's term in calling a *Schema*, which is on the
one side an object capable of being observed while on the other side
it is General. (Of course, I always use 'general' in the usual sense of
general as to its object. If I wish to say that a sign is general as to its
matter, I call it a Type, or Typical.) Now let us see how the Diagram
entrains its consequence. The Diagram sufficiently partakes of the
percussivity of a Percept to determine, as its Dynamic, or Middle,
Interpretant, a state [of] activity in the Interpreter, mingled with
curiosity. As usual, this mixture leads to Experimentation. It is the
normal Logical effect; that is to say, it not only happens in the cortex of
the human brain, but must plainly happen in every Quasi-mind in which
Signs of all kinds have a vitality of their own. Now, sometimes in one
way, sometimes in another, we need not pause to enumerate the ways,
certain modes of transformation of Diagrams of the system of diagrammatization used have become recognized as permissible. Very likely the
recognition descends from some former Induction, remarkably strong
owing to the cheapness of mere mental experimentation. Some circumstance connected with the purpose which first prompted the construction of the diagram contributes to the determination of the
permissible transformation that actually gets performed. The Schema
sees, as we may say, that the transformate Diagram is substantially
contained in the transformand Diagram, and in the significant features
to it, regardless of the accidents, — as, for example, the Existential
Graph that remains after a deletion from the Phemic Sheet is contained
in the Graph originally there, and would do so whatever colored ink
were employed. The transformate Diagram is the Eventual, or Rational,

Interpretant of the transformand Diagram, at the same time being a
new Diagram of which the Initial Interpretant, or signification, is the
Symbolic statement, or statement in general terms, of the Conclusion.
By this labyrinthine path, and by no other, is it possible to attain to
Evidence; and Evidence belongs to every Necessary Conclusion.

There are at least two other entirely different lines of argumentation
each very nearly, and perhaps quite, as conclusive as the above, though
less instructive, to prove that all necessary reasoning is by diagrams. One
of these shows that every step of such an argumentation can be represented, but usually much more analytically, by Existential Graphs. Now
to say that the graphical procedure is more analytical than another is to
say that it demonstrates what the other virtually assumes without proof.
Hence, the Graphical method, which is diagrammatic, is the sounder
form of the same argumentation. The other proof consists in taking up,
one by one, each form of necessary reasoning, and showing that the
diagrammatic exhibition of it does it perfect justice.

Let us now consider non-necessary reasoning. This divides itself,
according to the different ways in which it may be valid, into three
classes: probable deduction; experimental reasoning, which I now call
Induction; and processes of thought capable of producing no conclusion
more definite than a conjecture, which I now call Abduction. I examined
this subject in an essay in the volume of "Studies in Logic by Members of
the Johns Hopkins University," published in 1883; and have since made
three independent and laborious investigations of the question of
validity, and others connected with it. As my latest work has been
written out for the press and may sometime be printed, I will limit what
I say here as much as possible. The general principle of the validity of
Induction is correctly stated in the Johns Hopkins Essay, but is too
narrowly defined. All the forms of reasoning there principally considered
come under the class of Inductions, as I now define it. Much could now
be added to the Essay. The validity of Induction consists in the fact
that it proceeds according to a method which though it may give pro-
visional results that are incorrect will yet, if steadily pursued, eventually
correct any such error. The two propositions that all Induction possesses
this kind of validity, and that no Induction possesses any other kind
that is more than a further determination of this kind, are both susceptible of demonstration by necessary reasoning. The demonstrations are
given in my Johns Hopkins paper; and although the description of the
mode of validity there is too narrow, yet it covers the strongest inductions and most of the reasonings generally recognized as Inductions. It
is characteristic of the present state of logic that no attempt has been
made to refute the demonstrations, but the old talk conclusively refuted

by me goes on just the same. To say that the validity of Induction rests on Necessary Reasoning is as much as to say that Induction separated from the deduction of its validity does not make it evident that its conclusion has the kind of justification to which it lays claim. This being the case, it is not surprising that Induction, separated from the deduction of its validity, makes no essential use of diagrams. But instead of experimenting on Diagrams it experiments upon the very Objects concerning which it reasons. That is to say, it does so in an easily extended sense of the term "experiment"; the sense in which I commonly employ the word in the critical part of logic.

The third mode of non-necessary reasoning, if we are to count the deduction for probabilities as a class, though it ought not to be reckoned such, is Abduction. Abduction is no more nor less than guessing, a faculty attributed to Yankees.* (*In point of fact, the three most remarkable, because most apparently unfounded, guesses I know of were made by Englishmen. They were Bacon's guess that heat was a mode of motion, Dalton's of chemical atoms, and Young's (or was it Wallaston's) that violet, *green* (and not yellow, as the painters said) and red were the fundamental colors.) Such validity as this has consists in the generalization that no new truth is ever otherwise reached while some new truths are thus reached. This is a result of Induction; and therefore in a remote way Abduction rests upon diagrammatic reasoning.

The System of Existential Graphs the development of which has only been begun by a solitary student, furnishes already the best diagram of the contents of the logical Quasi-mind that has ever yet been found and promises much future perfectionment. Let us call the collective whole of all that could ever be present to the mind in any way or in any sense, the *Phaneron*. Then the substance of every Thought (and of much beside Thought proper) will be a Consistituent of the Phaneron. The Phaneron being itself far too elusive for direct observation, there can be no better method of studying it than through the Diagram of it which the System of Existential Graphs puts at our disposition. We have already tasted the first-fruits of this method, we shall soon gather more, and I, for my part, am in confident hope that by-and-by (not in my brief time) a rich harvest may be garnered by this means.

What, in a general way, does the Diagram of Existential Graphs represent the mode of structure of the Phaneron to be like? The question calls for a comparison, and in answering it a little flight of fancy will be in order. It represents the structure of the Phaneron to be quite like that of a chemical compound. In the imagined representation of the

Phaneron (for we shall not, as yet, undertake actually to construct such a Graph), in place of the ordinary spots, which are Graphs not *represented* as compound, we shall have Instances of the absolutely Indecomposable Elements of the Phaneron (supposing it has any ultimate constituents, which, of course, remains to be seen, until we come to the question of their Matter; and as long as we are, as at present, discoursing only of their possible Forms, their being may be presumed), which [are] close enough analogues of the Atoms in the Chemical Graph of "Rational Formula." Each Elementary Graph, like each chemical element, has its definite Valency, — the number of Pegs on the periphery of its Instance, — and the Lines of Identity (which never branch) will be quite analogous to the chemical bonds. This is resemblance enough. It is true that in Existential Graphs we have the Cuts, to which nothing in the chemical Graph corresponds. Not yet, at any rate. We are now just beginning to rend away the veil that has hitherto enshrouded the constitution of the proteid bodies; but whatever I may conjecture as to those vast supermolecules, some containing fifteen thousand molecules, whether it seems probable on chemical grounds, or not, that they contain groups of opposite polarity from the residues outside those groups, and whether or not similar polar submolecules appear within the complex inorganic acids, it is certainly too early to take those into account in helping the exposition of the constitution of the phaneron. Were such ideas as solid as they are, in fact, vaporous, they ought to be laid aside until we have first thoroughly learned all the lessons of that analogy between the consitution of the phaneron and that of chemical bodies which consists in both the one and the other being composed of elements of definite valency.

In all natural classifications, without exception, distinctions of form, once recognized, take precedence over differences of matter. Who would now throw Iron, with its valency, perhaps, of eight, as used to be done, into the same class with Manganese, of valency seven, Chromium, with its valency of six (though these three all belong to the even fourth series), and Aluminum with valency three and in the odd series three, rather than with Nickel and Cobalt, and even along with Ruthenium, Rhodium, and Palladium of the sixth series, and with the tenth-series Osmium, Iridium, and Platinum? Or who would for one instant liken ordinary alcohol to methyl ether (which has the same Material composition) instead of with the alcoholates? The same precedence of Form over Matter is seen in the classification of psychical products. Some of Rafael's greatest pictures, — the Christ bearing the cross, for example, — are suffused with a brick red tinge, intended, I doubt not, to correct for the violet blueness of the deep shade of the chapels in which they were meant to be hung. But who would classify Rafael's paintings according to their predominant tinges instead of according to the nature of the composition, or the stages of Rafael's development? There is no need of insisting upon a matter so obvious. Besides there is a rational explanation of the precedence of Form over Matter in natural classifications. For such classifications are intended to render the composition of the entire classified collection rationally intelligible,—no matter what else they may be intended to show; and Form is something that the mind can "take in," assimilate, and comprehend; while Matter is always foreign to it, and though recognizable, is incomprehensible. The reason of this, again, is plain enough: Matter is that by virtue of which an object gains Existence, a fact known only by an Index, which is connected with the object only by brute force; while Form, being that by which the object is such as it is, is comprehensible. It follows that, assuming that there are any indecomposable constituents of the Phaneron, since each of these has a definite Valency, or number of Pegs to its Graph-Instance, this is the only Form, or, at any rate, the only intelligible Form, the Element of the Phaneron can have, the Classification of Elements of the Phaneron must, in the first place, be classified according to their Valency, just as are the chemical elements.

We call a Spot a Medad, Monad, Dyad, Triad, Tetrad, or by some other such name, according as its Valency, or the number of its Pegs, is 0, 1, 2, 3, 4, etc. It is to be remarked that a Graph not only has attachments to other Graphs through its Pegs and through Lines of Identity, but is also attached to the Area on which it is scribed, this Area being a Sign of a logical Universe. But it is not the same kind of attachment, since the Entire Graph of the Area is after a fashion *predicated* of that Universe, while the Lines of Identity represent Individual Subjects of which the two connected Spots are predicated either being regarded as determining the other. There would therefore be a confusion of thought in adding one to the number of Pegs and calling the sum the Valency. It would rather be the sum of two different categories of Valency. But in the case of the Medad, where there is no Peg, the possibility of scribing the Graph upon an Area is the only Valency the Spot has,—the only circumstance that brings it and other thoughts together. For this reason, we can, without other than a Verbal inconsistency, due to the incompleteness of our Terminology, speak of a Medad as a Monad. For some purposes, it is indispensable so to regard it.

I am now going to make a few notes which may be useful to a person in reflecting upon this subject, even if I am not led to make here any further remarks upon them.

It is likely to prove convenient to have at one's disposition a certain formula which follows as a Corollary from Listing's Census-Theorem. The formula is, $2(K + S - X) = V - L$; or, in words, for any Graph which is separated from others, twice the sum of the Cyclosis, K, added to the excess of the number of spots, S, over the chorisis, X, is equal to the excess of the sum of the Valencies of all the Spots in the Graph, V, over the number of Loose Ends, L. The *Chorisis*, X, is the number of separate pieces which go to make up the Graph. The *Cyclosis*, K, is the number of Lines of Identity in the Graph which might be severed without increasing the Chorisis. A *Loose End* is an extremity of a Line of Identity not abutting upon any Spot. Such is the end of a Line of Identity on the Area of a Cut which abuts upon the Cut, itself. Since the Reader may not be familiar with the Census Theorem, I will give an immediate demonstration of the truth of the formula. Taking any Graph whatever, let the Capital letters, K, S, X, V, L, denote respectively its Cyclosis, number of Spots, Chorisis, Total Valency of its Spots, and number of Loose Ends. Sever every Line of Identity in the Graph at one point. (But it will not be necessary to cut any that has a loose end. Observe, however, that we *assume without proof* that the number of Lines of Identity is Finite. Observe, too, that a point of Teridentity *must* be regarded as a Spot.) Let the little letters k, s, x, v, l, refer to the Graph resulting from this operation, each letter having the same predicative signification as the corresponding capital. There will now be no Cyclosis: $k = 0$. The Spots will be the same as before: $s = S$. The Chorisis will be the same as the number of Spots: $x = S$. The Total Valency will not have been altered: $v = V$. There will be a Loose End for every Peg: $l = V$. Thus, both members of the proposed formula will vanish: namely, both $2(k + s - x) = 2(0 + S - S)$; and $v - l = V - V$; and thus the formula verifies itself for this state of extreme dissection. Now restore the original Graph by bringing together the two loose ends that have resulted from severing each Line of Identity. Mend the lines *one by one*, in any order of succession you like. Each such act of mending will leave the Total Valency unaltered, but will diminish the number of Loose Ends by 2; thus increasing the value of the second member of the formula by 2. If the two loose ends brought together belong, at the time when they are brought together, to separate pieces of the Graph, the mending cannot affect the Cyclosis, nor the number of Spots, but will diminish the Chorisis by 1, thus increasing the first member of the equation, $2(K + S - X)$, by 2. If, however, the loose ends brought together do *not* belong to separate pieces, at the time they are brought together, they belong to the same piece; and the mending will increase the Cyclosis by 1, while leaving the number of Spots and the

Chorisis unchanged; thus again increasing the first member of the equation by 2. Thus, both members of the equation are increased by the same amount at each step of the operation, and the equation remains as true after the step as before. Hence, it remains as true after all the mendings as it was in its extreme dissection. But it was then true; and is therefore true at the end. At the end it is the original Graph again; which was *any Graph we pleased*. Hence, the formula is true of Any Graph we please, so long as the Lines of Identity are finite in number.

The formula does not teach one much; but perhaps it will help to keep one in mind of what sort of work a really scientific research into the Phaneron must be. It must be a work of diagrammatic thinking, first and last.

Logic requires great subtlety of thought, throughout; and especially in distinguishing those characters which belong to the diagram with which one works, but which are not significant features of it considered as the Diagram it is taken for, from those that testify as to the Form represented. For not only may a Diagram have features that are not significant at all, such as its being drawn upon "laid" or upon "wove" paper; not only may it have features that are significant but are not diagrammatically so; but one and the same construction may be, when regarded in two different ways, two altogether different diagrams; and that to which it testifies in the one capacity, it must not be considered as testifying to in the other capacity. For example, the Entire Existential Graph of a Phemic Sheet, in any state of it, is a Diagram of the logical Universe, as it is also a Diagram of a Quasi-mind; but it must not, on *that* account, be considered as testifying to the identity of those two. It is like a telescope eye piece which at one focus exhibits a star at which the instrument is pointed, and at another exhibits all the faults of the objective lens.

Among Existential Graphs there are two that are remarkable for being truly *continuous* both in their Matter and in their corresponding Signification. There would be nothing remarkable in their being continuous in either, or in both respects; but that the continuity of the Matter should correspond to that of the Signification is sufficiently remarkable to limit these Graphs to two; the Graph of Identity represented by the Line of Identity, and the Graph of coëxistence, represented by the Blank. Here, Reader, I overhear you asking what I mean by True Continuity. If I miss-hear, it is because I am expecting you to ask the logical questions, — for *questions* may logically follow, as well as Assertions.

Well, Reader, I reply, in asking me that question, 'What do I mean by True Continuity?' you are asking one of the most difficult questions of logic. We know very well that the Continuity of the Theory of Functions, which I call Pseudo-Continuity, is a certain order among the individual members of a Collection whose multitude is the same as that of the Collection of all possible collections of integer numbers. But between any two points of what I call a Truly Continuous Line there is room for any multitude of points whatsoever, and therefore of an endless series of multitudes all infinitely greater than the total multitude of points of which the linear Pseudo-continuum consists. Now logicians have always rightly said that no collection of individuals whatever is adequate to presenting all the possible variations of a general term; and consequently the points of a True Linear Continuum cannot be actual constituent parts of it. Its only parts, as Kant says, are homogeneous (in respect to those qualities which belong to all the parts) with the whole, and those homogeneous parts are indeterminate, in that each may end and the next begin where you will. This is why every continuum may be regarded as the actualization of a generalized relation having the form of the relation of three (or four?) points upon a line. But it is quite evident that Kant is right (though his nominalism made the truth appear to him more psychological than logical, as it truly is) in making the primitive relation to be of the form of the relation of two instants of time, or what is the same thing as the relation between a logical antecedent and consequent. The reason that in order to define the relation of a point upon a line to another point it is necessary to speak of a third, if not also of a fourth, point is that on the line one does not distinguish, as in the sequence of time and in that of logic, one direction from the other. But here we come upon a disputed question among exact logicians; namely, Which is the more primitive (or fundamental, or simple) form of relation, that of an Equiparance (i.e. a reciprocal relation), or that of a Disquiparance? I say that it is the Disquiparance, or rather, it is the *Opponency*, or relation of which a specialization may be a Disquiparance. All the arguments in favor of the primitivity of the Equiparance will be found upon analysis to amount substantially and in principle to this: though "is a cousin" of" and "is a companion of" are both equiparances, yet "is a cousin of a companion of" is a disquiparance; and thus a disquiparance is a compound of two equiparances. Mr. Kempe in his great memoir published in the *Philosophical Transactions* for 1886 has a System of Graphs in which the Spots have no definite Valency, and there is usually but one kind of line, which signifies whatever Equiparance may have been agreed upon. Now he places on such a line two spots of different colors, as shown in Fig. 1 and remarking that this compounded line signifies a Disquiparance (as it manifestly usually will), he regards that as proof that an undirected line is simpler than a directed

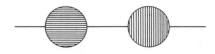

Fig. 1

line. But I propose to show that if this has any semblance of a sound argument, it must be so understood as to be a mere variant of the argument about the cousin of a companion. For representing Fig. 1 in an Existential Graph, and putting, *l*, for the equiparant relation signified by Kempe's plain line, *g* for the gules spot, and *z* for the azure spot, the Graph represented must be one or other of the three of Fig. 2, or else some other to which the same remarks will apply. The first and third of these Graph-instances can be severed in the middle so as to separate each into two equiparants similar to "is cousin of" and "is companion of."

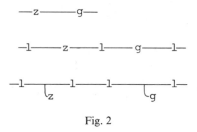

Fig. 2

But the second cannot be separated into two equiparants and therefore affords no semblance of an argument; and the same is true of the third, if it be cut elsewhere than in the middle. For an Equiparant is a general description of relation which, if it describes the relation of any individual object, *A*, to another, *B*, in every case also describes the relation of *B* to *A*. Let *l* be such a relation. Then the effect of joining *z* to it in the manner shown in the third graph is to make the relation inapplicable to the relation of *A* to *B* unless *B* happens to be described by *z*, which of course cannot always be the case, if *z* has any definite signification. If this will (or may) leave the description applicable to the relation of *B* to *A* yet inapplicable to that of *A* to *B*. In short this fragment of the Graph will (or may) signify a disquiparant relation; and there will be no semblance of argument. I will now refute the argument from the cousin of a companion in two distinct ways. In the first place, when a relation forms the predicate of a proposition, it is, in a certain sense, specialized. That is to say it is only a special case of the relation that has any relevancy to the two Subjects of the proposition; and the relation as it is in the proposition only has that limited extension. An Equiparance, however, is such only in its full generality, and usually ceases to be equiparant when limited to a special relate and correlate. That obviously is the reason why (to make use, for a moment, of the General Algebra of Logic) the relative product of two equiparants such as $A:B \psi B:A$ and $A:C \psi C:A$, gives in one order $B:C$ and in the other order $C:B$. That is, in the first order, the $A:B$ of one and the $C:A$ of the other are irrelevant, while in the other order, the $A:C$ of one and the $B:A$ of the other are irrelevant, and might as well be absent. Rectify this by making every Index of an individual that occurs in either occur also in the other, as in

$$(A:B \psi B:A \psi C:D \psi D:C)(A:C \psi C:A \psi B:D \psi D:B)$$

and the product will be equiparant: $A:D \psi B:C \psi C:B \psi D:A$. That is one answer to the argument. In the second place, it cannot be the fact that the relative product of two equiparants *may be* a disquipant, —and it is not always so, nor *ever* so if either of the equiparants is a concurrent, —cannot suffice to prove that Equiparance is more primitive, fundamental, or simple than Disquiparance, inasmuch as any disquiparant *whatsoever* relatively multiplied into its own converse, which is equally disquipant, will give an equiparant product; and that *without any specialization at all*; so that the method of argumentation that my opponents have adopted much rather tends to prove my contention than theirs. When I speak of them as opponents, I mean they are accidentally so, as regards some particular questions. They are Exact Logicians, toiling in the honest and sincere scientific way. That and their great genius commands my respect. I think them somewhat incautious and liable to fallacious thinking; but of all of us logicians that is the peculiar danger, owing to the nature of our subject of thought. In the discussion of this particular question the method mentioned comprises their entire armory of reason. They are given to applying it in most involved forms, notwithstanding the manifest danger of fallacy's lurking complicated argumentation to prove any proposition that relates solely to extremely simple constituents of thought. That method of theirs makes decidedly in my favour. Nevertheless, I do not altogether approve of it. It seems to me to involve doubtful assumptions. I do not, to begin with, think that the distinction between Equiparance and Disquiparance has any just claim to primacy among divisions of Dyadic Relative Terms. If I were quite sure that any formal division of them could lay such a claim, I should unhesitatingly give my vote for the distinction between *Opponents* and *Concurrents*. Opponents are terms expressing relations in which one individual object can stand to another. Concurrents are mere specializations of identity. Ordinary adjectives are concurrents. In the next place, I conceive the question of the most important division of Signs of Dyadic Relation to be subordinate to the question of the Forms of Dyadism, or Twoness, itself. Now Dyadism itself has no generality. We should come down to the most specialized possible Relatives. These are manifestly the relation of a single designate individual to another and the relation of such an individual to itself, $(A:B)$ and $(A:A)$. . The latter form I hold to be *degenerate*. That is to say, it is eviscerated of the kernel of Twoness, and is a mere empty shell of Twoness. It is, therefore, a derived idea. But I do not see that it is, on that account, necessarily composite, in any proper and usual sense. Yet since $(A:A)$ is the matter of Oneness masquerading under the guise of Twoness, while $(A:B)$ is simply Twoness in its own proper guise, I conceive the latter to be more direct and in the only sense in which either is composite, to be decidedly the simpler. Since Disquiparants are assimilated to $(A:B)$, and Equiparants to $(A:A)$, the Disquiparant appears to me to be the primitive, and the Equiparant to be the degenerate, Form. I have thus given a distant hint, and no more, of the way in which, as it seems to me, this question ought to be treated. At this stage of our study of the Phaneron, I could not present the Method, as it really is.

This question being settled the nature of continuity may be regarded as sufficiently understood for our purpose. Let us return then to the two continuous Graphs, which are the Blank and the Line of Identity. I will begin with the latter. The Immediate Interpretant of Identity is, I think, simple. If it were the so-called "numerical" identity only that the line signified, perhaps it might not be so, since that sort of Identity implies Existence and apparently something more. But the line of Identity is not confined to metallic areas: it is also scribed on Color. Now a Possibility, not having Existence, cannot be a subject of Numerical Identity. Nor can we say that the Identity signified by the Line of Identity is the most special agreement possible; for we should not hesitate to employ the line to express that the *same* man who fought the battle of Leipzig fought the battle of Waterloo. Now years having elapsed, it was certainly not in entire strictness the same individual; for an individual is determinate in all respects, and therefore in date. The Eventual Interpretant of the assertion "A is *identical* with B" is, "A will serve all purposes instead of B," or in other words "Whatever is true of A is true of B," and the "all purposes" and "whatever is true" refer to limited logical universes of purposes or (what is the same at bottom) of predicates. In Form, this is the statement of a Disquiparance. It is Equiparant only because the denial of an assertion is itself an assertion and fitness for one purpose is unfitness for the reverse purpose. Hereupon you will remark that if the relation to be expressed is thus Disquiparant in Form but rendered Equiparant by its Matter, then a perfect Diagram of it should have its two extremities unlike in Form, yet like in Matter, — if any meaning can be attached to that; and you will ask how I make out that to be true of the Line of Identity. I answer that I am not hired as an advocate of existential Graphs. I suppose, like some other human inventions of which antiquaries can tell us, that it has its imperfections; and I am desirous of finding them out and exposing them to the comment of all my dear neighbours. Should you say, 'if the System is as imperfect as that, its inventor, who has spent so many years upon it, its inventor must be very nearly an idiot,' your *consequence* will be very wrong, but nowhere in this world could you find more heartfelt assent to your *consequent* than in my solitary study. By no means accept anything that inventor says about Logic, unless you see for yourself that it is true. Yet let me tell you that, fool as he is, he has important truth to communicate that is not quite smothered in blunders. A Line of Identity that abuts upon a Cut, whether on its Area or on its Place may *look* alike at its two ends; but an essential part of every diagram is the Conventions by which it is interpreted; and the principle that Graphs are Endoporeutic in interpretation, as they naturally will be in the process of scribing, confers a definite *sens*, as the French say, a definite way of facing, a definite front and back, to the Line. If a Line of Identity does not abut upon a Cut, then that extremity of it from which the motion of the Graphist's pencil starts will be its hinder end, while the extremity at which the motion ceases will be the forward end. But since the Interpreter is at liberty to take it the other way, it would be a grave logical fault to add any barb or other mark to show which way the line faced, because it would be introducing a rhetorical element into what is designed to be a purely logical diagram. If you ask how I make out that the line faces one way in form but the matter obliterates the distinction, I ask you to recall the definitions of Matter and Form that go back to Aristotle (though it is hard to believe they are not earlier; and the metaphysical application of $\H{v}\lambda\eta$ sounds to me like some late Ionic philosopher, and not a bit like Aristotle, whom it would also have been more like to claim it, if it were his). Form is that which makes anything such as it is, while matter makes it to be. From this pair of beautiful generalizations are born a numerous family of harmonious and interresemblant acceptions of the two words. In speaking of Graphs we may well call the Principles of their Interpretation (such as the Endoporeutic Principle) the Form; the way of shaping and scribing them (such as leaving the Line without barbs) the Matter. Nothing could be in better accord with the general definitions of Form and of Matter.

I have already, in a former chapter, shown how a continuous Line of some thickness necessarily signifies Identity in the System of Existential Graphs. The necessary character of this interpretation may win a pardon for any slight imperfection in the Diagrammatization of Identity by such a Line, should we detect any such imperfection of Diagrammatization in it. Kant, in one of his most characteristic familiarly famed chapters, beginning S.642 of the "Critik[2] der reinen Vernunft," the well-known "*Anhang zur transcendentalen Dialektik*, which treats *Von dem regulativen Gebrauch der Ideen der reinen Vernunft*," sets up a sharp distinction between the constitutive and the regulative application of concepts, and lays down, as regulative principles, three laws, of which one, the *Gesetz der Affinität*, becomes highly pertinent to our present question, provided, in the first place, we understand his "continuirliche Uebergang" from one concept to another in the sense of True Continuity, as we should, and, in the second place, if we recognize, as we must, that Kant's distinction is not absolute, inasmuch as all the so-called "constitutive" applications of principles are, at bottom, regulative. The reader will find means, I hope, to admit the latter condition problematically. He cannot yet be expected to grant its truth, inasmuch as it is almost an exact definition of Pragmaticism; but in the sense of a hypothesis, as a proposition that may possibly be true, it seems to me he virtually has granted it in consenting to read a defence of pragmatism. In order to illustrate what it would mean to say that Identity is a Continuous Relation,—that is, continuous in meaning,—we may compare it with another. To say that he who commanded the French in the battle of Leipzig commanded them in the final battle of Waterloo, is not merely a statement of Identity: it is a statement of Becoming. There is an existential continuity in time between the two events. But, so understood, the statement asserts no Significative Identity, inasmuch as the intervening continuum is a continuum of Assertoric Truths. Now upon a continuous line there are no points (where the line is continuous), there is only room for points,—possibilities of points. Yet it is through that continuum, that line of generalization of possibilities that the actual point at one extremity necessarily leads to the actual point at the other extremity. The actualization of the two extremities consists in the two facts that at the first, without any general reason the continuum there begins while at the last, equally without reason, it is brutally, i.e. irrationally but forcibly cut off.

---

[2]   In an existential oval, Peirce wrote "Spell with a C."

19

DETACHED IDEAS CONTINUED AND THE DISPUTE
BETWEEN NOMINALISTS AND REALISTS (439)

When in 1866, Gentlemen, I had clearly ascertained that the three types of reasoning were Induction, Deduction, and Retroduction, it seemed to me that I had come into possession of a pretty well-rounded system of Formal Logic. I had, it is true, a decided suspicion that there might be a logic of relations; but still I thought that the system I had already obtained ought to enable me to take the Kantian step of transferring the conceptions of logic to metaphysics. My formal logic was marked by triads in all its principal parts. There are three types of inference— Induction, Deduction, and Retroduction—each having three propositions and three terms. There are three types of logical forms, the term, the proposition, and the inference. Logic is itself a study of signs. Now a sign is a thing which represents a second thing to a third thing, the interpreting thought. There are three ways in which signs can be studied, first as to the general conditions of their having any meaning, which is the *Grammatica Speculativa* of Duns Scotus, second as to the conditions of their truth, which is logic, and thirdly, as to the conditions of their transferring their meaning to other signs. The Sign, in general, is the third member of a triad; first a thing as thing, second a thing as reacting with another thing; and third a thing as representing another to a third. Upon a careful analysis, I found that all these triads embody the same three conceptions, which I call after Kant, my Categories. I first named them Quality, Relation, and Representation. I cannot tell you with what earnest and long continued toil I have repeatedly endeavored to convince myself that my notion that these three ideas are of fundamental importance in philosophy was a mere deformity of my individual mind. It is impossible; the truth of the principle has ever reappeared clearer and clearer. In using the word *relation*, I was not aware that there are relations which cannot be analyzed into relations between pairs of objects. Had I been aware of it, I should have preferred the word *Reaction*. It was also perhaps injudicious to stretch the meaning of the word Representation so far beyond all recognition as I did. However, the words Quality, Reaction, Representation might well enough serve to name the conceptions. The *names* are of little consequence; the point is to apprehend the conceptions. And in order to avoid all false associations, I think it far the best plan to form entirely new scientific names for them. I therefore prefer to designate them as *Firstness*, *Secondness*, and *Thirdness*. I will endeavour to convey to you some idea of these conceptions. They are ideas so excessively general, so much more general than ordinary philosophical terms, that when you first come to them they must seem to you vague.

Firstness may be defined as follows: It is the mode in which anything would be for itself, irrespective of anything else, so that it would not make any difference though nothing else existed, or ever had existed, or could exist. Now this mode of being can only be apprehended as a mode of feeling. For there is no other mode of being which we can conceive as having no relation to the possibility of anything else. In the second place, the First must be without parts. For a part of an object is something other than the object itself. Remembering these points, you will perceive that any color, say *magenta*, has and is a positive mode of feeling, irrespective of every other. Because, firstness is all that it is, for itself, irrespective of anything else, when viewed *from without* (and therefore no longer in the original fullness of firstness) the firstnesses are all the different possible sense-qualities, embracing endless varieties of which all we can feel are but minute fragments. Each of these is just as simple as any other. It is impossible for a sense quality to be otherwise than absolutely simple. It is only complex to the eye of comparison, not in itself.

A *Secondness* may be defined as a modification of the being of one subject, which modification is *ipso facto* a mode of being of quite a distinct subject, or, more accurately, secondness is that in each of two absolutely severed and remote subjects which pairs it with the other, not for my mind nor for, or by, any mediating subject or circumstance whatsoever, but in those two subjects alone; so that it would be just the same if nothing else existed, or ever had existed, or could exist. You see that this Secondness in each subject must be secondary to the inward Firstness of that subject and does not supersede that firstness in the least. For were it to do so, the two subjects would, insofar, become one. Now it is precisely their twoness all the time that is most essential to their secondness. But though the secondness is secondary to the firstness, it constitutes no limitation upon the firstness. The two subjects are in no degree one; nor does the secondness belong to them taken together. There are two Secondnesses, one for each subject; but these are only aspects of one Pairedness which belongs to one subject in one way and to the other in another way. But this pairedness is nothing different from the secondness. It is not mediated or brought about; and consequently it is not of a comprehensible nature, but is absolutely blind. The aspect of it present to each subject has no possible *rationale*. In their *essence*, the two subjects are not paired; for in its essence anything is what it is, while its secondness is that of it which is another. The secondness, therefore, is an accidental circumstance. It is that a blind reaction takes place between the two subjects. It is that which we experience when our will meets with resistance, or when something obtrudes itself upon sense. Imagine a magenta color to feel itself and nothing else. Now while it slumbers in its magenta-ness let it suddenly be metamorphosed into pea green. Its experience at the moment of transformation will be secondness.

The idea of Thirdness is more readily understood. It is a modification of the being of one subject which is a mode of a second so far as it is a modification of a third. It might be called an inherent reason. That dormitive power of opium by virtue of which the patient sleeps is more than a mere word. It denotes, however indistinctly, some reason or regularity by virtue of which opium acts so. Every law, or general rule, expresses a thirdness; because it induces one fact to cause another. Now such a proposition as, Enoch is a man, expresses a firstness. There is no reason for it; such is Enoch's nature, — that is all. On the other hand the result that Enoch dies like other men, as result or effect, expresses a Secondness. The necessity of the conclusion is just the brute force of this Secondness. In Deduction, then, Firstness by the operation of Thirdness brings forth Secondness. Next consider an Induction. The people born in the last census-year may be considered as a sample of Americans. That *these* objects should be Americans has no reason except that that was the condition of my taking them into consideration. There is Firstness. Now the Census tells me that about half those people were males. And that this was a necessary result is almost guaranteed by the number of persons included in the sample. There, then, I assume to be Secondness. Hence we infer the *reason* to be that there is some virtue, or occult regularity, operating to make one half of all American births male. There is Thirdness. Thus, Firstness and Secondness following have risen to Thirdness.

There are my three categories. I do not ask you to think highly of them. It would be marvellous if young students in philosophy should be able to distinguish these from a flotsam and jetsam of the sea of thought that is common enough. Besides, I do not ask to have them distinguished. All thought both correct and incorrect is so penetrated with this triad, that there is nothing novel about it, and [there being] no merit in having extracted it I do not at present make any definite assertion about these conceptions. I only say, here are three ideas, lying upon the beach of the mysterious ocean. They are worth taking home, and polishing up, and seeing what they are good for.

I will only say this. There is a class of minds whom I know more intimately probably than many of you do, in whose thought, if it can be called thought, Firstness has a relative predominance. It is not that they are particularly given to hypothetic inference, though it is true that they are so given; but that all their conceptions are relatively detached and sensuous. Then there are the minds whom we commonly meet in the world, who cannot at all conceive that there is anything more to be desired than power. They care very little for inductions, as such. They are nominalists. They care for the things with which they react. They do reason, so far as they see any use for it; and they know it is useful to read. But when it comes to a passage in which the reasoning employs the letters A, B, C, they skip that. Now the letters A, B, C are pronouns indispensable to thinking about Thirdness; so that the mind who is repelled by that sort of thought, is simply a mind in which the element of Thirdness is feeble. Finally, there is the geometrical mind, who is quite willing that others should snatch the power and the glory so long as he can be obedient to that great world-vitality which is bringing out a cosmos of ideas, which is the end toward which all the forces and all the feelings in the world are tending. These are the minds to whom I offer my three Categories as containing something valuable for their purpose.

These Categories manifest themselves in every department of thought; but the advantage of studying them in formal logic is, that there we have a subject which is very simple and perfectly free from all doubt about its premises, and yet is not like pure mathematics confined entirely to purely hypothetical premises. It is the most abstract and simple of all the positive sciences, and the correct theory of it is quite indispensable to any true metaphysics.

Having in 1867 made out the three categories, various facts proved to me beyond a doubt that my scheme of formal logic was still incomplete. For instance, I found it quite impossible to represent in syllogisms any course of reasoning in geometry or even any reasoning in algebra except in Boole's logical algebra. I had already ascertained that Boole's algebra was inadequate to the representation of ordinary syllogisms of the third figure; and though I had invented a slight enlargement of it to remedy this defect, yet it was of a make-shift character, plainly foreshadowing something yet unseen more organically connected with the part of the algebra discovered by Boole. In other directions, Boole's system strongly suggested its own imperfection. Pondering over these things, my note-book[1] shows that I soon came out upon the logic of relatives, and had so complemented the Boolian algebra as to give it power to deal with all dyadic relations. I learned from the great memoir of Prof. DeMorgan, dated [1859],[2] which he now sent me, that the same ground had already been considerably explored by him, though by a different route, starting from his own line of logical thought as expounded in his work *Formal Logic*.

It is now time to explain to you this Logic of Relatives. I will first give a chronology of the most important papers. I shall not mention any that are not quite fundamental. Relation was recognized as a part of the subject matter of logic by Aristotle and all ancient and medieval logicians. There is a tractate *DeRelativis* probably dating from the 14th century appended to the *Summulae of Petrus Hispanus*. Ockham and Paulus Venetus treat of it in their extended treatises on logic. Leslie Ellis made a single obvious remark on the application of Algebra to it which Professor Halsted thinks makes him the author of the logic of relatives. The remark is really fundamental. Yet it was exceedingly obvious; and was not followed out. Next came DeMorgan's Memoir. Then in 1870 [came] my first mode of extending Boole's Logical algebra to relatives. In 1883 I gave what I call the *Algebra of Dyadic Relatives*, which Schröder has fallen in love with. In the same volume O. H. Mitchell, in one of the most suggestive chapters that the whole history of logic can show, gave a method of treating a logical universe of several dimensions, which I soon after showed amounted to a new algebraic method of treating relatives generally. I call it the *General Algebra of Logic*. I think this the best of the algebraic methods. In 1890,[3] Mr. A. B. Kempe published an extended memoir of *Mathematical Form* which is really an important contribution to the logic of relatives. Schröder's third volume treats of the subject at great length, but in the interest of algebra rather than in that of logic. Finally about two years ago, I developed two intimately connected graphical methods which I call Entitative and Existential Graphs.

I shall here treat the subject by means of Existential Graphs, which is the easiest method for the unmathematical. Still, I shall not attempt to set this forth at all. I shall not even trouble you with the statement of its nine fundamental rules; far less with the score of others with which it

---

[1]   The note-book is MS. 339.
[2]   Peirce probably refers to "On the Syllogism IV" which appeared in the tenth volume of the *Cambridge Philosophical Transactions* in 1859. See *Collected Papers*, 3.643.
[3]   A memoir on "The Theory of Mathematical Form, Part I" appeared in the *Philosophical Transactions* of the Royal Society in 1886. It had been read on 18 June 1885.

is necessary to familiarize oneself in order to practice the method. But I shall describe in a confused, illogical fashion every essential feature of the system.

Let us pretend to assert anything we write down on the black board. As long as the board remains blank, whatever we may opine, we assert nothing. If we write down

> You are a good girl

we assert that. If we write

> You are a good girl
> You obey mamma

we assert both. This is therefore a copulative proposition. When we wish to assert something about a proposition without asserting the proposition itself, we will enclose it in a slightly drawn oval, which is supposed to fence it off from the field of assertions. Thus

> (You are a good girl) is much to be wished.

and again

> (You are a good girl) is false

This last assertion that a proposition is false is a *logical* statement about it; and therefore in a logical system deserves special treatment. It is also by far the commonest thing we have occasion to say of propositions without asserting them. For those reasons, let it be understood that if a proposition is merely fenced off from the field of assertion without any assertion being explicitly made concerning it, this shall be an elliptical way of saying that it is false.

> (You are a good girl)

Accordingly

> (You are a good girl)
> You obey mamma

is the copulative proposition "you are not a good girl, but you obey mamma." The denial of a copulative proposition is a *disjunctive* proposition. Thus

> (You are a good girl
> You obey mamma)

which denies that you are both a good girl and obey mamma, asserts that you are either not a good girl *or* you do not obey mamma. So

> ((You are a good girl)
> You obey mamma)

asserts that you are either a good girl or do not obey mamma, that is, "If you obey mamma you are a good girl." This is a *conditional* proposition. It is a species of disjunctive proposition. Copulative and disjunctive propositions are the two kinds of *Hypothetical* Propositions; and are generally recognized as such. Kant had a purpose in endeavoring to wrench this plain division into another form, or thought he had; and Kantian Logicians have been too feebleminded to dispute their master.

Now there remains only one little bit of a feature to complete the description of the system. But this little feature is everything. Namely, we will use a heavy line to *assert the identity of its extremities*. Thus

shall mean there *exists* something that is a good girl and is identical with something that obeys mamma. That is Some existing good girl obeys mamma. Or we can better express this thus

That is, Some existing good girl obeys the mamma of *her*. Now as long as there is but one such line of identity, whether it branches or not, the forms of inference are just the same as if there were no such line. But the moment there are two or more such lines, new forms of inference, unknown to ordinary logic, become possible. For example, let us write

That means, there is something which whoever loves obeys.

For it says there is something of which it is not true that somebody loves it and yet does not obey it. From this we can infer, by a very simple principle which means "whoever loves everything obeys something."

For it denies that there is something of which it is not true

that there is something that it does not love and yet of which it is not true that there is something that it obeys. You will find it impossible to express this simple inference syllogistically.

There is all the mathematics with which I intend to burden your poor heads. I hope you will suffer from the effects of it. I will try to give you a rest now, by passing to something easier. But mind, you certainly miss very important revelations concerning the question of nominalism and realism by being shut out from that great world of logic of which I have just given you one momentary glimpse as I whisk you by it in our railway journey through the forms of reasoning.

Any part of a graph which only needs to have lines of identity attached to it to become a complete graph, signifying an assertion, I call a *verb*. The places at which lines of identity can be attached to the verb I call its *blank subjects*. I distinguish verbs according to the numbers of their subject blanks, as *medads, monads, dyads, triads*, etc. A *medad*, or impersonal verb, is a complete assertion, like "It rains," "you are a good girl." A *monad*, or neuter verb, needs only one subject to make it a complete assertion, as

> —obeys mamma
> you obey—

A *dyad*, or simple active verb, needs just two subjects to complete the assertion as

> —obeys—
> ————   or   —is identical with—

A *triad* needs just three subjects as

> —gives—ₜto—
> —obeys both—ₐand—

Now I call your attention to a remarkable theorem. Every polyad higher than a triad can be analyzed into triads, though not every triad can be analyzed into dyads [Fig. 1].[4] Thus,

> —sells—to—for the price—

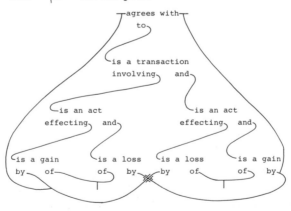

Fig. 1

---

[4]   The accompanying diagram was given with a statement that Peirce eliminated in the manuscript [Fig. 1.]

From this theorem we see that our list of categories is complete. I do not say there is no conception of Fourthness. We know that the tetrad was most sacred with the Pythagoreans. Of that, however, I believe I have the secret. If so, it argues nothing. But the fourthness in melody for example is indisputable. It contains however no peculiar intellectual element not found in Firstness, Secondness, and Thirdness. Let me say again that the connection of my categories with the numbers 1, 2, 3, although it affords a convenient designation of them, is a very trivial circumstance.

In the system of graphs may be remarked three kinds of signs of very different natures. First, there are the verbs, of endless variety. Among these is the line signifying identity. But, second the ends of the line of identity (and *every* verb ought to [be] conceived as having such loose ends) are signs of a totally different kind. They are demonstrative pronouns, indicating existing objects, not necessarily material things, for they may be *events*, or even *qualities*, but still objects, merely designated as *this* or *that*. In the third place the writing of verbs side by side, and the ovals enclosing graphs not asserted but subjects of assertion, which last is continually used in mathematics and makes one of the great difficulties of mathematics, constitute a third, entirely different kind of sign. Signs of the first kind represent objects in their firstness, and give the significations of the terms. Signs of the second kind represent objects as existing, — and therefore as reacting, — and also in their reactions. They contribute the *assertive* character to the graph. Signs of the third kind represent objects as representative, that is in their thirdness, and upon them turn all the inferential processes. In point of fact, it was considerations about the categories which taught me how to construct the system of graphs.

Although I am debarred from showing anything in detail about the logic of relatives, yet this I may remark, that where ordinary logic considers only a single, special kind of relation, that of similarity, — a relation, too, of a particularly featureless and insignificant kind, the logic of relatives imagines a relation in general to be placed. Consequently, in place of the *class*, which is composed of a number of individual objects or facts brought together in ordinary logic by means of their relation of similarity, the logic of relatives considers the *system*, which is composed of objects brought together by any kind of relations whatsoever. For instance, ordinary logic recognizes such reasoning as this:

> *A* does not possess any shade of Grey
> *B* does possess a shade of Grey
> ∴ *A* is not *B*

But the logic of relatives sees that the quality Grey here plays no part that any object might not play. Nor does the relation of possessing as a character differ in its logical powers from any other dyadic relation. It therefore looks upon that inference as but a special case of one like the following

> *A* loves everybody of the name of Grey
> *B* has a servant of the name of Grey
> ∴. *A* loves a servant of *B*.

But there *are* inferences in the logic of relatives which have not so much resemblance to anything in ordinary logic. Among the classes which ordinary logic recognizes, *general* classes are particularly important. These general classes are composed, not of real objects, but of possibilities, and hence it is that the nominalist for whom (though he be so mild a nominalist as Hegel was) a mere possibility which is not realized is nothing but what they call an "abstraction," and little better, if at all, than a fiction. It will be instructive, therefore, to inquire what it is in the logic of relatives which takes the logical position occupied in ordinary logic by "generality," or in medieval language by the *universal*.

Let us see, then, what it is that, in the logic of relatives, corresponds to generality in ordinary logic. From the point of view of Secondness, which is the pertinent point of view, the most radical difference between systems is in their multitudes. For systems of the same multitudes can be transformed into one another by mere change of Thirdness, which is not true of systems of different multitude. A system of multitude *zero* is no system, at all. That much must be granted to the nominalists. It is not even a quality, but only the abstract and germinal possibility which antecedes quality. It is at most being *per se*. A system of multitude unity is a mere First. A system of the multitude of two is like the system of truth and falsity in necessary logic. A system of the multitude of three is the lowest perfect system. The finite multitudes are all marked by this character that if there be a relation in which every individual in such a system stands to some other but in which no third stands to that other, then to every individual of the system some other individual stands in that relation. We next come to the multitude of all possible different finite multitudes, that is to the multitude of the whole numbers. A system of this multitude, which I call the *denumeral* multitude, is characterized by this, that though finite, yet its individuals have what I call *generative relations*. These are dyadic relations of relate to correlate such that, taking any one of them, whatever character belongs to the correlate of that relation whenever it belongs to the relate, and which also belongs to a certain individual of the system, which may be called the *origin* of the relation, belongs to every individual of the system. For example, in the system of cardinal numbers from *zero* up, the relation of being next lower in the order of magnitude is a generative relation. For every character which is such that if it belongs to the number next lower than any number also belongs to the number itself, and which also belongs to the number *zero* belongs to every number of the system. The next multitude is that of all possible collections of different finite collections. This is the multitude of irrational quantities. I term it the *first abnumeral* multitude. The next multitude is that of all possible collections of collections of finite multitudes. I call it the *second abnumeral multitude*. The next is the multitude of all possible collections of collections of collections of finite multitudes. There will be a denumeral series of such abnumeral multitudes. I prove that these are all different multitudes in the following way. In the first place, I say, that taking any such collection, which we may designate as the collection of *A*s, if each individual *A* has an identity distinct from all other *A*s, then it is manifestly true that each collection of *A*s has an identity distinct from all other collections of *A*s; for it is rendered distinct by containing distinctly different individuals. But the individuals of a denumeral collection, such as all the whole numbers, have distinct identities. Hence, it follows that the same is true of all the abnumeral multitudes. But this does not prove that those multitudes are all different from one another. In order to prove that, I begin by defining, after Dr. Georg Cantor, what is meant by saying that one collection of distinct objects, say the *B*s, is greater in multitude than another multitude of distinct objects, say the *A*s. Namely, what is meant is that while it is possible that every *A* should have a distinct *B* assigned to it exclusively and not to any other *A*, yet it is not possible that every *B* should have a distinct *A* assigned to it exclusively and not to any other *B*. Now then suppose the *A*s to form a collection of any abnumeral multitude, then all possible collections of different *A*s will form a collection of the next higher abnumeral multitude. It is evidently possible to assign to each *A* a distinct collection of *A*s for we may assign to each the collection of all the other *A*s. But I say that it is impossible to assign to every collection of *A*s a distinct *A*. For let there be any distribution of collections of *A*s which shall assign only one to each *A*, and I will designate a collection of *A*s which will not have been assigned to any *A* whatever. For the *A*s may be divided into two classes, the first containing every *A* which is assigned to a collection containing itself, the second containing every *A* to which is assigned a collection not containing itself. Now, I say that collection of *A*s which is composed of all the *A*s of the second class and none of the first has no *A* assigned to it. It has none of the first class assigned to it for each *A* of that class is assigned to but one collection which contains it while this collection does not contain it. It has none of the second class assigned to it for each A of this class is assigned only to one collection which does not contain it, while this collection does contain it. It is therefore absurd to suppose that any collection of distinct individuals, as all collections of abnumeral multitudes are, can have a multitude as great as that of the collection of possible collections of its individual members.

But now let us consider a collection containing an individual for every individual of a collection of collections comprising a collection of every abnumeral multitude. That is, this collection shall consist of all finite multitudes together with all possible collections of those multitudes, together with all possible collections of collections of those multitudes, together with all possible collections of collections of collections of those multitudes, and so on *ad infinitum*. This collection is evidently of a multitude as great as that of all possible collections of its members. But we have just seen that this cannot be true of any collection whose individuals are distinct from one another. We, therefore, find that we have now reached a multitude so vast that the individuals of such a collection melt into one another and lose their distinct identities. Such a collection is *continuous*.

Consider a line which returns into itself,—a ring [Fig. 2]. That line is a collection of points. For if a particle occupying at any one instant a single point, moves until it returns to its first position, it describes such a line, which consists only of the points that particle occupied during that time. But no point in this line has any distinct identity absolutely discriminated from every other. For let a point upon that line be marked [as in Fig. 3]. Now this mark is a discontinuity; and therefore I grant you, that this point is made by the marking distinctly different from all other points. Yet cut the line at that point [Fig. 4], and where is that marked point now? It has become two points. And if those two ends were joined together so as to show the place,—they would become one single point. But if the junction ceased to have any distinguishing character, that is any discontinuity, there would not be any distinct point there. If *we* could not distinguish the junction it would not appear distinct. But the line is a mere conception, as it is nothing but that which it can show; and therefore it follows that if there were no discontinuity there would *be* no distinct point there,—that is, no point absolutely distinct in its being from all others. Again going back to the line with two

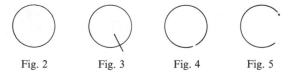

Fig. 2   Fig. 3   Fig. 4   Fig. 5

ends, let the last point of one end burst away [Fig. 5]. Still there is a point at the end still, and if the isolated point were put back, they would be one point. The end of a line might burst into any discrete multitude of points whatever, and they would all have been one point before the explosion. Points might fly off, in multitude and order like all the real irrational quantities from 0 to 1; and they might all have had that order of succession in the line and yet all have been at one point. Men will say this is self-contradictory. It is not so. If it be so prove it. The apparatus of the logic of relatives is a perfect means of demonstrating anything to be self-contradictory that really is so; but that apparatus not only absolutely refuses to pronounce this self-contradictory but it demonstrates, on the contrary, that it is not so. Of course, I cannot carry you through that demonstration. But it is no matter of *opinion*. It is a matter of plain demonstration. Even although I should have fallen into some subtle fallacy about the series of abnumeral multitudes, which I must admit possible in the sense in which it is possible that a man might add up a column of five figures in all its 120 different orders and always get the same result, and yet that result might be wrong, yet I say, although all my conclusions about abnumerals were brought to ruin, what I now say about continuity would stand firm. Namely, a continuum is a collection of so vast a multitude that in the whole universe of possibility there is not room for them to retain their distinct identities; but they become welded into one another. Thus the continuum is all that is possible, in whatever dimension it be continuous. But the general or universal of ordinary logic also comprises whatever of a certain description is possible. And thus the *continuum* is that which the logic of relatives shows the *true* universal to be. I say the *true* universal; for no realist is so foolish as to maintain that *no* universal is a fiction.

Thus,[5] the question of nominalism and realism has taken this shape: Are any continua real? Now Kant, like the faithful nominalist, that Dr. Abbot has shown him to be, says no. The continuity of Time and Space are merely subjective. There is nothing of the sort in the real thing-in-itself. We are therefore not quite to the end of the controversy yet; though I think very near it.

What is reality? Perhaps there isn't any such thing at all. As I have repeatedly insisted, it is but a retroduction, a working hypothesis which we try, our one desperate forlorn hope of knowing anything. Again it may be, and it would seem very bold to hope for anything better, that the hypothesis of reality though it answers pretty well, does not perfectly correspond to what is. But if there is any reality, then, so far as there is any reality, what that reality consists in is this: that

[5]   A marginal notation signed W. J. says "This is too abrupt along here. Should be more mediated to the common mind."

there is in the being of things something which corresponds to the process of reasoning, that the world *lives*, and *moves*, and *has its being*, in a logic of events. We all think of nature as syllogizing. Even the mechanical philosopher, who is as nominalistic as a scientific man can be, does that. The immutable mechanical law together with the laws of attraction and repulsion form the major premise, the instantaneous relative positions and velocities of all the particles whether it be "at the end of the sixth day of creation," —put back to an infinitely remote past if you like, though that does not lessen the miracle,—or whether it be at any other instant of time is the minor premise, the resulting accelerations form the conclusion. That is the very way the mechanical philosopher conceives the universe to operate.

I have not succeeded in persuading my contemporaries to believe that Nature also makes inductions and retroductions. They seem to think that her mind is in the infantile stage of the Aristotelians and Stoic philosophers. I point out that Evolution wherever it takes place is one vast succession of generalizations, by which matter is becoming subjected to ever higher and higher laws; and I point to the infinite variety of nature as testifying to her Originality or power of Retroduction. But so far, the old ideas are too ingrained. Very few accept my message.

I will submit for your consideration the following metaphysical principle which is of the nature of a retroduction: Whatever unanalyzable element *sui generis seems* to be in nature, although it be not really where it seems to be, yet must really be [in] nature somewhere, since nothing else could have produced even the false appearance of such an element *sui generis*. For example, I may be in a dream at this moment, and while I think I am talking and you are trying to listen, I may all the time be snugly tucked up in bed and sound asleep. Yes, that may be; but still the very semblance of my feeling a reaction against my will and against my senses, suffices to prove that there really is, though not in this dream, yet somewhere, a reaction between the inward and outward worlds of my life.

In the same way, the very fact that there seems to be Thirdness in the world, even though it be not where it seems to be, proves that real Thirdness there must somewhere be. If the continuity of our inward and outward sense be not real, still it proves that continuity there really is, for how else should sense have the power of creating it?

Some people say that the sense of time is not in truth continuous, that we only imagine it to be so. If that be so, it strengthens my argument immensely. For how should the mind of every rustic and of every brute find it simpler to imagine time as continuous, in the very teeth of the appearances,—to connect it with by far the most difficult of all the conceptions which philosophers have ever thought out,—unless there were something in their real being which endowed such an idea with a simplicity which is certainly in the utmost contrast to its character in itself. But this something must be something in some sense *like* continuity. Now nothing can be like an element so peculiar except that very same element itself.

Of all the hypotheses which metaphysicians have ever broached, there is none which quarrels with the facts at every turn, so hopelessly, as does their favorite theory that continuity is a fiction. The only thing that makes them persist in it is their notion that continuity is self-contradictory, and *that* the logic of relatives when you study it in detail will explode forever. I *have* refuted it before you, in showing you how a multitude carried to its greatest possibility necessarily becomes continuous. Detailed study will furnish fuller and more satisfying refutations.

The extraordinary disposition of the human mind to think of everything under the difficult and almost incomprehensible form of a continuum can only be explained by supposing that each one of us is in his own real nature a continuum. I will not trouble you with any disquisition on the extreme form of realism which I myself entertain that every true universal, every continuum, is a living and conscious being, but I will content myself with saying that the only things valuable, even here in this life, are the continuities.

The *zero* collection is bare, abstract, germinal possibility. The continuum is concrete, developed possibility. The whole universe of true and real possibilities forms a continuum, upon which this Universe of Actual Existence is, by virtue of the essential Secondness of Existence, a discontinuous mark—like a line figure drawn on the area of the blackboard. There is room in the world of possibility for any multitude of such universes of Existence. Even in this transitory life, the only value of all the arbitrary arrangements which mark actuality, whether they were introduced once for all "at the end of the sixth day of creation" or whether, as I believe, they spring out on every hand and all the time, as the act of creation goes on, their only value is to be shaped into a continuous delineation under the creative hand, and at any rate their only use for us is to hold us down to learning one lesson at a time, so that we may make the generalization of intellect and the more important generalizations of sentiment which make the value of this world. Whether when we pass away, we shall be lost at once in the boundless universe of possibilities, or whether we shall only pass into a world of which this one is the superficies and which itself is discontinuity of higher dimensions, we must wait and see. Only if we make no rational working hypothesis about it we shall neglect a department of logical activity proper for both intellect and sentiment.

Endeavors to effectuate continuity have been the great task of the Nineteenth Century. To bind together ideas, to bind together facts, to bind together knowledge, to bind together sentiment, to bind together the purposes of men, to bind together industry, to bind together great works, to bind together power, to bind together nations into great natural, living, and enduring systems was the business that lay before our great grandfathers to commence and which we now see just about to pass into a second and more advanced stage of achievement. Such a work will not be aided by regarding continuity as an unreal figment, it cannot but he helped by regarding it as the really possible eternal order of things to which we are trying to make our arbitrariness conform.

As to detached ideas, they are of value only so far as, directly or indirectly, they can be made conducive to the development of systems of ideas. There is no such thing as an absolutely detached idea. It would be no idea at all. For an idea is itself a continuous system. But of ideas those are most suggestive which detached though they seem are in fact fragments broken from great systems.

Generalization, the spilling out of continuous systems, in thought, in sentiment, in deed, is the true end of life. Every educated man who is thrown into business ought to pursue an avocation, a side-study, although it may be well to choose one not too remote from the subject of his work. It must be suited to his personal taste and liking, but whatever it is, it ought, unless his reasoning power is decidedly feeble, to involve some acquaintance with modern mathematics, at least with modern geometry, including topology, and the theory of functions. For in those studies there is such a wealth of forms of conception as he will seek elsewhere in vain. In addition to that, these studies will inculcate a strong dislike and contempt for all sham-reasoning, for all thinking made easy, for all attempts to reason without clothing conceptions in diagrammatic forms.

20

[THE PROBLEM OF MAP-COLORING] (154)[1]

He [Cantor] thereby treats a line as if it were composed of points as its ultimate parts. Doubtless he would grant that these points only constitute, as metaphysicians would say, the matter of the line, and that its form consists in their order. But I propose to show that a line does not consist of points at all.

In order to do this I have to enter into an inquiry which is perhaps not mathematical, although it is of a kind that every mathematician has to engage in. A purely mathematical memoir generally begins with positing a hypothesis and proceeds to trace out its consequences. What led the mathematician to this hypothesis he does not trouble himself to tell us. Yet the production of such hypotheses as imaginary quantity, infinitesimals, functions of various kinds, and the like has certainly been a great part of the achievements of mathematics. There is some advantage in defining the different sciences so as to make them cover what the different classes of inquirers have to do. Now almost every problem first presents itself to the mathematician in the shape of a confused state of facts out of which he has by a laborious analysis to extract his hypothesis.

However this may be, the immediate business which has first to engage our attention is the examination of that indistinct, yet clear, notion of continuity which is familiar to us all, and the identification of it with a perfectly distinct definition, such as can be employed in demonstrative reasoning. As a first step toward this, I wish to show that a continuum is not a collection, and exactly wherein it differs from every collection. For that purpose, I have to define collection. I do not aim in defining it to pin myself down to the usage of language. I am to define that unitary and scientific idea which coincides in the main with what we ordinarily mean by a collection.

[1] This paper is a variant in development of 8b in Volume 3. The opening five paragraphs of that paper are not repeated here.

We may set out from the statement that a *collection* is a definite whole of ultimate integrant parts, a statement roughly expressing what the word means, but not affording any adequate analysis of the idea, since it leaves "definite," "ultimate," "integrant," and "part" as unanalyzed notions. I know no concept of logic more difficult to analyze; and in coming to this task I think it will be useful first to make a few general logical remarks, and then to call attention to several essential features of all collections of which an analytic definition ought to take account.

Under the head of general logical remarks, I would first invite attention to certain phenomena observable in all experience. It need not excite surprise that I should call such a remark logical, since logic, although it is the most intimately allied to mathematics of any of the other sciences, is nevertheless not, like mathematics, a mere science of hypotheses but is a positive science founded on observation of facts. They are not, it is true, recondite facts demanding special observational means to bring them within the range of observation. On the contrary, they are facts the whole difficulty of remarking which arises from their being so universally present that there is no contrasting background to set them out. The particular phenomena I have in mind are that there are in experience three categories of elements. The first comprises qualities of sensation or feeling. Our attention is called to them by contrasts; but they are as they are in themselves independent of contrasts. They are not existences but mere qualities. They are what is left of experience after all relations and all actual reactions of sense are left out of account. They are not in themselves general, because generality consists in relation to particulars; but they are capable of generalization, that is, of being recognized by reflection in particulars.

The second category of elements of experience I term *reactions*. By a reaction I mean a direct consciousness of acting and of being acted on here and now. It is prominent in the sense of effort. It is impossible to have a sense of effort without a sense [of] resistance. Indeed, they are one and the same two-sided consciousness. Many psychologists hold that the sense of physical effort is nothing but a muscular sensation. Whether this be true or not, it seems to me we can detect the same sense of acting and being acted on here and now whenever we are conscious of receiving a sensation. I believe the consciousness of being acted upon and that of resisting are one and the same. But ordinary sensations in their first access are sub-conscious, deep in consciousness, and only come to our notice after a percept has been formed out of them; so that while we infer we have been acted on we do not recall the direct consciousness of it, and nothing brings the other aspect of that consciousness to our attention. In the case of a violent sensation, or shock, however, the dual character of the consciousness is marked. Our notion of this sort of consciousness is transferred by us to occurrences we see without us. We say that one body "acts" upon another; and the very word we use proclaims that we are comparing it to an exertion of the will. This consciousness is the root of all our conceptions of opposition and duality. A pair of objects is one thing "with" another; and this word "with" originally meant acting against. The same mode of consciousness is the chief ingredient of the ideas of existence and fact. The proverb that facts are stubborn things obscurely recognizes this. A fact, reality, or truth is that idea which insists upon ultimate acceptance, and finally bears down every resisting prejudice. There is no more philosophical definition of reality. The existence of an atom is not only manifested by, but consists in, its reactions against other atoms. A reaction is not merely, like qualities of feeling, not in itself general, but unlike them it is not capable of being generalized. It may, undoubtedly, submit to a general law, but it cannot itself become applicable to many things or be adequately distinguished in general terms. I term this character of reaction its anti-generality. It is an event here and now, and its essential character disappears upon generalization. I may resolve to act for general reasons; but in the action itself reason has no part and brute force, as the expressive phrase is, takes the field. The anti-generality of reaction is of great consequence in logic. We have seen that existence consists in reaction against the body of the universe. That existence is no general character is a point of great importance, insisted on by Kant and by Scotus. It furnishes the answer to the argument that God must exist because God is, by definition, a perfect being, and existence is an attribute of perfection. Anything that exists is a logical individual, which no general description can distinguish and no general law can explain, but which stands without a reason and defies the universe to annul it. Thus individuality consists in reaction. Demonstrative and relative pronouns denote their objects not by the intermediation of a description but by directly forcing attention to them. They are reactionary signs. Because fact involves reaction, no fact can be stated without the use of a reactionary word, or some sign that acts in the same way. A man walking along a road meets another who says, "There is a house on fire," and points with his finger (a reactionary sign) to where the house is visible. The first man continues his tramp and at the end of the day comes to a village where he says, "There was a house on fire." His hearer asks "Where?"; for the remark by itself hardly conveys any meaning, although the circumstances that the speaker has not dropped out of the sky but has come on foot, and that he has the air of saying something of local interest, are reactionary signs that the fire was at no very great distance. To the question "where?" the speaker may reply, "About a mile east of Bolton." This reply involves three reactionary words. A "mile" is 1760 yards; and a yard is the length between two lines engraved on a certain individual bar, which as individual cannot be distinguished by any general description, but must be recognized as a reactionary sign. "East" is doubly reactionary. For it is the direction 90° to the right of north. Now right is distinguished from left only by a difference of reaction, not in any general respect; and north is the direction of that pole of the heavens which is above the horizon in a certain part of the globe which, being individual, cannot be adequately distinguished by any general description but only again by a reactionary sign. Finally, "Bolton," being a proper name, is not a general description. It conveys a meaning only to the man in whom it forcibly awakens memories of an experience, direct or indirect of that town. It may be remarked, by the way, that right and left appear to be examples of the fact that reactions may be distinguished into classes which are not characterized by any general differences. Another example is the "local signs" by which the excitation of one nerve is distinguished from the entirely similar excitation of another nerve. But this is an obscure subject, upon which I avoid expressing a positive opinion. The force of reaction is capable of degrees. These degrees form a continuous series. This continuous serial order is itself neither a quality of feeling nor a reaction, but belongs to the third category of elements of experience. In order, however, that we may recognize that degrees of reaction do fall into series, we must have first been able to distinguish them from one another as reactions. I see before my eyes a dull red color. I compare it with my memory of a bright red. The dull red and the bright red are different feelings. The latter is the more intense feeling. But the dull red as actually before my eyes is incomparably more vivid than the bright red I only imagine. This vividness consists in the superior force with which it reacts upon me. It is not a difference in the feelings themselves. We may imagine a very low state of consciousness in which there is nothing but an unanalyzed feeling,—say a feeling of a certain purple,—without any analysis or other comparison, and without any experience of reaction. Still, that positive tone of purple would be there. It would have no particular grade of vividness, because vividness is entirely relative. This observation of a feeling shows that qualities of feeling are modes of consciousness which may in their own nature exist unmixed; although in our developed minds to which the distinction of Me and Not-me is ever present, this is impossible. A reaction, on the other hand, is an event in which one state of feeling is suddenly succeeded by another. It is absolutely impossible that such a break should take place without a feeling to be broken off, and another to take its place.

The third category of elements of experience consists in its *forms*, or combining elements. They are the intelligible, rational, order-bringing elements of experience. They are realized in nature without us and within us in laws and in regularities such as space, time, consistency of thought and purpose. But in themselves, apart from the reactions that constitute existence, they have only a mode of being consisting of what might be. Thus, an absolute vacuum void even of ether, would have a being consisting only in the fact that it afforded room into which things might be brought. An empty place is real in the sense that it is not a mere figment of mine or yours. Every rational inquirer will be compelled to admit that atoms might move into that place. That conditional compulsion constitutes its reality. But that compulsion is a rational compulsion utterly different from the blind reaction with which an atom forces itself into existence. Among forms, the most prominent in logic are generals. Let us start with any proposition, say, "There is a house that is on fire." This proposition if it is true represents a fact, a portion of experience. If it is not true it represents a similar fragment of a fictitious world of our own creation. From this proposition we strike out all re-actionary elements and we have a blank form " —— is a house that is on fire." What this represents in a *general*. Most logicians would remove the verb, and say the general is "a house on fire." They are certainly so far right, that the reactionary element of existence must be eliminated from the word "is," which in the blank form must be understood as express-ing the need that the description "a house on fire" should be applied to something before it can gain a place even in a fictitious world. The generality of a form consists in this *need*, and not merely in the *possibility* of description being applied to a limitless variety of singulars. Therein it differs from the generalizability of feeling. A quality of feeling, say scarlet, may be connected with different reactionary occasions. It may be compared with the feeling of red, with the color of magenta, or of crystals of mercuric iodide. In that sense, I call it *generalizable*. But on the other hand, it may be conceived to compose, in its simplicity, an entire conscious life, from which all reaction should be absent. A true general on the other hand is not fully what it tends to be until the blanks of a blank form are filled up. Nominalists say that a general is a mere name or conception. Realists say that it has a mode of being, not exist-ence, but only *in posse*, and that it may be real, like a law of nature, or may be a fiction. I wish here to avoid dispute on this point; but I will call attention to the circumstance that a name or conception is something that we actually do make, while a general consists not in our actually thinking something, but in the eternal possibility that it should be thought. It is an *ens rationis*. The blank form " —— is a house that is on fire," retains elements of feeling. Let these be deleted and we have " —— is a —— that is ——." This is a pure general form.

Thus feeling is the matter of experience, reaction is the breaks in it, form is the cement that welds it together. This is substantially what Hegel finds; which is natural, since he too, in his *Phänomenologie*, sets out to observe the elements of experience. But instead of recognizing these as independent elements of experience, he seeks to give an account of matters as if the third were the only one, thus on the one hand making the world too purely rational and on the other not recognizing that there is anything but the actual.

Another general logical remark which will prove pertinent is that there is another important kind of form besides generals, namely abstractions. The abstract and the general are often looked upon as much the same thing; but that is inexact logic. The abstract is a form, since its being consists in the possibility of something else. But no reactional determination will convert it into an ordinary concrete existence, for it is something added to the universe of concretes. For example, we know that if we take any three colors a fourth can be found such that no mixture of any three of the four will produce the fourth; but given these four and any fifth, there is some four of the five whose mixture will give the fifth. This is as much as to say that all colors form a three-dimensional system. This system is not itself a color, nor a class of colors. It is something over and above color whose being consists in the ways in which colors can differ. This is an abstraction; but it is not a pure abstraction, since it involves an element of feeling. An abstraction may be general, as a general character or relation, or it may be singular, like space and time. An abstraction not involving a reactionary element can only be distinct from another by differing in some general respect.

21

HOW TO REASON: A CRITICK OF ARGUMENTS (397)

ADVERTISEMENT

This work is distinguished from other logics, 1st, by the way it makes the nature of inquiry into real facts illuminate that of demonstration from fixed assumptions, and *vice versa*; 2nd, by drawing, not from any "cannot-help-thinking," but from an accurate analysis of inference, as its unavoidable consequences, rules that resolve the most obstinate logical doubts; and 3rd, by accepting (here is the upshot of the whole discussion) the principle of continuity for *lucerna pedibus* in all the dark paths of scientific and philosophical exploration.

After expounding the formal logic of tradition, together with its chief modern variations, the author points out that all reasoning of the degree of intricacy of elementary geometry (and *a fortiori* of political economy) involves the logic of relations, so that without that logic the gist of the reasoning cannot be stated. Now this study puts quite a new face upon syllogism, and shows that its air of being a process of abstract thought is merely due to the circumstance that in it the illative process is so rudimentary that an essential feature has widely escaped notice. That feature, obtrusive enough in reasoning about relations, is that in all reasoning there must be something amounting to a diagram before the mind's eye, and that the act of inference consists in *observing* a relation between parts of that diagram that had not entered into the design of its construction. In ordinary syllogism the formal statement of the two premises will serve for diagram; as

Every *S* is *M*; Every *M* is *P*.

Certainly, the conclusion cannot be drawn (unless, indeed, this diagram be translated into some other, say the Eulerian, for which something analogous will be true) without we notice, that is, make the imaginative *observation*, that the predicate of one premise is the subject of the other, so that the *nota notae* applies. Well may it tax the reader's trust, to be told that a considerable proportion of those who teach the art of reasoning should think it a sufficient answer to this to say that the general formal doctrine of relations is *extralogical*. It never was thought to be so, until its complexity was pointed out. Granting, however, for the sake of argument, that it is so; in what manner that affects the question a Philadelphia lawyer might be puzzled to show. Observe the author's line of argument. We find among inferences (not at all relating to geometry) cases in which the necessity for observation of diagrams forces itself upon our recognition; for without a very sensible exertion of the observing faculty we are impotent to draw the conclusion. Now the act of observation, having once been fully acknowledged in such a case, we easily trace it through a succession of simpler and simpler cases differing little from one another, until we are brought to perceive (what is so unobtrusive that at first we could not make sure of it) the existence of a perfectly analogous act of observation even in an ordinary syllogism. This is borrowing an instrument of thought from the naturalist, who constantly makes use of such series.

Other significant features of syllogism are brought to light in the same way; and it is further shown that syllogism does not, after all, represent the entire operation, nor the most arduous and important part of the operation of reasoning, be that reasoning as simple as syllogism can be; while the logic of relations, on the other hand, exhibits in its entirety the purely ratiocinative proceeding, — as distinguished from positive, or probable, inference. For example, logical machines have actually been constructed which will grind out relatively complicated syllogistic reasonings, and even dilemmas; though they stand one grade higher. But it is very easily shown that no machine not endowed with the power of arbitrary choice could possibly work out *the* conclusion from the simplest single premise of the relative sort; because in every such case an endless series of different conclusions are deducible from the same premise. Nor can these conclusions be summed up in one, except "implicitly," that is, in the sense in which the premise itself may be conceived to assert its most recondite and unexpected consequences. Now the curious thing is that having made this remark, which is rather obvious in the logic of relations, we turn back to ordinary non-relative logic and find, what nobody had suspected, essentially the same phenomenon there, though only *in embrio*. So that we produce a very simple inference that the logical machines cannot grind out. True they will give the conclusion (because it is logically necessary; whence the phenomenon is only *in embrio*), but they cannot show its relation to the premiss.

Again, the logic of relatives shows, for instance, that all the premisses of the theory of numbers may be compounded into one proposition, of no great complexity. And from this proposition as a premiss, by the application of stated rules,—though certainly not without observation and ingenuity in generous measure,—all the theorems of higher arithmetic may be deduced. Now, the old non-relative logic afforded so little insight into the nature of reasoning that, although "Every state of things in which a premise is true" may be the subject and "a state of things in which the conclusion is true" may be the predicate of a proposition, that logic permitted the giant Kant to erect his titanic structure upon the remark that a proposition whose falsity would involve a contradiction is one in which the premise is thought "versteckter Weise" in its subject. That he and his must maintain, or the bottom of his philosophy, I mean his view of the nature of the distinction between "analytical and synthetical judgments," falls out. The same defective logic permitted as subtle a thinker as J. S. Mill to hold that a mathematical demonstration is "a series of inductive inferences." The logic of relatives refutes this once for all, while preserving the truth that it embodies, namely that observation and ingenuity are involved in the reasoning process. For it leads us to perceive that purely deductive reasonings involve discovery as truly as does the experimentation of the chemist; only the discovery here is of the secrets of the mind within, instead of those of Nature's mind. Now the distinction between the Inward and the Outward, great and decisive as it is, is, after all, only a matter of degree. Inward phenomena are more amenable to our will, that is all.

The general idea of Hegel's Doctrine of Being forms a part of the logic of relatives, although it was thought necessary to exclude it from this volume. From this point of view a method for the investigation of those categories emerges of which Hegel's battledore-and-shuttlecock movement is one of the lower forms. This method, owing to the exclusion referred to, is in this volume only meagrely illustrated by showing how from the general idea of inference with its essential limitation to a scale of two,—the *true* and the *false*,—which is its only measure of propositions, is developed the conception of an unlimited scale of quantity. This finally leads to the conception of continuity, for which Dr. Georg Cantor's analysis is accepted with some correction. After respectful consideration of the different opinions that have found able defenders, the conclusion is reached that the conception of continuity involves no contradiction, and cannot be dispensed with. Two kinds of infinity are admitted; and it is shown that there is nothing contradictory, nor inconceivable, nor even very difficult about either. From this discussion flows the irresistible consequence that infinitesimals exist wherever there is continuity. The logic of the differential calculus is set forth from this point of view. The doctrine of limits is not denied. On the contrary, its value in theory and in special cases, perhaps even in practice is recognized. But it is shown to be an unnecessarily roundabout way.

The idea of quantity being developed is next applied in logic itself; and the statistical numbers of relative terms are considered, of which *probability* is a special case. This idea is found inapplicable to the imaginary world of mathematics; and what goes by the name of local probability is not, properly speaking, probability, at all. For insofar as things are imaginary, they are as we choose them to be; while probability supposes an order of experience independent of our will. In reference to probability propositions are of three classes.

The first consists of those which assert logical possibility or impossibility.

The second consists of assertions about the course of experience, in the long run.

The third consists of those which assert that something has been selected by a principle which determines that it shall belong to a genus while leaving accident to decide what one of that genus it shall be.

The philosophy of probability is considered. The rules of the calculation of chances are deduced and illustrated. The principal classes of problems are treated. The law of high numbers, the probability curve, and the theory of errors are particularly emphasized. This leads to statistical syllogism.

Using lower case letters for relative terms, and Greek miniscules for numerical ratios (which may have any positive values) the Aristotelian Figures are as follows:

FIGURES OF STATISTICAL SYLLOGISM

First Figure (Deduction)

In the long run, on the average, each M will be k of v Ps;
By the conditions of selection, on the average, among ρ Ss there is an l of an M;
∴ *Probably and approximately*, among ρ Ss there are ls of ks of v Ps.

To explain precisely the words "probably and approximately" would exceed present limits. Suffice it to say they imply that small deviations from ρ and v are probable, large deviations improbable; and further that upon indefinitely increasing the number of instances the conclusion will be justified, its error being indefinitely diminished.

Second Figure (Abduction)

In the long run, on the average, each M will be k of v Ps;
By observation, we find among ρ Ss are ls of ks of v Ps;
∴. *Provisionally and approximately, by the conditions of selection*, among ρ Ss is an l of an M.

"Provisionally," here, means that upon indefinitely increasing the number of instances the conclusion will be *modified* so that its error will be indefinitely diminished. "Probably" is omitted, because it is held (against the doctrine of Inverse Probabilities) that no definite value of the probability exists.

Third Figure (Induction)

By the conditions of selection, among ρ Ss is an l of an M;
By observation, we find among ρ Ss are ls of ks of v Ps;
∴. *Provisionally and approximately, in the long run*, each M will be k of v Ps.

Here is a theory which makes induction, hypothesis, and all positive synthetic inference to be nothing but *indirect probable syllogism*. It is essentially the same as that given by the author in 1867. If it be admitted, the rules of such inferences follow with demonstrative certainty. There is one of these rules which has not usually been sufficiently insisted upon; and examples are given to show how able men have been led into fallacious conclusions by neglect of it. That rule is that instances that have gone to suggest a theory cannot also be counted on as availing to raise the theory above the rank of a mere suggestion; or, to state the rule more clearly, a theory can only be supported by verified predictions, or what are virtually predictions.

Another consequence of the theory is a doctrine which, though by no means novel, is of no little importance, namely, that absolute certainty is absolutely unattainable. Even propositions that seem logically necessary are not quite certain; though here the defect is trifling. But all results of induction and similar reasoning are really ratios and have a sort of probable error, such as all the exactest observational sciences acknowledge in all their determinations. That is, there is really no reason to hold any such conclusions for absolutely exact, but there are overwhelming arguments to the contrary, excepting only cases where there is reason to think the ratio inferred cannot vary continuously. Felix Klein, one of the greatest of living mathematicians, closes a lecture on axioms with these words: "thus we are led in geometry to a certain modesty, such as is always in place in the physical sciences."

This is sound doctrine: but to listen to Büchnerite lecturing about Law and Energy, one would not dream that "a certain modesty" had any part in science. The importance of this species of acatalepsy for philosophy is plain, especially where scientific results generalized beyond all bounds clash to all appearance with common experience.

If, however, the field of possibility is not continuous, it may be that asbolutely exact conclusions *are* warranted. For this reason, among others, it is proper to consider the evidences for the reality of continuity. That we have a perfectly consistent *conception* of it has been shown. But what evidence is there that it is real? The author maintains that continuity is a datum of direct presentation. In this he is confirmed by the psychological studies of Professor W. James, to which he adds sundry arguments of his own. Besides, even if continuity be not given intuitively, its reality answers the logical conditions of a good theory.

The reality of continuity once admitted, the next question is what are we to regard as continuous and what as discontinuous? It is shown that to say anything is continuous is to leave possibilities open which are closed by asserting it is discontinuous. Accordingly, a regulative principle of logic requires us to hold each thing as continuous until it is proved to be discontinuous. But absolute discontinuity cannot be proved to be real, nor can any good reason for believing it so be alleged. We thus reach the conclusion that, as a regulative principle, at least, ultimate continuity ought to be presumed everywhere.

The reality of continuity appears most clearly in reference to mental phenomena; and it is shown that every general concept is, in reference to its invididuals, strictly a continuum. This (though asserted by Kant and others) did not appear quite evident as long as the doctrine of generals was restricted to non-relative terms. But in the light of the logic of relatives, the general is seen to be precisely the continuous. Thus, the doctrine of the reality of continuity coincides with that opinion the schoolmen called realism; and though as they held it, it was a crude notion enough, yet as Dr. F. E. Abbot has proved, in another dress it is the doctrine of all modern science.

This point reached, a massive foundation has been laid for a philosophy which shall not take for its first axiom a principle utterly irreconciliable with all spiritual truth. And so, with some lighter matter the volume is brought to its L'Envoy.

22

ON PHYSICAL GEOMETRY (257)

INTRODUCTION

Geometry has two parts. One frames an idea of space, whether suggested by real knowledge or not, develops it, generalizes upon it, and traces out consequences. This is a work of pure mathematics. The other part of geometry is an inquiry into the properties of real space. Being an investigation of the inorganic world, it is a branch of physics. There is no assurance of its conclusions being correct, except so far as they are based upon observations; and like other conclusions from observations they are affected by a probable error.

No doubt, our natural ideas of space are very nearly true; perhaps, considering their original vagueness, we may say that as far as they go they are quite true. But neither that nor any other facts in our possession, prove or tend to prove that those natural ideas, extended so as to be rigidly precise, are to be trusted implicitly as quite free from error, without the most thorough observational verification.

Modern advances in logic, in psychology, and in the history of thought have strengthened very much the general conception of physical science which was set forth by Whewell in 1837, namely that such science results from an evolution or growth of innate ideas, under the corrective, and suggestive, and supplementing influence of observation. The difference in the mode of development of geometry and of other branches of mechanics seems to be merely one of degree. Without *il lume naturale*, to which he appeals, Galileo could not have taken the first steps in dynamics, while the first suggestions from within have frequently been wrong. In like manner, our natural ideas of space are marvellously close to the truth; but the balance of observational evidence now is that they are not quite correct.

I shall not, in this paper, stop to examine the arguments of those who demur to any inquiry into physical geometry, but shall proceed, at once, to the inquiry itself.

I divide the properties of space into three classes. The first is made up of its *intrinsic properties* as a homogeneous continuum of three dimensions having a certain sort of connectivity. The second embraces its *optical properties*, consisting in the intersections of rays and planes. The third is composed of its metrical *properties*, such as that measurable angles are finite fractions of a whole turn and that the sum of the three angles of a finite triangle differs little from 180°.

PART I. INTRINSIC PROPERTIES OF SPACE

First Property. Space is supposed to be continuous. It is a matter of no small difficulty to make out precisely what that means, to analyze the conception of continuity. It is best to begin by considering continuity in one dimension. Time is likewise said to be continuous. An *instant* is a time that has no parts.* (*To reply that an instant is not a time would be to urge a mere verbal objection.) One element of the continuity of time is that there is no instant *next* after any instant. That is, there are always between two instants a series of instants without a first nor a last. This property is that of the infinite divisibility of time. The writer has proposed to term it *Kanticity*, because Kant gave this as the essence of continuity, of which it is only one element.

Another element of continuity may be described as follows: Let there be an endless series of instants, that is, a series such that there are instants of the series later than any given instant of the series. Such, for example, are the instants at which Achilles reaches the spot at which the tortoise was the previous instant of the series. Let there also be a beginningless series of instants, that is, a series having instants previous to each instant of the series. Let all the instants of the beginningless series be subsequent to all the instants of the endless series. Then, the property in question is that there is always at least one instant between those two series of instants. I will call this the *limit-property*; and these two together I will say constitute *quasi-continuity*.

It is quite evident from a logical point of view that the two propositions which make up quasi-continuity do not conflict. It also becomes familiar to us, when we consider that the first is true of the system [of] rational numbers and both of the system of all real numbers. But that those two together do not quite answer to the conception of continuity seems to be shown by the fact that both are also true of the system of numbers which remains after excising from the system of real numbers all which are greater than any one number and not greater than any other. Thus if the numbers continue from $-\infty$ to 1 inclusive and then all those above 1 up to 2 inclusive are omitted, and all above 2 remain, this system possesses both the two properties in question.

The two properties will not become contradictory if instead of instants or points in the enunciation of them anything else be substituted not in itself contradictory. This is logically evident. We may therefore substitute for points in that enunciation series of points similar to the whole series; and may enunciate without contradiction the following third property. If there be a series of systems each of innumerable points; and if the series of systems be such that after any system there is a system, that is, all the points of the one after all the points of the other; and if subsequent to that series of systems there be another series of systems, every system of this series having another such system before it, then there shall always be some system of innumerable points subsequent to the first series of systems and previous to the second series of systems. This property involves no contradiction; but it is not true of the system of real numbers, unless we admit infinitesimal differences. But if we admit infinitesimals in the sense of finite numbers divided by infinite numbers, that is, if we admit that it is possible for equidistant points on a line to be as many as the points of a line without being in contact, then the property in question is possessed by such a line of points.

If the points on a line possess this third property, the system of real numbers is inadequate to discriminating them. If A represents the distance of any point from the origin and $A + i$ the distance of another point sufficiently near $A + i = A$, that is, the same number will attach to both; and in that sense $i = 0$. But a new system of numbers may be affixed to the points between the points A and $A + i$, and according to that way of measuring i and 0 will cease to be equal.

If a series of points possesses this third property I shall say that it is *continuous*.

Let us now see what evidence exists or can be expected that space possesses any or all of these properties.

Of course, none of them can be directly observed. They must be inferred to account for something we do observe. The credibility of a hypothesis is greatly increased when we can satisfy ourselves that the property supposed belongs to *anything*, especially if that thing seems to be of the same general nature as the subject of the hypothesis. Accordingly, it is very important to inquire whether Time is continuous.

Now a fact that goes to show that time is continuous is that our consciousness seems to flow in time. If we suppose that we are immediately sensible of time, the origin of the idea is explained; but if not, then we must cast about for some other way of accounting for our having the idea. Now there are great difficulties in the way of supposing that we are immediately conscious of time, and therefore of the past and future, unless we suppose it to possess the third property of continuity, so that we can be immediately conscious of all that is within an infinitesimal interval from any instant of which we are immediately conscious. . . .

23

METHODS OF REASONING (748)

FIRST METHOD. THE SIMPLE CONSEQUENCE

By a proposition, is meant, in logic, anything which can be held for true, or which can be supposed to be so held. Thus, "all men are free and equal," is a proposition, and it is at the same time composed of two propositions, that all men are free and that all men are equal. It is plain that any compound of propositions forms a proposition.

Reasoning is accepting a proposition as true, while recognizing some other proposition as the reason for it. By recognizing a proposition as the reason for another, is meant recognizing that the belief in the former causes the belief in the latter in a way in which true propositions will not (at least usually) produce belief in such as are false.

A proposition accepted on account of a reason is called a conclusion, and is said to be inferred from that reason. The reason, itself a proposition, is generally conceived as composed of several propositions, and these are termed the premises of the inference.

The reasoner always conceives, however vaguely, that there is some general rule by which he passes from premises to conclusion; otherwise he would say that the premises were followed by the conclusion, but not that they caused or determined the latter. If this rule be conceived to be one which will never lead from true premises to a false conclusion, the reasoning is said to be necessary; but if the rule is conceived to be one which will only lead to the truth on the whole, or in the long run, while it may occasionally lead to error, the reasoning is said to be probable. For the present, we shall confine our attention to necessary inferences.

The rule connecting the premises with the conclusion of a necessary inference may always be stated in this form: that if propositions of a certain description are true, then a proposition related to them in a certain way will also always be true. This rule, so stated, is called the leading principle of the inference.

An inference whose leading principle is true is said to be a valid inference; its reason is called sound, and its conclusion is said to really follow from the premises.

All truth is ascertained by observation; but a proposition the truth of which can easily be ascertained by observing the parts of a diagram, or something of the sort, which we can construct at pleasure, is said to be *evident*. Thus, it is evident that a triangular pyramid of cannonballs having two balls on a side consists of four balls in all. A leading principle that is evident is called a logical principle, the inference to which it belongs is termed complete, and its conclusion is said to follow logically from the premises. For example, suppose a silver dollar fails to ring when thrown upon a table: the necessary conclusion is that it is counterfeit. This is a valid, but not a complete inference, for its leading principle is that every dollar, not counterfeit, rings when thrown upon the table; which is true, but not evident. To complete the reasoning, we add this leading principle as a premise, whereupon we obtain the following inference: "Every good dollar rings, etc.: this dollar does not ring, etc.: hence, this dollar is not good." The leading principle of this is that nothing possesses any character which is never found in conjunction with another character which it is known to possess: and this is evidently true.

The rudest kind of reasoning consists in drawing a conclusion from a single premise. The premise, in this case, is called the antecedent, the conclusion the consequent, while the inference itself is called a consequence, which name is likewise given to the leading principle. I reason in this way when I hear a person make a statement, and jump at once to the conclusion that what he says is true, without stopping to consider whether he is a veracious witness or not; or when I see a magician put a pocket-handkerchief in a box, and am unreflectingly led to believe it remains in the box five minutes later, without asking myself whether this is what generally happens in magicians' tricks or not. Such reasoning, however sound, is plainly incomplete, and represents an entirely unreflecting and uncritical state of mind. There are, however, simple consequences that are logically complete: these are usually termed immediate inferences. Such, for example, is the inference from "There is a number smaller than any other," to "There is but one number than which no other is smaller." The principle that having once accepted a proposition we are right in adhering to it, until the matter is reexamined, may be represented by the formula, "A, therefore A." This is called the *identical inference*.

It may here be remarked that it will be found particularly convenient, in discussing methods of reasoning, to denote propositions by letters of the alphabet; so that "If A, then B," will be a compendious way of saying that if a certain proposition, A, is true, than a certain proposition, B, is true.

SECOND METHOD. THE *MODUS PONENS*

The moment that a person who has made the incomplete inference, "A, therefore B," is led to reflect upon or criticize his procedure in the slightest degree, he will recognize the leading principle "If A, then B," as a premise and thus reform his inference, as follows:

> If A, then B,
> But A:
> Hence, B.

The form of inference is called the *modus ponens*.

A proposition consisting of two clauses connected by an "if," so that one proposition is said to follow from the other, called in grammar as it formerly was in logic, a conditional sentence, has in our times been more commonly termed hypothetical by the logicians. That which, in grammar, is called s simple sentence, is in logic termed a categorical proposition. Such a proposition is put by logicians into the standard form, "S is P," where S and P represent two names. Thus, the dog runs is stated in logic in the form, the dog is running. This often violates the usage of language, but logic has its own forms of expression, and the important point in this case is to show that the simple sentence is equivalent to saying that objects to which one name applies another name applies. Propositions asserting that if A is B is, whenever A is B is, wherever A is B is, and whatever A is B is, are all of the same general nature, and any fact which can be stated in the form of a hypothetical proposition can also be stated in the form of a categorical proposition. Hypothetical propositions usually assert nothing with regard to the actual state of things and relate only to what is possible. This range of possibility sometimes includes all that could without self-contradiction be supposed to be true, sometimes all that is in accord with physical laws, and generally all that in some supposable state of knowledge would not be known to be false. Take, for example, the proposition "If this patient has yellow fever his temperature will shortly rise." Here, we are supposed not to know whether the patient has yellow fever or not, but we are supposed to know the different courses that the disease takes in patients like the one in hand, and the hypothetical proposition is equivalent to this categorical one: Every yellow fever patient like the one in hand shortly experiences a rise in temperature. To take another example, suppose If I am well tomorrow, I shall go fishing. We do not know whether I shall be well or not; we further do not know but I shall be well and in good humor, we do not know but I shall be well and in bad humor, etc. But the statement is that every such case is a case in which I should go fishing. It is easy to see that every hypothetical proposition . . . In many cases the state of ignorance supposed is fictitious, which is indicated in grammar by the use of past tenses. "If the witness were telling the truth, he would not blush." We here go back to an imaginary state of knowledge in which we do not know whether the witness blushes or tells the truth but among all the possibilities that open, all those in which he tells the truth are cases in which he does not blush. In like manner, the meaning of every hypothetical proposition can be precisely expressed in categorical form, the only difficulty being to settle the exact meaning which the hypothetical proposition, usually a vague mode of speech, bears. The following are offered as exercises.

If the population of the United States increases, from 1880 to 1890, as fast as that of the state of Ohio did from 1870 to 1880, and if the population of France increases at the same rate as it did from 1872 to 1881, then at the end of the century, the population of the United States will be triple that of France.

At the end of a century, the population of the United States will be as dense as that of Europe is now, unless some catastrophe should prevent.

If an inhabitant of the planet Jupiter can climb a mountain on that planet as easily as a man can climb a proportionally high mountain on the earth, the former must have 30 times the strength of the latter as compared with his mass: and if, further, the inhabitant of Jupiter is as much larger than a man as Jupiter is than the earth, and is no more likely to break his bones by muscular contraction than a man is, the strength of the material of his bones must be 300 times as great as that of ours per unit of mass.

If Juan Perez de Manchena had not been the confessor of Queen Isabella, Columbus would never have discovered America.

If Newton had not discovered the law of gravitation, some other man would.

If Sir Philip Francis wrote the letters of Junius, he was a singular instance of a vain and ambitious man never showing his real ability. . . .

Every hypothetical proposition describes a state of things in its antecedent, and then in its consequent assigns another description thereto; and almost every categorical proposition, though it be not equivalent to a hypothetical one, amounts to the assertion that something it describes may be otherwise described. These two descriptions may [be] expressible in general terms, and the latter, at least, usually is so; but it is an important theorem of logic that no proposition whatever can be completely and fully expressed in general terms alone. For let us consider what is meant, precisely, by a general term or general description. Such a description is a single word or phrase or series of phrases, or something equivalent to that,—say, a signal from a code. A description is thus a conventional sign of some kind. It may have some resemblance to the thing signified, like an onomatopoeic word, and that resemblance may have had something to do with its selection as a sign. Still, once selected, it has become conventional. By this, I do not mean that it was established by any treaty, but only that it is significant of its object by virtue of a mental habit associating together the word and thing. A habit is a general rule operative within the organism, and hence a conventional sign is naturally general. Besides, an idea has no individual identity,—two ideas exactly alike are the same idea. Accordingly, if a conventional sign is not general, it is not purely conventional. Now an object can always be imagined, or at least supposed, which shall reunite any two descriptions that are not absolutely contradictory, so that a proposition that merely says that among supposable objects there is one of a given description to which another given description is applicable might as well be left unsaid. But though descriptions apply to any supposable objects of the sort described, propositions are usually restricted to objects now existing, or to those which have existed or will exist, or to such as can exist in conformity with physical laws and other conditions, or to those that are met with in some realm of fiction, etc. This restriction, an essential part of the meaning of the proposition, cannot, for the reason just given, be expressed by any general description. It is usually indicated by a reference to an avenue of sense or source of information by which we are placed in real relation with the world of objects to which the proposition refers. Very often, it is only the tone of the discourse which gives us to understand whether what is said is to be taken as history, physical possibility, or fiction. In other cases, phrases such as "the fact is," "according to the nature of things," and the like are employed, and these no doubt partake of the nature of conventional signs. Yet so far as they refer us to some living experience or to something with which we have been made familiar by its action on us and ours on it, they signify their objects, not by virtue of habitual association merely, but by the force of a real causal connection. In short, though all words are to some extent conventional, yet some of them do not possess that generality which is the distinguishing mark of purely conventional signs. "Here," "now," "this," are rather like finger-pointings which forcibly direct the mind to the object denoted. They resemble the letters of a geometrical diagram in merely serving to bring the mind back to the same identical object which has previously come before it; and differ from conventional signs in being made to signify their objects by being actually attached to them. When I say I mean discourse to refer to the real world, the world "real" does not describe what kind of a world it is: it only serves to bring the mind of my hearer back to that world which he knows so well by sight, hearing, and touch, and of which these sensations are themselves indices of the same kind. Such a purely demonstrative sign is a necessary appendage to a proposition, to show what world of objects, or as the logicians say what "universe of discourse" it has in view.

Inferences like the following are to be referred to the modus ponens:

Whenever A is B is,	Wherever A is B is,	Whatever A is B is,
Now A is,	Here is A,	This is A,
Hence, now B is.	Hence, here is B.	Hence, this is B.

Forms like these are the simplest examples of logical inferential formulae. They represent, in a diagrammatical way, the relations of the parts of an inference. The premises of every logical inference state that certain relations subsist between certain objects, and to draw the conclusion, we have to contemplate these relations and to see that where these relations subsist something else is true. It is indispensable that an act of observation should be performed. For instance, in reasoning by the *modus ponens*, it is necessary actually to notice that the proposition stated in the second premise is the same as the antecedent of the first premise. It will not, therefore, be sufficient to state the relations; it is necessary actually to exhibit them, or to represent them by signs the parts of which shall have analogous relationships. This principle, of fundamental importance in logic, will be made more clear when we come to more difficult forms of reasoning. It explains why diagrammatic representation of inferences are to be preferred to general descriptions.

It will be seen that we have to do in logic with three kinds of signs, 1st, diagrams, which stand for their objects by virtue of being like them, 2nd, indices, which stand for their objects by virtue of being connected with them, and 3rd, descriptions, which stand for their objects by virtue of being mentally associated with them.

THIRD METHOD. BARBARA

A slightly more advanced method of reasoning is shown in the following schema:

> If A, then B;
> If B, then C.
> Hence, if A, then C.

This mode of reasoning (stated in categorical form) is known to logicians as the syllogism in Barbara. Several of the rules current in logic are only forms of the statement that Barbara is valid reasoning. Such is the maxim "Nota notae est nota rei ipsius," the mark of a mark is a mark of the thing itself, where the mark of a mark must be interpreted as a description applicable to everything to which another description applies. Such too is the dictum de omni, "whatever is asserted of the whole of a class is asserted of every part thereof." Another way is to say with DeMorgan that the relation of antecedent to consequent is a transitive one, that is, if A is in this relation to B, and B to C, then A is in this relation to C.

A description which applies to everything to which another description applies is said to be wider than the latter and to contain the latter under it. The validity of Barbara may therefore be stated in the form, Whatever contains something that contains a third itself contains that third. This conception is used in the technical names of the parts of the syllogism. Stating Barbara in the categorical form, we have

> Every M is P,
> Every S is M,
> Hence, every S is P.

The description M, which occurs in both premises but is eliminated from the conclusion, is called the middle term, S and P the extremes. The description S, which is contained under the middle term, is called the minor extreme; the description P, under which the middle term is contained, is called the major extreme; and the same adjectives are applied to the premises in which S and P are respectively found.

The major premise of Barbara is a Rule, the minor premise is the statement that a certain Case comes under that rule, while the conclusion gives the Result of the rule in that case. Barbara, thus consists in the direct application of a rule; and every proceeding of that sort is of the same nature as the inference in Barbara. When, in consequence of receiving a certain sensation, we act in a certain way by force of a habit, which is a rule operative within the organism, our action cannot be said to be an inference, but it conforms to the formula of Barbara. So when an effect follows upon its cause by virtue of a law of nature, the operation of causation takes place in Barbara. These instances give us an inkling that logic is far more than an art of reasoning: its forms have psychological and metaphysical importance.

FOURTH METHOD. INDIRECT INFERENCE

To say that from two premises, P and Q, a certain conclusion R follows, is the same as to say that from either of the premises, say P, we may conclude "If Q then R," that is to say, P being once granted, R follows from Q. Applying this transformation to the syllogism in Barbara,

     S is M, M is P: therefore, S is P.

we obtain two forms of immediate inference, viz.:

     M is P:
     Hence, If S is M, S is P.

and

     S is M:
     Hence, If M is P, S is P.

Now the following is a regular inference in Barbara:

     If S is M, S is P;
     If S is P, then X.
     Hence, If S is M, then X.

But the minor premise here (the first one) is the conclusion of the first of our two immediate inferences; consequently, the following inference is valid:

     M is P,
     If S is P, then X;
     Hence, If S is M, then X.

This form of inference is called the minor indirect syllogism. The following is a concrete example:

     All men are mortal,
     If Enoch and Elijah were mortal, the Bible errs;
     Hence, if Enoch and Elijah were men the Bible errs.

Again, we may start with this syllogism in Barbara:

     If M is P, then S is P;
     If S is P, then X;
     Hence, If M is P, then X.

The first premise of this is the conclusion of the second immediate in-ference given above. Substituting for it the premise of that inference, we have the following form, which is called the major indirect syllogism:

S is M,
If S is P, then X:
Hence, If M is P, then X.

Example:

All patriarchs are men,
If all patriarchs are mortal, the Bible errs;
Hence, if all men are mortal the Bible errs.

We have already seen that the assertion that a thing exists is not of the nature of a general description. It may now be added that the asser-tion that any sort of thing does not exist is of that nature. In point of fact, every general description amounts to the statement that some sort of thing does not exist. To say that all crows are black is the same as to say that non-black crows do not exist. Propositions are thus of two essen-tially different kinds; those which assert that one thing follows from another, being thus equivalent to hypotheticals, and these amount to denying the existence of something: and secondly those which affirm the existence of something and thus amount to denying that one thing follows from another. "The cockatrice has a serpentine tail," means that if anything is a cockatrice it has a serpentine tail, and amounts to denying the existence of a cockatrice without such a tail. It may be noticed by the way that though the categorical is thus put into hypo-thetical form, yet the relation of subject to predicate is not shown to be nothing more than what is involved in the relation of antecedent to con-sequent, but that the former relation is a special form of the latter. Propositions of the first class are termed universal, those of the second particular. A compound proposition may be partly of one kind and partly of the other. Thus, if I say "all the mammals of Australia carry their young in pouches," I affirm that there are mammals in Australia, and at the same time state that if there are any they carry their young in pouches. There are other propositions, which in their main import belong distinctly to one of the two classes, and yet imply the truth of a proposition of the other class. Take this example: "Some men would not be terrified by any ghost." This is distinctly the affirmation of the existence of something; yet it implies that if there be a ghost, there will be men whom it cannot terrify. Again, "If all men are sinners, some men are fools," is distincly hypothetical, yet it also implies the existence of some men.

To deny a proposition, then, is to apply to it a general description, that of being false. If now in the second of the two forms of immediate inference given above we replace the P by the special description "false," and put the premise into hypothetical form, we get

If S, then M.
Hence, if M is false, S is false.

This is the principle of the mode of inference called the reductio ad absurdum, by which we show that if a proposition, S, were true, another proposition, M, would necessarily be true, but that M is not true, so that S cannot be true. This inference takes the name of the modus tollens when put in the form

If S then M,
M is false,
Hence, S is false.

If in the formula given for the minor indirect syllogism, we put "false" for X, we obtain the following:

M is P,
That S is P is false,
Hence, that S is M is false.

This mood is called Bocardo. Making the same substitution in the major indirect syllogism, we get

S is M,
That S is P is false,
Hence, that M is P is false.

This mood is called Baroko. Baroko and Bocardo are simple cases of the reductio ad absurdum. Baroko is useful for establishing distinctions between things, as when we reason,

All fishes are propagated from eggs,
But it is not true that whales are propagated from eggs,
Therefore, it is not true that whales are fishes.

Bocardo is useful for establishing exceptions to rules, as thus:

The dingo is a mammal of Australia,
But it is not true that dingos carry their young in pouches,
Therefore, it is not true that all the mammals of Australia carry their young in pouches.

There are many other modes of syllogism derivable in similar ways, but they are of no great importance, and will be passed by for the moment.

FIFTH METHOD. THE PRINCIPLE OF EXCLUDED MIDDLE

We have thus far considered but a single property of negation, namely that denying a proposition is applying a general to it. The formula of the modus ponens gives us by transformation this form of immediate inference:

    S is true,

    Hence, if from S P follows, then P is true.

Substitute for P the special description "false" and we have the immediate inference

    S is true,

    Hence, the denial of S is false.

The statement of the validity of this general inference is termed the principle of contradiction. The converse of this principle, namely that from the denial of the denial of a proposition the truth of that proposition follows, is termed the principle of excluded middle. This principle constitutes a distinct axiom concerning negation. It is not involved in any of the traditional forms of inference, but it is involved in certain modes introduced into logic by DeMorgan. I do not mean that reasoning involving this principle had never been used before, but only that such forms had escaped recognition as distinct from others, and indeed DeMorgan himself does not see that they are essentially different from the rest. The inference from "If A is true, B is false," to "If B is true, A is false," depends merely on the principle of contradiction, that is on "false" being a general predicate. For "false" put any predicate, C, and it will be equally true that if it be admitted that A being granted, C follows from B, we must also admit that B being granted, C follows from A. On the other hand, the inference from "If A is false, B is true," to "If B is false, A is true," is of an entirely different nature, and depends on the principle of excluded middle.

24

SKETCH OF A NEW PHILOSOPHY (928)

DEVELOPMENT OF THE METHOD

1. It is not a historical fact that the best thinking has been done by words, or aural images. It has been performed by means of visual images and muscular imaginations. In reasoning of the best kind, an imaginary experiment is performed. The result is inwardly observed, and is as unexpected as that of a physical experiment. On the other hand, the success of outward experimentation depends on there being a reason in nature. Thus, reasoning and experimentation are essentially analogous.

2. According to what law are fruitful conceptions developed? Their first germs present themselves in concrete and confused forms. The human mind, without being able to draw certain truth from its own depths, has nevertheless a natural bias toward true ideas of force and of human nature. It finds such ideas simple, easy, natural. Natural selection may be supposed to account for this to some extent. Yet the first origins of fruitful ideas can only be referred to chance. They promptly sink into oblivion if the mind is unprepared for them. If they meet allied ideas, a welding process takes place. This is the great law of association, the one law of intellectual development. It is very different from a mechanical law, in that it is only a gentle force. If ideas once together were rigidly associated, intellectual development would be frustrated. Association is external and internal, two grand divisions. These act in two ways, first to carry ideas up and make them broader, second to carry them out in detail.

3. Development of the theory of assurance, mainly as in my previous writings.

4. Philosophy seeks to explain the universe at large, and show what there is intelligible or reasonable in it. It is thus committed to the notion (a postulate which however may not be completely true) that the process of nature and the process of thought are alike.

Analysis of the logical process. (Much of this in my previous writings.) The conceptions of First, Second, Third, rule all logic, and are therefore to be looked for in nature. Chance and the law of high numbers. The process of stirring up a bag of beans preparatory to taking out a sample handful analogous to the welding of ideas.

The only thing ever inferrable is a ratio of frequency. This truth compels the introduction of the conception of continuity throughout logic.

Logic teaches that Chance, Law, and Continuity must be the great elements of the explanation of the universe.

5. The philosopher must regard opinions as so many vivisection-subjects, to be studied for their natural history affinities. He must even take this attitude to his own opinion. Opinion has a regular growth, though it may get stunted or deformed. To take the next step in philosophy vigorously and promptly, we must study our own historical position.

The drama of the last three centuries of struggling thought, in politics and sociology, in science, in mathematics, in philosophy, briefly narrated.

The ruling ideas of today.

6. How large numbers bring about regular statistics in social matters. The peculiar reasoning of political economy; the Riccardian inference. Analogy between the laws of political economy and those of intellectual development. This teaches the necessity of similar ideas in philosophy. Darwin and Adam Smith.

7. Mathematicians have exploded axioms. Metaphysics was always an ape of mathematics, and the metaphysical axioms are doomed. The regularity of the universe cannot be reasonably supposed to be perfect. Absolute chance was believed by the ancients.

8. The present deadlock in molecular physics. If we are to cast about at random for theories of matter, the number of such theories must be at least ten million, and it must take the race a century to test each one. Hence the chances are that there will be five million centuries before any substantial advance.

Hence, the only hope is to get some notion what laws and forces are naturally to be expected. We must have a natural history of laws of nature. The only way to attain this is to explain these laws, and the only explanation is to show how they came about, how they have grown. But if they are growing, they are not absolutely rigid. Errors of observation and real chance departures from law.

9. The Darwinian hypothesis stated in skeleton form, or a Darwinian skeleton key to philosophy. Its elements are *Sporting* or accidental varia-

tion, *heredity* not absolute but a gentle force, and adaptation, which means reproductivity. This key opens a theory of evolution applicable to the inorganic world also. The Lamarkian principle is limited, the Darwinian general.

10. The modern psychology and the law of association. Habit and breaking up of habit. Feeling sinks in habit. Application to philosophy.

11. The monism of the modern psychologists is really materialism. The unreasonableness of it. The idea of supposing a particular kind of machine feels is repugnant to good sense and to scientific logic. "Ultimates" cannot be admitted. The only possible way of explaining the connection of body and soul is to make matter effete mind, or mind which has become thoroughly under the dominion of habit, till consciousness and spontaneity are almost extinct.

12. The Absolute in metaphysics fulfills the same function as the absolute in geometry. According as we suppose the infinitely distant beginning and end of the universe are *distinct*, *identical*, or *nonexistent*, we have three kinds of philosophy. What should determine our choice of these? Observed facts. These are all in favor of the first.

13. Resumé of all these principles of the method of philosophy. Method relaxed for this sketch.

APPLICATION OF THIS METHOD

14. The process, the beginning, the end.
15. The law of assimilation. Disturbance and its propagation.

16. Development of Time. How to conceive of time being developed. How its different properties came about.

17. Development of space and of the laws of matter and motion. They could not be otherwise.

18. Gravitation and molecular forces.

19. The chemical elements.

20. Protoplasm.

21. Consciousness. Development of God.

22. The end of things.

Chapter I. Method. The close and essential analogy between fruitful thinking and experimentation. On the one hand, the success of experimentation depends on there being a reason in Nature. Proof of this. On the other hand, reasoning of much power has, as a historical fact, never been performed by means of words, or other sounds, nor even to any great extent by means of pure retinal sensations, but by means of muscular sensations and visual images which have in the imagination been put in motion, so that a sort of imaginary experiment is made; and the result has been observed inwardly, as that of a physical experiment is outwardly; and this result is as unexpected and had been as occult as that of a physical experiment.

The main proposition of this chapter depends mainly on the analysis of historical facts about the great reasonings of science, mathematics, etc. aided by results of modern psychology.

Chapter II. Method, continued. The law of the development of fruitful conceptions. In what manner their first germs present themselves. There is absolutely no account to be given of the truly germinal ideas in their first origin, except that they occur by chance. (Although it is to be remarked that the mind has some bias toward true ideas.) But if the mind is unprepared, they immediately go into oblivion. If they find other ideas to which they can attach themselves, a welding process takes place. This welding is the only law of mental development. Its peculiar character, a gentle force. It acts in causing the growing together of ideas to form higher ideas,—a process roughly called generalization; and also to assimilate one idea into another, so as to fill out the detail of this latter.

This chapter draws upon the history of science, upon psychology, and upon mathematics.

Chapter III. Method, continued. Nature of assurance. This chapter is founded upon various previous publications of mine; but gives an improved view of the subject.

Chapter IV. Method, continued. What is the general nature of the problem of philosophy? To show what can be found of intelligible and reasonable in the universe at large. This supposes the process of nature and the process of thought to be identical; at least, so far as philosophy can succeed in its attempt. The first thing to be done therefore is to re-examine the logical process, to dissect it and find its principal elements, with a view of endeavoring to trace these in nature.

The logical process restated. The elementary and fundamental ideas contained in logic.

Chapter V. Method, continued.   If there is a regular growth about philosophy, we want to place ourselves historically, and take the inevitable next step in the most vigorous and prompt way. A coldly critical attitude necessary even toward our own opinions; we must fairly look all round.

The drama of the last three centuries of struggling thought, in politics and sociology, in science, in mathematics, in philosophy.

The ideas of today.

Chapter VI. Method, continued.   The present condition of society. Its lesson for philosophy. Significance of statistics.

Chapter VII. Method, continued.   The present attitude of mathematics. Axioms exploded. The lesson for philosophy.

Chapter VIII. Method, continued.   The present state of molecular physics. Prospects on the existing system very blue. Only hope is a natural history of laws and forces. How is this to be attained? Lesson for philosophy.

Chapter IX. Method, continued.   Modern psychology. The law of association. Its lesson for philosophy.

Chapter X. Method, continued.   The monism of the modern psychologists is mere materialism. The only possible explanation of the connection of body and soul.

Chapter XI. Method, continued.   The elliptic, parabolic and hyperbolic philosophy.

25

[CONCEPTIONS OF MODERN MATHEMATICS] (from 950)

Certain other conceptions of modern mathematics are indispensable
to a philosophy which is to be upon the intellectual level of our age.
On the first place, there is the conception of a space of more than three
dimensions, ordinarily regarded as highly mysterious, but really easy
enough. We can have no visual image of a space of three dimensions; we
can only see its projections upon surfaces. A perspective view, or picture on a plane surface, represents all we can see at any one time. It
would, therefore, be unreasonable to ask how a space of four or five
dimensions would look, in any other than a projective sense. Take one
of those glass paperweights cut into the form of a polyhedron. Photo-
graphs of such a body of three dimensions taken from one point of view
before and after giving it one turn are sufficient to determine how it must
look however it be turned. Now, if it had four dimensions instead of
three, the difference would be that by a certain peculiar effort we
could turn it so as to give a different perspective form from any that the
first two pictures would account for. Now, the reader does not [know]
all about the geometry of three or even of that of two dimensions; therefore, he cannot ask to have a complete idea of space of four dimensions;
but the property that I have just mentioned, that in such a space by a
peculiar effort a body could be turned so as to look in a way that ordinary
perspective would not account for, this gives a sufficient idea of space
of four dimensions. From this, all the other properties of that space
could be deduced.

The principal use that philosophy has to make of the conception of
n-dimensional space is in explaining why the dimensions of real space
are three in number. We see from this study that it is the restriction
in the number of dimensions which constitutes the fact to be explained.
No explanation of why space has more than two dimensions is called for,
because that is a mere indeterminacy. But why bodies should be restricted to move in three dimensions is one of the problems which
philosophy has to solve.

Another mathematical conception to be studied is that of imaginary quantities. Several illusory accounts of this conception have been given; yet I believe the true account is the most usual. If one man can lift a barrel of flour, how many men can just lift a bushel? The answer one fourth of a man is absurd, because we are dealing with a kind of quantity which does not admit of fractions. But a similar solution in continuous quantity would be correct. So, there is a kind of quantity which admits of no negative values. Now, as the scheme of quantity with negatives is an extension of that of positive quantity, and as the scheme of positive quantity is an extension of that of discrete quantity, so the scheme [of] imaginary quantity is an extension of that of real quantity. To determine the position of a point upon a plane requires two numbers (like latitude and longitude) and if we choose to use a single letter to denote a position on a plane and choose to call what that letter signifies a quantity, then that quantity is one which can only be expressed by two numbers. Any point on the plane taken arbitrarily is called zero, and any other is called one. Then, the point which is just as far on the other side of zero is, of course, -1; for the mean of 1 and -1 is zero. Take any three points A, B, C, forming a triangle, and find another point, D, such that the triangle ACD is similar to the triangle ABC. Then, we naturally write $(B - A):(C - A) = (C - A):(D - A)$. Apply this to the case where A is the zero point, B the point 1, and D the point -1. Then C will be the point at unit distance from A, but at right angles to AB and this point will represent a quantity i such that $i^2 = -1$ [Fig. 1].

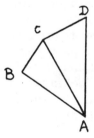
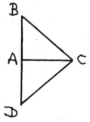

Fig. 1

Imaginary quantities are put to two very different uses in mathematics. In some cases, as in the theory of functions, by considering imaginary quantities, and not limiting ourselves to real quantity (which is but a special case of imaginary quantity) we are able to form important generalizations and bind together different doctrines in a manner which leads us to great advances of the most practical kind. In other cases, as in geometry, we use imaginary quantity, because the problems to be solved are too difficult in the case of real quantity. It is very easy to say how many inflections a curve of a certain description will have, if imaginary inflexions are included, but very difficult if we are restricted to real inflexions. Here imaginaries serve only a temporary purpose, and will one of these days give place to a more perfect doctrine.

To give a single illustration of the generalizing power of imaginaries, take this problem. Two circles have their centres at the distance D, their radii being R_1 and R_2. Let a straight line be drawn between their two points of intersection, at what distances will the two centres be from this line? Let these distances be x_1 and x_2, so that $x_1 + x_2 = D$. Then the square of the distance from one of the intersections of the circle to the line through their centres will be, by the Pythagorean proposition $R_1^2 - x_1^2 = R_2^2 - x_2^2$. These two equations give

$$x_1 = \frac{1}{2}D + \frac{R_1^2 - R_2^2}{2D}$$

$$x_2 = \frac{1}{2}D - \frac{R_1^2 - R_2^2}{2D}$$

Now the two circles may not really intersect at all, and yet x_1 and x_2 continue to be real. In other words, there is a real line between two imaginary intersections whose distance from the line of centres is

$$\frac{1}{2D}\sqrt{4R_1^2R_2^2 - (D^2 - R_1^2 - R_2^2)^2}$$

I do not know whether the theory of imaginaries will find any direct application in philosophy or not. But, at any rate, it is needed for the full comprehension of the mathematical doctrine of the absolute. For this purpose we must first explain the mathematical extension of the theory of perspective* (*In the main given in Brooke Taylor's Perspective, 1715). In the figure [Fig. 2][1] let O be the eye, or centre of projection, let

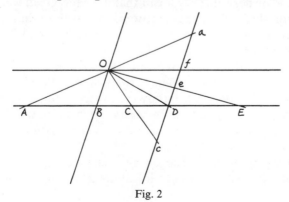

Fig. 2

[1]   The same conception with a similar diagram is to be found in MS. s-5.

the line *afeDc* represent the plane* (*The reader need not be informed that a [plain] is not a *plane*. It is flat but need not be level. Thus, the vertical wall of a room is a *plane*.) of projection seen edgewise, and let the line *ABCDE* represent a natural plane seen edgewise. Any straight line, as *OE*, being drawn from the eye to any point on the natural plane, will cut the plane of the picture or plane of projection in *e*, the point which represents the point *E*. The mathematician (sometimes the artist, too) extends this to the case where *C*, the natural point is nearer the eye than the corresponding point of the picture. He also extends the same rule to the case where *A*, the natural point, and *a*, the point of the picture, are on opposite sides of the eye. Here is the whole principle of geometrical projection. Suppose now that three points in nature, say *P*, *Q*, *R*, really lie on a straight line. Then the three lines *OP*, *OQ*, *OR*, from these points to the eye, will lie on one plane. This plane will be cut by the plane of the picture in a straight line (because the intersection of any two planes is a straight line). Hence, the points *p*, *q*, *r* which are the representations in the picture of *P*, *Q*, *R*, also lie in a straight line, and in general every straight line in nature is represented by a straight line in the picture and every straight line in the picture representing a straight line in a plane not containing the eye represents a straight line. Now according to the doctrine of Euclid, that the sum of the angles of a triangle is 180°, the parts of a natural plane at an infinite distance are also represented by a straight line in the picture, called the vanishing line of that plane. In the figure *f* is the vanishing line (seen edgewise) of the plane *ABCDE*. Note how the passage from *e* through *f* to *a* corresponds to a passage from *E* off to infinity and back from infinity on the other side to *A*. Euclid or no Euclid, the geometer is forced by the principles of perspective to conceive the plane as joined on to itself through infinity. Geometers do not mean that there is any continuity through infinity; such an idea would be absurd. They mean that if a cannonball were to move at a continuously accelerated rate toward the north so that its perspective representation should move continuously, it would have to pass through infinity and reappear at the south. There would really be a *saltus* at infinity and not motion proper. Persons absorbed in the study of projective geometry almost come to think there really is in every plane a line at infinity. But those who study the theory of functions regard the parts at infinity as a point. Both views are fictions which severally answer the purposes of the two branches of mathematics in which they are employed.

As I was saying, if the Euclidean geometry be true and the sum of the angles of a triangle equal 180°, it follows that the parts of any plane at infinity are represented by a right line in perspective. If that proposition be not true, still the perspective representation of everything remains exactly the same. Could some power suddenly change the properties of space so that the Euclidean doctrine should cease to be true, all things would *look* exactly as they did before. Only when you come to measure the real differences between objects would you find those distances, especially the long distances, essentially altered. There would be two possible cases. . . .

KEY TO GREEK TERMS

Professor Ralph L. Ward of Hunter College of City University of New York undertook the task of transcription and transliteration for the production of this appendix which has been compiled by him. He notes that there are numerous etymologies of Latin, Greek and English words in Peirce's work which are faulty from the present day standpoint but which have been left unaltered in this edition. Moreover no attempt has been made in this edition to insert the usual asterisk before purely reconstructed words adduced by Peirce; nor has there been an attempt to insert macrons over vowels of Latin words cited in the text.

ἐν ἀριθμῷ	hen arithmōi	p. 7
Καὶ πεπερασμένην εὐθεῖαν	Kai peperasmenēn eutheian	p. 42
κατὰ τὸ συνεχὲς ἐπ'	kata to syneches ep'	
εὐθείας ἐκβαλεῖν	eutheias ekbalein	
Ἡιτήσθω ἀπὸ παντὸς	Ēitēsthō apo pantos	p. 44
σημείου ἐπὶ πᾶν σημεῖον	sēmeiou epi pan sēmeion	
εὐθεῖαν γραμμὴν ἀγαγεῖν	eutheian grammēn agagein	
τῶν ἐντὸς τοῦ σχήματος	tōn entos tou schēmatos	p. 72
κειμένων	keimenōn	
τῶν ἐντὸς τοῦ σχήματος	tōn entos tou schēmatos	p. 72
κειμένων	keimenōn	
τὸ ὅλον τοῦ μέρους μεῖζον	to holon tou merous meizon	p. 72
τὸ ὅλον τοῦ μέρους μεῖζον	to holon tou merous meizon	p. 72
ποῦ στῶ	pou stō	p. 174
ἐπαγωγή	epagōgē	p. 182
ἡ ἀπὸ τῶν καθ' ἕκαστον	hē apo tōn kath' hekaston	p. 182
ἐπὶ τὰ καθόλου ἔφοδος	epi ta katholou ephodos	p. 182
ἀπαγωγή	apagōgē	p. 183
ἀπαγωγή	apagōgē	p. 183
παράδειγμα	paradeigma	p. 184
Στοιχεῖα	Stoicheia	p. 201
Εὐθεῖα γραμμή ἐστιν, ἥτις	Eutheia grammē estin, hētis	p. 202
ἐξ ἴσου ἐφ' ἑαυτῆς	ex isou eph' heautēs	p. 202
σημείοις κεῖται	sēmeiois keitai	
γωνία	gōnia	p. 202
κλίσις	klisis	
ἐξ ἴσου	ex isou	
κοιναὶ ἔννοιαι	koinai ennoiai	p. 203
πορίσματα	porismata	p. 214
Καινὰ στοιχεῖα	Kaina stoicheia	p. 235
ὅπερ ἔδει δεῖξαι	hoper edei deixai	p. 238
μηδενός + -άδα	mēdenos + -ada	P. 278
κοιναὶ ἔννοιαι	koinai ennoiai	p. 287
κοινὴ ἔννοια	koinē ennoia	p. 287
κοινή	koinē	p. 287
ὅπερ ἔδει δεῖξαι	hoper edei deixai	p. 291
ἐνέργεια	energeia	p. 309
ὕλη	hylē	p. 329

INDEX OF NAMES

Abbot, F.E., 343, 358
Agassiz, L., ix, 64, 66
American Journal of Mathematics, 50, 95
Aristotle, vi, vii, 19, 51, 68, 167, 184, 202, 203, 269, 329

Bacon, R., 102, 320
Baldwin, J.M., 141
Berkeley, G., 44, 45
Boethius, 205
Boltzmann, L., 37
Boole, G., 1, 106, 117, 119, 124
Bolzano, B., 117
Brugsch, K., 171

Cantor, G., xxii, xxiii, 7, 50, 117, 125, 347, 355
Carnegie Institution, 13
Carus, P., 25
Cauchy, A.L., 157
Cayley, A., 46, 50, 124, 151
Century Dictionary, 55
Chrystal, G., 271
Clifford, W.K., 269
Comte, A., 15, 272

Dalton, J., 320
Darwin, C., 142
Dedekind, J.W.R., 34, 50, 159
De Morgan, A., 76, 124, 152, 176, 241, 269
    Formal Logic, 335
Descartes, R., 87, 99, 102, 167, 195
Dickens, C., 48
Diodorus, 169
Dirichlet, P.G.L., 87

Encyclopedia Britannica, 271
Epicurus, 167
Euclid, 43, 44, 45, 49, 89, 94, 201, 205, 236, 291, 384
Euler, L., 46, 69

Fermat, P. de, 28, 87
Ferrero, A., 125
Franklin, C. Ladd-, 174

Galileo, 102, 359
Galton, F., ix, 34, 125
Goethe, J.W. von
    Theory of colors, 44
Gray, A., ix

Hamilton, W.R., 269
Harvey, W., 102
Hegel, G.W.F., 30, 51, 103, 138, 167, 355
    Phänomenologie, 19
Heiberg, J.H., 214, 286
Herbart, J.F., 61
Hobbes, T., 45
Hooke, R., 69
Hume, D., 70, 71
Huygens, C., 69, 87

Johns Hopkins Studies in Logic, 22, 174, 319
Journal of Speculative Philosophy, 19

Kant, I., 1, 9, 19, 51, 61, 72, 84, 152, 167, 176, 253, 258, 325, 330, 343, 349
Kempe, A.B., 125, 325
    Memoir on Mathematical Form, xix, 335
Kepler, J., 102
Keyser, C.J., 78
Klein, F., 357

Lamark, J.B., 142
Lambert, J.H., 61

Langley, S.P., 71
Lebesque, V.A., 87
Legendre, A.M., 46, 87
Leibniz, G.W., vi, 61, 151
Listing, J.B.
  Census Theorem, 46, 323
Locke, J., 167
Lockyer, J.N., 227

Mach, E., 143
Maxwell, C.,
  Electricity and Magnetism, 128
Mill, J.S., 32, 70, 71, 103, 117, 158, 241, 252
Mitchell, O.H., 174
Monist, 23
Muir, T., 151

Newcomb, S., 32, 159

Ockham, William of, 335

Palmer, E.H., viii
Paulus Venetus, 335
Pearson, K., 25, 34, 125
Peirce, B., i, 270,
  Analytic Mechanics, 53, 55
  Geometry, xii
Peirce, C.S., 152, 335
  Algebra of Dyadic Relatives, 1883
Petrie, F., 67
Petrus Hispanus
  Summulae, 335
Pherecydes, 140
Philo, 169

Photometric Researches, ix
Plato, 68, 286
Playfair, J., 215
Poincaré, H., 32, 37
Popular Science Monthly, 249
Port-Royalists, 241
Proclus, 205
Pythagoras, 68, 167

Ramus, P., 102
Renouf, LePage, 171
Riemann, B., 85, 124
Royce, J., 141

Schröder, E., 34, 61, 150, 152, 159, 335
Scotus, Duns, vi, 167, 169, 276, 349
  Grammatica Speculativa, 331
Socrates, 68, 247
Spencer, H., xviii, 141
Sylvester, J.J., 151

Thomson and Tait,
  Natural Philosophy, xv

Venn, J., 25
Von Staudt, 124

Wallaston, 320
Wallis, J., 87
Weierstrass, K., 55, 124
Whewell, W., 359
Wolff, C., 286

Young, T., 69, 320

INDEX OF SUBJECTS

abduction, 25, 37, 320
absolute, 377
abstraction, 11, 49, 50, 160-163, 212
   corollarial, 56
   theorematic, 56
aleph, 50
antecedentals, 278
argument, 22

barbara, 369
barycentric calculus, 47, 51
Begriffe, 30
belief, 39, 40, 41
boolian, 96

calendar, xviii
categories, 18, 19, 30, 51, 307, 331-34
cause, 296
chance, 140
classification of sciences, 15, 64-68, 227,
   272-75
collection, x, 56, 164
congruence, xiii
continuity, xi, xxiv, 266, 324, 346
   pseudo-, 325
continuum, 50, 134, 135, 342, 343
correlate, xxii
counting, 89

deduction, 37
   theorematic, 1
   corollarial, 1, 215
definitions
   affirmation, 249, 250
   axiom, 237, 287
   belief, 249, 250
   convention, 287
   corollary, 237
   diagram, 237
   dyadics, 191
   icon, 237

   judgment, 249, 250
   letter, 237
   logic, 20, 54, 248
   mathematics, 19, 199, 265, 268
   postulate, 202, 237
   premisses, 58
   problem, 291
   proposition, 248
   quantity, 265
   scholium, 237
   theorem, 237, 289
determinants, 151
diagrams, Eulerian, 150
dichotomic mathematics, 151, 165, 285
dichotomous divisions, 301
discrimination, 186
disquiparants, 327
doctrine of chances, 34, 103
doubt, 40, 41
dynamics, xix

economics of research, 26, 28, 62, 63,
   184
ecthesis, 290
entelechy, 294, 300
equiparants, 327
esthetics, 19, 192, 197
ethics, 19, 197
evidence, 3, 7
evolution, 140
exactitude, x

fallacy, 25, 70
   3 kinds, 61
fermatian inference, 79
filament, 163
form, 297
function, 120

generalization, 346
geometry

physical, 269
    topical, 50, xxii
graphs, 324
    existential, 150, 320, 335
group, 229

habit, 140, 143
history of science, 68, 69, 143, 378

icon, 242, 256
imaginaries, 382
index, 242, 256
induction, 37, 158, 320
    validity of, 25
inference (6 modes), 279-81
infinity, 77
    mathematical, 266
infinitesimal, xi, 55, 56, 266, 355

law, 296
leading principle, 363
least squares, 60
logic, 197
    critical, 24
    divisions, 21
    of chance, 24
    of number, 34

machines, calculating, 10
map coloring, 347
mathematical concepts, 18
matter, 293, 296
mean, 120
metaphysics, 192
methodeutic, 26, 62, 63
miracles, 70
Modus Ponens, 365, 368
multitude, xxii, 266
    denumeral, 340
    fundamental theorem, xxii
    orders of, 341

nominal, 285
nominalism, 295
normative sciences, 197

ontology, 192

parameter, 120
particle, 163
pedagogy, 186

phenomenology, 192
physics, 192
physiognosy, 29
point, Boscovichian, 53, 55
pons asinorum, 20, 207, 289
postulate, 286
potential, 128
pragmaticism, 330
principle
    excluded middle, xiii, xx, 373
    identity of indiscernibles, xx
    leading, 175, 176, 363
    pragmatic, xii, 162
probability, 25, 55, 59, 125, 251
problem, 4-color, 46, 47
proof, 46
propositions, 22
psychics, 192
psychognosy, 29, 192
psychology, 192, 248

questions, twenty, xii
quoddam, 294

reactions, 348
real, 285
reality, 162, 343, 358
    of time, 32
    of space, 32
reasoning, 314, 315
    ampliative, 1
    analogy, 38
    corollarial, 1, 38
    deductive, 38
    explicatory, 1
    of mathematics, 47
    probable, 198
    retroductive, 183
    theorematic, 1, 38, 41, 42, 49
relate, xxii
relations
    dyadic, xxii, 165
    generative, 340
    triadic, 165

science, 188
    branch, 16
    class, 16
    family, 16
    order, 16
    species, 16

varieties, 16
semiotic, 20, 54
sign, xxi, 20, 239-47, 297, 299, 314
space
    elliptic, 288
    n-dimensional, 381
    properties of, 360
statistical ratio, 59
stechiologic, 21
surface, Riemannian, 268
syllogism of transposed quantity, 76, 77
symbol, 243, 259, 266
synechism, xvi

tabula rasa, 262

testimonies, 257
trees, 128

uniformities, 250
    of nature, 31
universal, 340
Urtheil, 39, 247
utility, scientific, 27

valency, 322
variable, 120
vortex, 298

yard, 258